THIS BOOK
IS THE PROPERTY OF
The Board of Education of Baltimore County

LOCH RAVEN SR.

NO. OF BOOK_____

DATE OF PURCHASE_____

RULES

1. The pupil to whom this book is loaned will be held responsible for its careful use and will be expected to return it or pay for it before he is given credit for the subject.
2. Any writing in or marking upon, or otherwise defacing it will be considered a material injury, for which such book must be replaced or paid for.
3. Teachers shall inspect all books at least once a month and report their condition at the end of each year.

PUPIL'S NAME	DATE LOANED
Steve Ingram	
Brad Blanche	

Mr. Psenieska

Fundamentals of College Algebra

FOURTH EDITION

EARL W. SWOKOWSKI
Marquette University

Prindle, Weber & Schmidt, Incorporated
Boston, Massachusetts

Prindle, Weber & Schmidt is a division of Wadsworth, Inc.

Printed in the United States of America.
Second printing: December 1978
Third printing: September 1979

Cover image "Emergence" ©Copyright 1977 by Jan Ehrenworth.
All rights reserved. Used by permission of the artist.

Library of Congress Cataloging in Publication Data
Swokowski, Earl William

   Fundamentals of college algebra.

   Includes index.
   1. Algebra. I. Title.
QA154.2.S96 1978   512.9   77-26256
ISBN 0-87150-253-4

Preface

The fourth edition of *Fundamentals of College Algebra* reflects the continuing change in the needs and abilities of students who enroll in precalculus mathematics courses. The goal of this new edition is to maintain the mathematical soundness of earlier editions, but to make the tone of the book less formal by means of rewriting, placing more emphasis on graphing, and adding many new exercises, applied problems, examples, and figures. Optional exercises for students who use hand-held calculators are included in appropriate sections. The use of a second color to highlight figures and important statements will further enhance the appeal of the text to students.

This edition has greatly benefited from suggestions and comments of the following reviewers and survey respondents:

A. N. Aheart *(West Virginia State College)*, C. C. Alexander *(University of Mississippi)*, J. Baker *(Western Carolina University)*, P. Bauer *(University of Wisconsin, Marshfield)*, J. Bennett *(University of Evansville)*, B. P. Bockstege *(Broward Community College)*, R. Dana *(Lake City Community College)*, D. Deckard *(University of Michigan)*, B. C. Detwiler *(Western Kentucky University)*, J. Dewar *(Loyola Marymount University)*, F. Dodd *(University of South Alabama)*, W. Duncan *(McLennan Community College)*, C. V. Duplissey *(University of Arkansas-Little Rock)*, L. Estergard *(Brevard Community College)*, H. Fox *(University of Wisconsin, Waukesha)*, L. E. Fuller *(Kansas State University)*, R. Georing *(Phoenix College)*, H. E. Hall *(DeKalb Community College)*, R. Hamm *(College of Charleston)*, D. A. Happel *(Briar Cliff College)*, S. E. Hardy *(Georgia State College)*, W. Holstrom *(Elgin Community College)*, C. G. Hunkovsky *(Cochise College)*, G. Kolettis *(University of Notre Dame)*, J. R. Loughrey *(Canada College)*, L. J. Luey *(City College of San Francisco)*, R. D. McWilliams *(Florida State University)*, C. Miracle *(University of Minnesota)*, P. R. Montgomery *(University of Kansas)*, J. J. Morrell *(Ball State University)*, G. W. Nelson *(North Dakota State University)*, B. Partner *(Ball State University)*, J. W. Patterson *(Atlanta Junior College,* W. D. Popejoy *(University of Northern Colorado)*, M. W. Rennie *(Washington State University)*, W. Sanders *(Houston State University)*, D. Sherbert *(University of Illinois)*, P. Sherman *(University of Oregon)*, M. Shurlds *(Mississippi State University)*, L. Sons *(Northern Illinois University)*, D. R. Stocks *(University of Alabama-Birmingham)*, A. Sullenberger *(Tarrant County Junior College)*, W. R. Sunkman *(Bemidji State College)*, D. F. Thames *(Lamar University)*, J. L. Whitcomb *(University of North Dakota)*, S. Whitman *(University of Alabama)*, T. L. Williams *(Idaho State University)*, W. Wright *(Loyola Marymount University)*, K. Yanosko *(Florida State University)*.

In addition, I wish to single out as especially helpful the detailed reviews of various stages of the revised manuscript by Donald L. Dykes *(Kent State University)*, Mark P. Hale, Jr. *(University of Florida)*, Douglas W. Hall *(Michigan State University)*, A. J. Hulin *(University of New Orleans)*, Burnett Meyer *(University of Colorado)*, and Russell J. Rowlett *(University of Tennessee)*.

I am also grateful to the staff of Prindle, Weber & Schmidt, Inc., for their cooperation and valuable assistance. In particular, Elizabeth Thomson was very helpful in her role as production editor, and Executive Editor John Martindale was a constant source of advice and encouragement.

Special thanks are due to my wife Shirley and the members of our family: Mary, Mark, John, Steve, Paul, Tom, Bob, Nancy, and Judy. All have had an influence on the book — either directly, through working exercises, proofreading, or typing, or indirectly, through continued interest and moral support.

To all of the people named here and to the many unnamed students and teachers who have helped shape my views about precalculus mathematics, I express my sincere appreciation.

Earl W. Swokowski

Table
of
Contents

CHAPTER 1

Fundamental
Concepts
of Algebra

The material in this chapter is basic to the study of algebra. We begin by discussing properties of real numbers. Next we turn our attention to exponents and radicals, and how they may be used to simplify complicated algebraic expressions.

1 ALGEBRA: A POWERFUL LANGUAGE AND TOOL

A good foundation in algebra is essential for advanced courses in mathematics, the natural sciences, and engineering. It is also required for problems which arise in business, industry, statistics, and many other fields of endeavor. Indeed, every situation which makes use of numerical processes is a candidate for algebraic methods.

Algebra evolved from the operations and rules of arithmetic. The study of arithmetic begins with addition, multiplication, subtraction, and division of numbers, such as

$$4 + 7, \quad (37)(681), \quad 79 - 22 \quad \text{and} \quad 40 \div 8.$$

In *algebra* we introduce symbols or letters a, b, c, d, x, y, etc., to denote *arbitrary* numbers and, instead of special cases, we often consider *general* statements such as

$$a + b, \quad cd, \quad x - y \quad \text{and} \quad x \div a.$$

This *language of algebra* serves a two-fold purpose. First, it may be used as a shorthand, to abbreviate and simplify long or complicated statements. Second, it is a convenient means of generalizing many specific statements. To illustrate, at an early age, children learn that

$$2 + 3 = 3 + 2, \quad 4 + 7 = 7 + 4, \quad 5 + 9 = 9 + 5, \quad 1 + 8 = 8 + 1$$

and so on. In words, this property may be phrased "if two numbers are added, then the order of addition is immaterial; that is, the same result is obtained whether the second number is added to the first, or the first number is added to the second." This lengthy description can be shortened, and at the same time made easier to

1

understand, by means of the algebraic statement

$$a + b = b + a$$

where a and b denote arbitrary numbers.

Many illustrations of the generality of algebra may be found in formulas used in science and industry. For example, if an airplane flies at a constant rate of 300 mph (miles per hour) for two hours, then the distance it travels is given by

$$(300)(2), \quad \text{or } 600 \text{ miles.}$$

If the rate is 250 mph and the elapsed time is 3 hours, then the distance traveled is

$$(250)(3), \quad \text{or } 750 \text{ miles.}$$

If we introduce symbols, and let r denote the constant rate, t the elapsed time, and d the distance traveled, then the two illustrations we have given are special cases of the general algebraic formula

$$d = rt.$$

When specific numerical values for r and t are given, the distance d may be found readily by an appropriate substitution in the formula. Moreover, the formula may also be used to solve related problems. For example, suppose the distance between two cities is 645 miles, and we wish to find the constant rate which would enable an airplane to cover that distance in 2 hours and 30 minutes. Thus we are given

$$d = 645 \text{ miles}, \qquad t = 2.5 \text{ hours}$$

and the problem is to find r. Since $d = rt$ it follows that

$$r = \frac{d}{t}$$

and hence for our special case,

$$r = \frac{645}{2.5} = 258 \text{ mph.}$$

That is, if an airplane flies at a constant rate of 258 mph, then it will travel 645 miles in 2 hours and 30 minutes. In like manner, given r, the time t required to travel a distance d may be found by means of the formula

$$t = \frac{d}{r}.$$

The preceding example indicates how the introduction of a general algebraic formula not only allows us to solve special problems conveniently, but also to enlarge the scope of our knowledge by suggesting new problems that can be considered.

We have given only two elementary illustrations of the value of algebraic methods. There are an unlimited number of situations where a symbolic approach may lead to insights and solutions that would be impossible to obtain using only numerical processes. As you proceed through this text and go on to either more advanced courses in mathematics or fields which employ mathematics, you will become further aware of the importance and the power of algebraic techniques.

2   REAL NUMBERS

Real numbers are used considerably in all phases of mathematics and you are undoubtedly well acquainted with symbols which are used to represent them, such as

$$1, \quad 73, \quad -5, \quad \frac{49}{12}, \quad \sqrt{2}, \quad 0, \quad \sqrt[3]{-85}, \quad 0.33333..., \quad 596.25$$

and so on. The real numbers are said to be **closed** relative to operations of addition (denoted by $+$) and multiplication (denoted by $\cdot$). This means that to every pair a, b of real numbers there corresponds a unique real number $a + b$ called the **sum** of a and b and a unique real number $a \cdot b$ (also written ab) called the **product** of a and b. These operations have the following properties, where all lower-case letters denote arbitrary real numbers, and where 0 and 1 are special real numbers referred to as **zero** and **one**, respectively.

(1.1)   Commutative Properties

$$\boxed{a + b = b + a, \qquad ab = ba}$$

(1.2)   Associative Properties

$$\boxed{a + (b + c) = (a + b) + c, \qquad a(bc) = (ab)c}$$

(1.3)   Identities

$$\boxed{a + 0 = a = 0 + a, \qquad a \cdot 1 = a = 1 \cdot a}$$

(1.4)   Inverses
For every real number a, there is a real number denoted by $-a$ such that

$$\boxed{a + (-a) = 0 = (-a) + a}$$

For every real number $a \neq 0$, there is a real number denoted by $1/a$ such that

$$\boxed{a\left(\frac{1}{a}\right) = 1 = \left(\frac{1}{a}\right)a}$$

(1.5)   Distributive Properties

$$\boxed{a(b + c) = ab + ac, \qquad (a + b)c = ac + bc}$$

The equals sign, $=$, used in properties (1.1)–(1.5) means, of course, that the expressions immediately to the right and left of the sign represent the same real number. The real numbers 0 and 1 are sometimes referred to as the **additive identity** and **multiplicative identity**, respectively. We call $-a$ the **additive inverse** of a (or the **negative** of a). If $a \neq 0$, then $1/a$ is called the **multiplicative inverse** of a (or the **reciprocal** of a). The symbol a^{-1} is often used in place of $1/a$. Thus, by definition,

$$a^{-1} = \frac{1}{a}$$

Example 1   Verify the following special cases of properties (1.2) and (1.5).

    (a) $2 + (3 + 4) = (2 + 3) + 4$

    (b) $2 \cdot (3 \cdot 4) = (2 \cdot 3) \cdot 4$

    (c) $2 \cdot (3 + 4) = 2 \cdot 3 + 2 \cdot 4$

    (d) $(2 + 3) \cdot 4 = 2 \cdot 4 + 3 \cdot 4$

Solutions   To verify each of parts (a)–(d) we perform the operations indicated on opposite sides of the equals sign and observe that the resulting numbers are identical. Thus

(a)
$$2 + (3 + 4) = 2 + 7 = 9$$
$$(2 + 3) + 4 = 5 + 4 = 9$$

(b)
$$2 \cdot (3 \cdot 4) = 2 \cdot 12 = 24$$
$$(2 \cdot 3) \cdot 4 = 6 \cdot 4 = 24$$

(c)
$$2 \cdot (3 + 4) = 2 \cdot 7 = 14$$
$$2 \cdot 3 + 2 \cdot 4 = 6 + 8 = 14$$

(d)
$$(2 + 3) \cdot 4 = 5 \cdot 4 = 20$$
$$2 \cdot 4 + 3 \cdot 4 = 8 + 12 = 20$$

Since $a + (b + c)$ and $(a + b) + c$ are always equal we may, without ambiguity, use the symbol $a + b + c$ to denote the real number they represent. Similarly, the notation abc is used to represent either $a(bc)$ or $(ab)c$. An analogous situation exists if four real numbers a, b, c, and d are added. For example, we could consider

$$(a + b) + (c + d), \quad a + [(b + c) + d], \quad [(a + b) + c] + d,$$

and so on. It can be shown that regardless of how the four numbers are grouped, the same result is obtained, and consequently it is customary to write $a + b + c + d$ for any of these expressions. Furthermore, it follows from the Commutative Properties (1.1) that the numbers can be interchanged in any way. For example,

$$a + b + c + d = a + d + c + b = a + c + d + b.$$

We shall justify a manipulation of this type by referring to "commutative and associative properties of real numbers." A similar situation exists for multiplication, where the expression *abcd* is used to denote the product of four real numbers.

The Distributive Properties (1.5) are useful for finding products of many different types of expressions. The next example provides two illustrations. Others will be found in the exercises.

Example 2   If *a*, *b*, *c*, and *d* denote real numbers, show that

(a) $a(b + c + d) = ab + ac + ad$

(b) $(a + b)(c + d) = ac + bc + ad + bd$

Solutions   Each product may be found by using property (1.5) several times. The reader should supply reasons for each step in the following.

(a)
$$a(b + c + d) = a[(b + c) + d]$$
$$= a(b + c) + ad$$
$$= (ab + ac) + ad$$
$$= ab + ac + ad$$

(b)
$$(a + b)(c + d) = (a + b)c + (a + b)d$$
$$= (ac + bc) + (ad + bd)$$
$$= ac + bc + ad + bd$$

If $a = b$ and $c = d$, then since *a* and *b* are merely different names for the same real number, and likewise for *c* and *d*, it follows that $a + c = b + d$ and $ac = bd$. This is often called the **substitution principle**, since we may think of replacing *a* by *b* and *c* by *d* in the expressions $a + c$ and ac. As a special case, using the fact that $c = c$ gives us the following rules.

(1.6)

> If $a = b$, then $a + c = b + c$.
>
> If $a = b$, then $ac = bc$.

We sometimes refer to the rules in (1.6) by the statements "any number *c* may be added to both sides of an equality" and "both sides of an equality may be multiplied by the same number *c*." These rules constitute two extremely important algebraic manipulations. We shall make heavy use of them in Chapter 2, in conjunction with solving equations.

Properties (1.1)–(1.5) can be used to prove the following results (see Exercises 36 and 37).

(1.7)

> $a \cdot 0 = 0$ for every real number *a*.
>
> If $ab = 0$, then either $a = 0$ or $b = 0$.

The statements in (1.7) imply that $ab = 0$ *if and only if* either $a = 0$ or $b = 0$. The phrase "if and only if," which is used throughout mathematics, always has a two-fold character. Here it means that if $ab = 0$, then $a = 0$ or $b = 0$ and, *conversely*, if

$a = 0$ or $b = 0$, then $ab = 0$. Consequently, if both $a \neq 0$ and $b \neq 0$, then $ab \neq 0$; that is, *the product of two nonzero real numbers is always nonzero.*

The following rules for negatives can also be proved directly from properties (1.1)–(1.5). (See Exercise 41.)

(1.8)

$$-(-a) = a$$
$$(-a)b = -(ab) = a(-b)$$
$$(-a)(-b) = ab$$
$$(-1)a = -a$$

The operation of **subtraction** (denoted by $-$) is defined by

(1.9)

$$a - b = a + (-b)$$

The next example indicates that the Distributive Properties hold for subtraction.

Example 3   If a, b, and c are real numbers, show that

$$a(b - c) = ab - ac.$$

Solution   We shall list reasons after each step as follows.

$$\begin{aligned}
a(b - c) &= a[b + (-c)] & (1.9) \\
&= ab + a(-c) & (1.5) \\
&= ab + [-(ac)] & (1.8) \\
&= ab - ac & (1.9)
\end{aligned}$$

If $b \neq 0$, then **division** (denoted by $\div$) is defined by

(1.10)

$$a \div b = a\left(\frac{1}{b}\right) = ab^{-1}$$

The symbol a/b is often used in place of $a \div b$, and we refer to it as the **quotient of a by b** or the **fraction a over b**. The numbers a and b are called the **numerator** and **denominator**, respectively, of the fraction. It is important to note that since 0 has no multiplicative inverse, a/b is not defined if $b = 0$; that is, *division by zero is not permissible*. Also note that

$$1 \div b = \frac{1}{b} = b^{-1}.$$

The following rules for quotients may be established, where all denominators are nonzero real numbers.

$$\frac{a}{b} = \frac{c}{d} \quad \text{if and only if } ad = bc$$

$$\frac{a}{b} = \frac{ad}{bd}$$

$$\frac{a}{-b} = \frac{-a}{b} = -\frac{a}{b}$$

(1.11)

$$\frac{a}{b} + \frac{c}{b} = \frac{a+c}{b}$$

$$\frac{a}{b} + \frac{c}{d} = \frac{ad+bc}{bd}$$

$$\frac{a}{b} \cdot \frac{c}{d} = \frac{ac}{bd}$$

$$\frac{a}{b} \div \frac{c}{d} = \frac{a}{b} \cdot \frac{d}{c} = \frac{ad}{bc}$$

Example 4 Find (a) $\dfrac{2}{3} + \dfrac{9}{5}$ (b) $\dfrac{2}{3} \cdot \dfrac{9}{5}$ (c) $\dfrac{2}{3} \div \dfrac{9}{5}$

Solutions Using (1.11) we have

(a) $$\frac{2}{3} + \frac{9}{5} = \frac{(2\cdot 5) + (3\cdot 9)}{3\cdot 5} = \frac{10 + 27}{15} = \frac{37}{15}$$

(b) $$\frac{2}{3} \cdot \frac{9}{5} = \frac{2\cdot 9}{3\cdot 5} = \frac{18}{15} = \frac{6\cdot 3}{5\cdot 3} = \frac{6}{5}$$

(c) $$\frac{2}{3} \div \frac{9}{5} = \frac{2}{3} \cdot \frac{5}{9} = \frac{2\cdot 5}{3\cdot 9} = \frac{10}{27}$$

The **positive integers** $1, 2, 3, 4, \ldots$ may be obtained by adding the real number 1 successively to itself. The negatives, $-1, -2, -3, -4, \ldots$, of the positive integers are referred to as **negative integers**. The **integers** consist of the totality of positive and negative integers together with the real number 0.

Observe that by the Distributive Properties, if a is a real number then

$$a + a = (1 + 1)a = 2a$$

and

$$a + a + a = (1 + 1 + 1)a = 3a.$$

Similarly the sum of four a's is $4a$, the sum of five a's is $5a$, and so on.

If $a, b,$ and c are integers and $c = ab$, then a and b are called **factors**, or **divisors**, of c. For example the integer 6 may be written as

$$6 = 2\cdot 3 = (-2)(-3) = 1\cdot 6 = (-1)(-6).$$

Hence $1, -1, 2, -2, 3, -3, 6,$ and -6 are factors of 6.

A positive integer p different from 1 is **prime** if its only positive factors are 1 and p. The first few primes are 2, 3, 5, 7, 11, 13, 17, and 19. One of the reasons for the importance of prime numbers is that every positive integer a different from 1 can be expressed in one and only one way (except for order of factors) as a product of primes. (The proof of this result will not be given in this book.) As examples, we have

$$12 = 2 \cdot 2 \cdot 3, \quad 126 = 2 \cdot 3 \cdot 3 \cdot 7, \quad 540 = 2 \cdot 2 \cdot 3 \cdot 3 \cdot 3 \cdot 5.$$

A real number is called a **rational number** if it can be written in the form a/b, where a and b are integers and $b \neq 0$. Real numbers that are not rational are called **irrational**. The ratio of the circumference of a circle to its diameter is an irrational real number and is denoted by π. It is often approximated by the decimal 3.1416 or by the rational number 22/7. We use the notation $\pi \approx 3.1416$ to indicate that π is *approximately equal* to 3.1416. To cite another example, a real number a such that $a^2 = 2$, where a^2 denotes $a \cdot a$, is not rational. There are two such irrational numbers denoted by the symbols $\sqrt{2}$ and $-\sqrt{2}$.

Real numbers may be represented by decimal expressions. Decimal representations for rational numbers either terminate or are nonterminating and repeating. For example, it can be shown by long division that a decimal representation for 7434/2310 is 3.2181818..., where the digits 1 and 8 repeat indefinitely. The rational number 5/4 has the terminating decimal representation 1.25. Decimal representations for irrational numbers may also be obtained; however, they are always nonterminating and nonrepeating. The process of finding decimal representations for irrational numbers is usually difficult. Often some method of successive approximation is employed. For example, the device learned in arithmetic for extracting square roots can be used to find a decimal representation for $\sqrt{2}$. Using this technique we successively obtain the approximations 1, 1.4, 1.41, 1.414, 1.4142, and so on.

Sometimes, it is convenient to use the notation and terminology of sets. A **set** may be thought of as a collection of objects of some type. The objects are called **elements** of the set. Capital letters $A, B, C, R, S, \ldots$ will often be used to denote sets. Lower-case letters $a, b, x, y, \ldots$ will represent elements of sets. Throughout our work $\mathbb{R}$ will denote the set of real numbers, and $\mathbb{Z}$ the set of integers. If every element of a set S is also an element of a set T, then S is called a **subset** of T. For example, $\mathbb{Z}$ is a subset of $\mathbb{R}$. Two sets S and T are said to be equal, written $S = T$, if S and T contain precisely the same elements. The notation $S \neq T$ means that S and T are not equal.

EXERCISES

In each of Exercises 1–10, justify the equality by stating only one of the properties (1.1)–(1.5).

1 $(4 \cdot 5) \cdot 4 = 4 \cdot (4 \cdot 5)$

2 $3 \cdot (4 + 5) = (4 + 5) \cdot 3$

3 $(4 \cdot 5) \cdot 4 = 4 \cdot (5 \cdot 4)$

4 $(4 + 5) \cdot 3 = 4 \cdot 3 + 5 \cdot 3$

5 $3 \cdot (5 + 0) = 3 \cdot 5$

6 $3 + (-3) = 0$

7 $1 \cdot (2 + 3) = 2 + 3$

8 $(1 + 2) + 1 = 1 + (1 + 2)$

9 $(1/4)4 = 1$

10 $0 \cdot 1 = 0$

Use properties (1.1)–(1.5) to find the products in Exercises 11–20, where all letters represent real numbers.

11 $a(b + 3) + 2(b + 3)$

12 $c(d + 1) + 5(d + 1)$

13 $(a + 2)(b + 3)$

14 $(c + 5)(d + 1)$

15 $2x(y + 2) - 3(y + 2)$

16 $2p(6q + 5) - 3(6q + 5)$

17 $(2x - 3)(y + 2)$

18 $(2p - 3)(6q + 5)$

19 $(4r + 5)(3s + 6)$

20 $(3u + 7)(4v - 1)$

In each of Exercises 21–30, write the expression as a rational number whose numerator is as small as possible.

21 $\dfrac{3}{4} + \dfrac{5}{3}$

22 $\dfrac{1}{5} + \dfrac{3}{2}$

23 $\dfrac{5}{6} \cdot \dfrac{2}{3}$

24 $\dfrac{4}{9}\left(\dfrac{1}{2} + \dfrac{3}{4}\right)$

25 $\dfrac{1}{2} + \dfrac{1}{4} + \dfrac{1}{6}$

26 $\dfrac{3}{-2} \cdot \dfrac{-5}{6}$

27 $\dfrac{13}{4} \div \left(\dfrac{3}{2} + 1\right)$

28 $\dfrac{2}{5} + \dfrac{7}{4} + \dfrac{3}{2}$

29 $\dfrac{5}{7}\left(\dfrac{3}{2} - \dfrac{5}{6}\right)$

30 $\dfrac{10}{11} \cdot \dfrac{11}{10}$

31 Show, by means of examples, that the operation of subtraction on $\mathbb{R}$ is neither commutative nor associative.

32 Show that the operation of division, as applied to nonzero real numbers, is neither commutative nor associative.

In Exercises 33–40, all letters denote real numbers and no denominators are zero.

33 Prove or disprove: $\dfrac{1}{a} + \dfrac{1}{b} = \dfrac{1}{a + b}$

34 Prove or disprove: $\dfrac{a}{b + c} = \dfrac{a}{b} + \dfrac{a}{c}$

Prove the rules in Exercises 35–40.

35 $a = -a$ if and only if $a = 0$.

36 $a \cdot 0 = 0$ (*Hint:* Write $a \cdot 0 = a \cdot (0 + 0) = a \cdot 0 + a \cdot 0$ and then add $-(a \cdot 0)$ to both sides.)

37 If $ab = 0$, then either $a = 0$ or $b = 0$.

38 $\dfrac{-a}{-b} = \dfrac{a}{b}$

39 $\left(\dfrac{a}{b}\right)^{-1} = \dfrac{b}{a}$

40 $\dfrac{a}{b} + \dfrac{c}{b} = \dfrac{a+c}{b}$ (*Hint:* Write $\dfrac{a}{b} + \dfrac{c}{b} = ab^{-1} + cb^{-1}$ and use the Distributive Properties.)

41 Prove (1.8).

3   COORDINATE LINES

It is possible to associate the set of real numbers with the set of all points on a line l in such a way that for each real number a there corresponds one and only one point, and conversely, to each point P on l there corresponds precisely one real number. Such an association between two sets is referred to as a **one-to-one correspondence**. We first choose an arbitrary point O, called the **origin**, and associate with it the real number 0. Points associated with the integers are then determined by laying off successive line segments of equal length on either side of O as illustrated in Figure 1.1. The points corresponding to rational numbers such as 23/5 and $-1/2$ are obtained by subdividing the equal line segments. Points associated with certain irrational numbers, such as $\sqrt{2}$, can be found by geometric construction. For other irrational numbers such as π, no construction is possible. However, the point corresponding to π can be approximated to any degree of accuracy by locating successively the points corresponding to 3, 3.1, 3.14, 3.141, 3.1415, 3.14159, and so on. It can be shown that to every irrational number there corresponds a unique point on l and, conversely, every point that is not associated with a rational number corresponds to an irrational number.

Figure 1.1

The number a that is associated with a point A on l is called the **coordinate** of A. An assignment of coordinates to points on l is called a **coordinate system** for l, and l is called a **coordinate line**, or a **real line**. A direction can be assigned to l by taking the **positive direction** along l to the right and the **negative direction** to the left. The positive direction is noted by placing an arrowhead on l as shown in Figure 1.1.

The numbers which correspond to points to the right of 0 in Figure 1.1 are called **positive real numbers**, whereas those which correspond to points to the left of 0 are **negative real numbers**. The real number 0 is neither positive nor negative. The set of positive real numbers is **closed** relative to addition and multiplication; that is, if a and b are positive, then so is the sum $a + b$ and the product ab.

Note that if a is positive, then $-a$ is negative. Similarly, if $-a$ is positive, then $-(-a) = a$ is negative. A common error is to think that $-a$ is always a negative number; however, this is not necessarily the case. For example, if

$a = -3$, then $-a = -(-3) = 3$, which is positive. Indeed, whenever a is negative the number $-a$ is positive.

If a and b are real numbers, and $a - b$ is positive, we say that *a is greater than b* and write $a > b$. An equivalent statement is *b is less than a*, written $b < a$. The symbols $>$ or $<$ are called **inequality signs** and expressions such as $a > b$ or $b < a$ are called **inequalities**. From the manner in which we constructed the coordinate line l in Figure 1.1, we see that if A and B are points with coordinates a and b, respectively, then $a > b$ (or $b < a$) *if and only if A lies to the right of B*. The following definition is stated for reference, where a and b denote real numbers.

(1.12)

> $a > b$ means $a - b$ is positive.
>
> $b < a$ means $a - b$ is positive.

As indicated above, $a > b$ and $b < a$ have exactly the same meaning.

As illustrations of (1.12) we may write

$$5 > 3 \text{ since } 5 - 3 = 2 \text{ is positive;}$$
$$-6 < -2 \text{ since } -2 - (-6) = -2 + 6 = 4 \text{ is positive;}$$
$$-\sqrt{2} < 1 \text{ since } 1 - (-\sqrt{2}) = 1 + \sqrt{2} \text{ is positive;}$$
$$2 > 0 \text{ since } 2 - 0 = 2 \text{ is positive;}$$
$$-5 < 0 \text{ since } 0 - (-5) = 5 \text{ is positive (or } -5 \text{ is negative).}$$

The last two illustrations are special cases of the following general properties.

> $a > 0$ if and only if a is positive.
>
> $a < 0$ if and only if a is negative.

These properties follow directly from (1.12).

It should also be clear from our discussion that if a and b are real numbers, then *one and only one of the following is true:*

$$a = b, \quad a > b, \quad a < b.$$

We sometimes refer to the **sign** of a real number as being positive or negative according as the number is positive or negative. The next result concerning signs can be proved using the rules in (1.8).

(1.13)   **Laws of Signs**

> (i)   The product of two negative real numbers is positive.
>
> (ii)   The product of a positive and a negative real number is negative.

As illustrations we see, from (1.8), that

$$(-2)(3) = -(2 \cdot 3) = -6$$
$$(2)(-3) = -(2 \cdot 3) = -6$$
$$(-2)(-3) = 2 \cdot 3 = 6.$$

There are several other useful symbols which involve inequality signs. In particular, $a \geq b$, which is read **a is greater than or equal to b**, means that either $a > b$ or $a = b$ (but not both). As an illustration we have

(1.14)

$$a^2 \geq 0 \text{ for every real number } a.$$

The symbol $a \leq b$ is read **a is less than or equal to b** and means that either $a < b$ or $a = b$. The expression $a < b < c$ means that both $a < b$ and $b < c$, in which case we say that **b is between a and c**. We may also write $c > b > a$. For instance,

$$1 < 5 < \frac{11}{2}, \quad -4 < \frac{2}{3} < \sqrt{2}, \quad 3 > -6 > -10.$$

Other variations of the inequality notation are used. For example, $a < b \leq c$ means both $a < b$ and $b \leq c$. Similarly, $a \leq b < c$ means both $a \leq b$ and $b < c$. Finally, $a \leq b \leq c$ means both $a \leq b$ and $b \leq c$.

If a is a real number, then it is the coordinate of some point A on a coordinate line l, and the symbol $|a|$ is used to denote the number of units (or distance) between A and the origin, without regard to direction. Referring to Figure 1.2 we see that for the point with coordinate -4 we have $|-4| = 4$. Similarly, $|4| = 4$. In general, if a is negative we change its sign to find $|a|$, whereas if a is nonnegative, then $|a| = a$; that is,

(1.15)

$$|a| = \begin{cases} a & \text{if } a \geq 0. \\ -a & \text{if } a < 0. \end{cases}$$

The nonnegative number $|a|$ is called the **absolute value** of a.

Figure 1.2

Example 1 Find $|3|$, $|-3|$, $|0|$, $|\sqrt{2} - 2|$, and $|2 - \sqrt{2}|$.

Solution Since 3, $2 - \sqrt{2}$, and 0 are nonnegative, we have by (1.15),

$$|3| = 3, \quad |2 - \sqrt{2}| = 2 - \sqrt{2}, \quad \text{and} \quad |0| = 0.$$

Since -3 and $\sqrt{2} - 2$ are negative, we use the formula $|a| = -a$ of (1.15) to obtain

$$|-3| = -(-3) = 3 \quad \text{and} \quad |\sqrt{2} - 2| = -(\sqrt{2} - 2) = 2 - \sqrt{2}.$$

Note that in Example 1, $|-3| = |3|$ and $|2 - \sqrt{2}| = |\sqrt{2} - 2|$. It can be shown in general that

$$|a| = |-a| \text{ for every real number } a.$$

We shall use the concept of absolute value to define the distance between any two points on a coordinate line. Let us begin by noting that the distance between the points with coordinates 2 and 7 shown in Figure 1.3 equals 5 units on *l*. This distance is the difference, 7 − 2, obtained by subtracting the smaller coordinate from the larger. If we employ absolute values, then since $|7 - 2| = |2 - 7|$, it is unnecessary to be concerned about the order of subtraction. We shall use this as our motivation for the following definition.

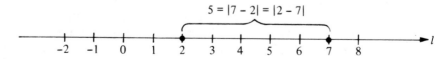

Figure 1.3

(1.16)

Let a and b be the coordinates of two points A and B, respectively, on a coordinate line *l*. The **distance between A and B**, denoted by $d(A, B)$, is defined by

$$d(A, B) = |b - a|$$

The number $d(A, B)$ in Definition (1.16) is also called the **length of the line segment AB**.

Observe that since $d(B, A) = |a - b|$ and $|b - a| = |a - b|$, we may write

$$d(A, B) = d(B, A).$$

Also note that the distance between the origin O and the point A is

$$d(O, A) = |a - 0| = |a|$$

which agrees with the geometric interpretation of absolute value illustrated in Figure 1.2. The formula for $d(A, B)$ in (1.16) is true regardless of the signs of a and b, as illustrated in the next example.

Example 2 Let A, B, C, and D have coordinates $-5, -3, 1$, and 6 respectively on a coordinate line *l* (see Figure 1.4). Find $d(A, B)$, $d(C, B)$, $d(O, A)$, and $d(C, D)$.

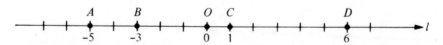

Figure 1.4

Solution   By (1.16),

$$d(A, B) = |-3 - (-5)| = |-3 + 5| = |2| = 2$$
$$d(C, B) = |-3 - 1| = |-4| = 4$$
$$d(O, A) = |-5 - 0| = |-5| = 5$$
$$d(C, D) = |6 - 1| = |5| = 5.$$

These answers can be checked visually by referring to Figure 1.4.

The concept of absolute value has uses other than that of finding distances between points. Generally, it is employed whenever we are interested in the "magnitude" or "numerical value" of a real number without regard to its sign.

EXERCISES

In Exercises 1–4, replace the □ with either $<$, $>$, or $=$.

1 (a) $-7 \square -4$ (b) $3 \square -1$ (c) $1 + 3 \square 6 - 2$

2 (a) $-3 \square -5$ (b) $-6 \square 2$ (c) $1/4 \square 0.25$

3 (a) $1/3 \square 0.33$ (b) $125/57 \square 2.193$ (c) $22/7 \square \pi$

4 (a) $1/7 \square 0.143$ (b) $(3/4) + (2/3) \square 19/12$ (c) $\sqrt{2} \square 1.4$

In each of Exercises 5–16, express the given statement in terms of inequalities.

5 -8 is less than -5.

6   2 is greater than 1.9.

7   0 is greater than -1.

8 $\sqrt{2}$ is less than π.

9 x is negative.

10 y is positive.

11 a is between 5 and 3.

12 b is between 1/10 and 1/3.

13 b is greater than or equal to 2.

14 x is less than or equal to -5.

15 c is not greater than 1.

16 d is nonnegative.

Rewrite the numbers in Exercises 17–20 without using symbols for absolute value.

17 (a) $|4 - 9|$ (b) $|-4| - |-9|$ (c) $|4| + |-9|$

18 (a) $|3 - 6|$ (b) $|0.2 - (1/5)|$ (c) $|-3| - |-4|$

19 (a) $3 - |-3|$ (b) $|\pi - 4|$ (c) $(-3)/|-3|$

20 (a) $|8 - 5|$ (b) $-5 + |-7|$ (c) $(-2)|-2|$

In each of Exercises 21–24, the given numbers are coordinates of three points A, B, and C (in that order) on a coordinate line l. For each, find:
(a) $d(A, B)$;   (b) $d(B, C)$;   (c) $d(C, B)$;   (d) $d(A, C)$.

21 -6, -2, 4

22   3, 7, -5

23   8, -4, -1

24 -9, 1, 10

25 Are the positive real numbers closed relative to:
(a) subtraction?   (b) division?   Explain.

26 Are the negative real numbers closed relative to
(a) addition?   (b) multiplication?   (c) subtraction?   Explain.

Rewrite each of Exercises 27–30 without using symbols for absolute value.

27 $|5 - x|$ if $x > 5$

28 $|x^2 + 1|$

29 $|-4 - x^2|$

30 $|a - b|$ if $a < b$

4 INTEGRAL EXPONENTS

Throughout the remainder of this chapter, all symbols for elements will denote real numbers. We have had occasion to use the notation a^2 for the real number $a \cdot a$. Similarly we write

$$a^3 = a \cdot a \cdot a, \quad a^4 = a \cdot a \cdot a \cdot a, \quad a^5 = a \cdot a \cdot a \cdot a \cdot a$$

and generally, if n is any positive integer,

(1.17)

$$a^n = \underbrace{a \cdot a \cdot \cdots \cdot a}_{n \text{ factors}}$$

where n factors, all equal to a, appear on the right-hand side of the equals sign. The positive integer n is called the **exponent** of a in the expression a^n, and a^n is read "a to the nth power" or simply "a to the n." Note that $a^1 = a$. Some numerical examples are

$$\left(\frac{1}{2}\right)^5 = \frac{1}{2} \cdot \frac{1}{2} \cdot \frac{1}{2} \cdot \frac{1}{2} \cdot \frac{1}{2} = \frac{1}{32}$$

$$(-3)^3 = (-3)(-3)(-3) = -27$$

$$(\sqrt{2})^4 = \sqrt{2}\sqrt{2}\sqrt{2}\sqrt{2} = (\sqrt{2})^2(\sqrt{2})^2 = (2)(2) = 4$$

It is important to remember that if n is a positive integer, then an expression such as $3a^n$ means $3(a^n)$ and not $(3a)^n$. The real number 3 is called the **coefficient** of a^n in the expression $3a^n$. Similarly, $-3a^n$ means $(-3)a^n$, not $(-3a)^n$. For example, we have

$$5 \cdot 2^3 = 5 \cdot 8 = 40 \quad \text{and} \quad -5 \cdot 2^3 = -5 \cdot 8 = -40.$$

Example 1 Express each of the following in the form a^n, where n is a positive integer.

(a) $a^3 a^4$ (b) $(a^4)^3$ (c) $\dfrac{a^5}{a^2}$, where $a \neq 0$

Solutions   Using (1.17) we have:

(a) $$a^3a^4 = (aaa)(aaaa) = a^7$$

(b) $$(a^4)^3 = (a^4)(a^4)(a^4) = (aaaa)(aaaa)(aaaa) = a^{12}$$

(c) $$\frac{a^5}{a^2} = \frac{aaaaa}{aa} = \frac{aa}{aa} \cdot \frac{aaa}{1} = aaa = a^3$$

Example 2   Express the following in forms which involve only powers of a and b:

(a) $(ab)^4$ (b) $(a/b)^3$, where $b \neq 0$.

Solutions   (a)   Using (1.17) and the Commutative and Associative Properties many times, we have

$$(ab)^4 = (ab)(ab)(ab)(ab)$$
$$= (aaaa)(bbbb) = a^4b^4$$

(b)   Using (1.17) and (1.11) we obtain

$$\left(\frac{a}{b}\right)^3 = \frac{a}{b} \cdot \frac{a}{b} \cdot \frac{a}{b} = \frac{aaa}{bbb} = \frac{a^3}{b^3}$$

The solutions of Examples 1 and 2 are tedious and boring, to say the least. If the exponents were larger, the methods of solution would be even more aggravating. Fortunately, the following rules make such tasks less painful.

(1.18)   Laws of Exponents

If a, b are real numbers and m, n are positive integers, then

(i) $a^m a^n = a^{m+n}$

(ii) $(a^m)^n = a^{mn}$

(iii) $(ab)^n = a^n b^n$

(iv) $\left(\dfrac{a}{b}\right)^n = \dfrac{a^n}{b^n}$, if $b \neq 0$

(v)   If $a \neq 0$, then

$$\frac{a^m}{a^n} = a^{m-n} \text{ for } m > n$$

$$\frac{a^m}{a^n} = \frac{1}{a^{n-m}} \text{ for } n > m$$

$$\frac{a^m}{a^n} = 1 \text{ for } m = n$$

Let us illustrate laws (i), (ii), and (v) by reworking Example 1. Thus

$$a^3 a^4 = a^{3+4} = a^7$$

$$(a^4)^3 = a^{4\cdot 3} = a^{12}$$

$$\frac{a^5}{a^2} = a^{5-2} = a^3$$

Example 2 could be solved in similar fashion using laws (iii) and (iv).

A complete proof of (1.18) requires the method of mathematical induction discussed in Chapter Eleven. However, if we are allowed to count an arbitrary number of factors, then it is easy to supply arguments which establish the laws. To prove (i) we have

$$a^m a^n = \underbrace{a\cdot a\cdots\cdots a}_{m \text{ factors}}\cdot\underbrace{a\cdot a\cdots\cdot a}_{n \text{ factors}}$$

Since the total number of factors a on the right is $m+n$, this expression is equal to a^{m+n}.

Similarly, we could write

$$(a^m)^n = \underbrace{a^m\cdot a^m\cdots\cdot a^m}_{n \text{ factors } a^m}$$

and count the number of times a appears as a factor on the right-hand side. Since $a^m = a\cdot a\cdots\cdot a$ where a occurs as a factor m times, and since the number of such groups of m factors is n, the total number of factors is $m\cdot n$. This gives us (ii) of (1.18).

Laws (iii) and (iv) of (1.18) can be obtained in similar fashion and are left to the reader. Law (v) is clear if $m = n$. For the case $m > n$, the integer $m - n$ is positive and we may write

$$\frac{a^m}{a^n} = \frac{a^n a^{m-n}}{a^n} = \frac{a^n}{a^n}\cdot a^{m-n} = 1\cdot a^{m-n} = a^{m-n}.$$

A similar argument can be used if $n > m$.

We can, of course, use symbols for real numbers other than a and b when applying (1.18), as in the following illustrations.

$$x^5 x^6 = x^{5+6} = x^{11} \qquad (y^5)^7 = y^{5\cdot 7} = y^{35}$$

$$(rs)^7 = r^7 s^7 \qquad \left(\frac{p}{q}\right)^{10} = \frac{p^{10}}{q^{10}}$$

$$\frac{c^8}{c^3} = c^{8-3} = c^5 \qquad \frac{u^3}{u^8} = \frac{1}{u^{8-3}} = \frac{1}{u^5}$$

The Laws of Exponents can be extended to rules such as $a^m a^n a^p = a^{m+n+p}$, $(abc)^n = a^n b^n c^n$, and so on. For convenience we shall also refer to such generalizations as (i) and (iii) of (1.18).

Example 3   Simplify each of the following:

(a) $(3x^3y^4)(4xy^5)$ (b) $(2a^2b^3c)^4$ (c) $\left(\dfrac{2r^3}{s}\right)^2\left(\dfrac{s}{r^3}\right)^3$

Solutions   We shall justify each step by referring to an appropriate property of real numbers.

(a) $(3x^3y^4)(4xy^5) = (3)(4)x^3xy^4y^5$ (commutative and associative properties)

$\qquad\qquad\qquad = 12x^4y^9$ (i) of (1.18)

(b) $\qquad\qquad (2a^2b^3c)^4 = 2^4(a^2)^4(b^3)^4c^4$ (iii) of (1.18)

$\qquad\qquad\qquad\qquad = 16a^8b^{12}c^4$ (ii) of (1.18)

(c) $\left(\dfrac{2r^3}{s}\right)^2\left(\dfrac{s}{r^3}\right)^3 = \left(\dfrac{2^2r^6}{s^2}\right)\left(\dfrac{s^3}{r^9}\right)$ (iv), (iii), and (ii) of (1.18)

$\qquad\qquad\qquad = 2^2\left(\dfrac{r^6}{r^9}\right)\left(\dfrac{s^3}{s^2}\right)$ (properties of quotients)

$\qquad\qquad\qquad = 4\left(\dfrac{1}{r^3}\right)(s)$ (v) of (1.18)

$\qquad\qquad\qquad = \dfrac{4s}{r^3}$ (properties of quotients)

It is possible to extend our work to exponents which are negative integers or 0. If we want (i) of (1.18) to be true for $n = 0$, then $a^m \cdot a^0 = a^{m+0} = a^m$ and, if $a \neq 0$, multiplication by $1/(a^m)$ leads to $a^0 = 1$. Thus, in order to be consistent with our previous development we introduce the following definition.

(1.19)   Definition

> If a is any nonzero real number, then $a^0 = 1$.

Applying (1.19) we see that

$$3^0 = 1, \quad (-14)^0 = 1, \quad (5 - \sqrt{2})^0 = 1$$

and so on. The symbol 0^0 will be left undefined. Thus, whenever we write $x^0 = 1$, it is always assumed that x represents a *nonzero* real number.

Let us next turn our attention to negative exponents. In Section 2 we used the notation $a^{-1} = 1/a$ if $a \neq 0$. If (i) of (1.18) is to be true in this situation, then we must have

$$a^{-2} = a^{-1} \cdot a^{-1} = \frac{1}{a} \cdot \frac{1}{a} = \frac{1}{a^2}$$

$$a^{-3} = a^{-2} \cdot a^{-1} = \frac{1}{a^2} \cdot \frac{1}{a} = \frac{1}{a^3}$$

and so on. It is natural, therefore, to define negative integral exponents as follows.

(1.20) **Definition**

> If a is a nonzero real number and n is a positive integer, then $a^{-n} = \dfrac{1}{a^n}$.

It is possible to show that the Laws of Exponents are valid for all integers m and n, whether positive, negative, or zero; however, we shall not discuss the proofs. If negative exponents are allowed, then (v) of (1.18) may be abbreviated to read

$$\frac{a^m}{a^n} = a^{m-n}, \text{ for } \textit{all} \text{ integers } m \text{ and } n$$

For example,

$$\frac{a^2}{a^5} = a^{2-5} = a^{-3} = \frac{1}{a^3} = \frac{1}{a^{5-2}}$$

In the future it will be assumed that (1.18) is true for all integral exponents and we shall justify steps in proofs by referring to (1.18) even though negative exponents are involved.

To simplify statements, we shall assume that in all problems involving exponents, symbols which appear in denominators represent *nonzero* real numbers.

Example 4 Eliminate negative exponents and simplify:

(a) $(a^{-2}b^3)^{-3}$ (b) $\dfrac{8x^3y^{-5}}{4x^{-1}y^2}$

Solutions (a)

$$(a^{-2}b^3)^{-3} = (a^{-2})^{-3}(b^3)^{-3} \qquad \text{(iii) of (1.18)}$$
$$= a^6 b^{-9} \qquad \text{(ii) of (1.18)}$$
$$= a^6 \cdot \frac{1}{b^9} \qquad \text{(1.20)}$$
$$= \frac{a^6}{b^9} \qquad \text{(properties of quotients)}$$

(b)

$$\frac{8x^3y^{-5}}{4x^{-1}y^2} = \frac{8}{4}\frac{x^3}{x^{-1}}\frac{y^{-5}}{y^2} \qquad \text{(properties of quotients)}$$
$$= 2x^{3-(-1)}y^{-5-2} \qquad \text{(v) of (1.18)}$$
$$= 2x^4y^{-7} \qquad \text{(simplifying)}$$
$$= \frac{2x^4}{y^7} \qquad \text{(1.20)}$$

Another method for simplifying (b) is to multiply numerator and denominator of the given fraction by y^5x, thereby eliminating negative exponents. Thus

$$\frac{8x^3y^{-5}}{4x^{-1}y^2} = \frac{8x^3y^{-5}}{4x^{-1}y^2} \cdot \frac{y^5x}{y^5x} \qquad \text{(identity element)}$$

$$= \frac{8}{4}\frac{x^4y^0}{x^0y^7} \qquad \text{(Why?)}$$

$$= 2\frac{x^4(1)}{(1)y^7} \qquad (1.20)$$

$$= \frac{2x^4}{y^7} \qquad \text{(identity element)}$$

EXERCISES

In Exercises 11–50, the word "simplify" means to replace the given expression by one in which letters representing real numbers appear only once, and no negative exponents occur.

Express the numbers in Exercises 1–10 in the form a/b, where a and b are integers.

1 $\left(-\frac{3}{5}\right)^3$

2 $(-2)^4$

3 $3^2 + (2 \cdot 3)$

4 $(-3)^2 - 2^3$

5 $\left(\frac{4}{5}\right)^{-2}\left(\frac{2}{5}\right)^3$

6 $\dfrac{5^{-2}}{2^{-5}}$

7 $3^0 + 0^3$

8 $2^{-4} + (-4)^2$

9 $\dfrac{3^2 \cdot 5^{-3}}{5 \cdot 3^{-1}}$

10 $(15,678)^0$

Simplify the expressions in Exercises 11–50.

11 $(4a^3)(3a^5)$

12 $(-3b^4)(4b^2)(\frac{1}{6}b^7)$

13 $\dfrac{(2c^7)(3c^2)}{12c^{12}}$

14 $\dfrac{18c^{11}}{(6c^4)^2}$

15 $(4x^2)^2(3x^4)$

16 $(5y^2)(-2y^3)^3$

17 $\dfrac{(4b^2)^3}{(2b^3)^2}$

18 $\dfrac{(6z^2)(2z^2)^4}{(3z^2)^5}$

19 $(8t^3u^5)(2^{-1}tu^{-3})$

20 $(m^4)(-4m^3)(3m^{-2})$

21 $(-2a^2bc^{-4})(5a^{-1}b^3c^5)$

22 $(8x^2y^3)(2x^4y^{-2})(\frac{1}{4}x^{-3}y)$

23 $(-4x^3)^2(8x^4)^{-1}$

24 $(3z^3y^4w^{-9})^0$

25 $\dfrac{(-6a^3)^2}{9a^4}$

26 $\dfrac{(4c^{-3})(-3c^7)}{6c^2}$

27 $(4a^2b^{-1})^3$

28 $\left(\dfrac{3x^2y^3}{4x^3y^2}\right)\left(\dfrac{2xy^4}{6x^5}\right)$

29 $(-2u^{-1}y^2)^3(5u^2y)^{-1}$ 30 $(3x^{-2}yz^3)^4$

31 $(-3ab^2)^3\left(\dfrac{a^5}{9b}\right)^2$ 32 $\dfrac{(10r^{-2}s)^3}{(50r^3s^{-1})^2}$

33 $\left(\dfrac{5t^2v}{4tv^3}\right)^2\left(\dfrac{2tv^3}{t^2v}\right)^4$ 34 $\left(\dfrac{3p}{q^3}\right)^{-2}\left(\dfrac{-q}{2p^2}\right)^3$

35 $\left(\dfrac{16a^2b^{-4}c}{a^5bc^7}\right)^0$ 36 $(a+b)^2(a+b)^{-2}$

37 $(3x^2y^{-3})^{-2}$ 38 $(-4a^3b^{-2})^{-2}$

39 $\dfrac{8u^2v^{-1}}{2u^{-1}v^3}\div\dfrac{u^3}{v^2}$ 40 $(7x^2y^{-5}+3x^{-1}y^2)^0$

41 $(3a+2b)^3(3a+2b)^{-2}$ 42 $(cd^{-1})^{-1}$

43 $[(x^2y^{-1})^2]^{-3}$ 44 $\dfrac{(x^2y^{-3}z)^{-2}}{(x^{-2}yz^3)^{-1}}$

45 $[(a^{-1})^{-1}]^{-1}$ 46 $(p^{-1}q^{-1})^{-1}$

47 $\dfrac{(-2rs^2vp^{-1})^3}{(5r^{-3}sv^2p)^{-2}}$ 48 $\left(\dfrac{y^{-1}}{2x^{-1}}\right)\left(\dfrac{2x}{y}\right)^{-1}$

49 $\dfrac{(xy)^{-1}}{x^{-1}y^{-1}}$ 50 $\left(\dfrac{a^{-2}}{b^{-2}}\right)\left(\dfrac{b}{a}\right)^2$

5   RADICALS

We have used the symbol $\sqrt{2}$ to denote the positive real number whose square is 2. In general, if $a > 0$ and $b > 0$, we write

$$\boxed{\sqrt{b} = a \qquad \text{if and only if } a^2 = b}$$

and call a the **(principal) square root of b.**

Example 1   Find $\sqrt{9}$, $\sqrt{16}$, and $\sqrt{25/49}$.

Solutions   We have

$$\sqrt{9} = 3, \qquad \sqrt{16} = 4, \qquad \text{and} \qquad \sqrt{25/49} = 5/7$$

since

$$3^2 = 9, \qquad 4^2 = 16 \qquad \text{and} \qquad (5/7)^2 = 25/49.$$

Note that if $b > 0$ and $a^2 = b$, then also $(-a)^2 = b$, and hence there are always two real numbers, one positive and one negative, that have b as their square. It is very important to note that $\sqrt{b}$ denotes only the *positive* real number whose square is b.

We also write

$$\sqrt[3]{b} = a \qquad \text{if and only if } a^3 = b$$

In this case if $b > 0$, then $a > 0$, whereas if $b < 0$, then $a < 0$. The number a is called the **principal cube root of b**, or simply the **cube root of b**.

Example 2 Find $\sqrt[3]{8}$, $\sqrt[3]{-8}$, and $\sqrt[3]{27/64}$.

Solutions We have

$$\sqrt[3]{8} = 2, \qquad \sqrt[3]{-8} = -2, \qquad \text{and} \qquad \sqrt[3]{27/64} = 3/4$$

since

$$2^3 = 8, \qquad (-2)^3 = -8, \qquad \text{and} \qquad (3/4)^3 = 27/64.$$

In general we write

(1.21)

$$\sqrt[n]{b} = a \qquad \text{if and only if } a^n = b$$

provided that both a and b are nonnegative real numbers and n is a positive integer, or that both a and b are negative and n is an odd positive integer. The number a is called the **principal nth root** of b. For brevity we usually drop the adjective "principal." Of course, if $n = 2$ we write $\sqrt[2]{b} = \sqrt{b}$.

Example 3 Find $\sqrt[5]{-243}$, $\sqrt[4]{625/81}$, and $\sqrt[6]{64}$.

Solutions By (1.21),

$$\sqrt[5]{-243} = -3, \qquad \sqrt[4]{625/81} = 5/3 \qquad \text{and} \qquad \sqrt[6]{64} = 2$$

since

$$(-3)^5 = -243, \qquad (5/3)^4 = 625/81 \qquad \text{and} \qquad 2^6 = 64.$$

Note that we have not defined $\sqrt[n]{b}$ if $b < 0$ and n is an *even* positive integer. The reason is that if n is even, then $a^n \geq 0$ for every real number a. We shall use the terminology "$\sqrt[n]{b}$ exists" if there is a real number a such that $a^n = b$. In the future, whenever we use the symbols $\sqrt[n]{b}$, $\sqrt[m]{c}$, $\sqrt[p]{x}$, and so on, it is assumed that these roots exist. To complete our terminology, the symbol $\sqrt[n]{b}$, is called a **radical**, the number b is called the **radicand**, and n is the **index** of the radical. The symbol $\sqrt{}$ is called a **radical sign**.

If $\sqrt{b} = a$, then $a^2 = b$; that is, $(\sqrt{b})^2 = b$. Similarly, if $\sqrt[3]{b} = a$, then $a^3 = b$, or $(\sqrt[3]{b})^3 = b$. In general, if n is any positive integer, then

(1.22)
$$(\sqrt[n]{b})^n = b$$

It also follows from (1.21) that if $b > 0$, or if $b < 0$ and n is an odd positive integer, then

(1.23)
$$\sqrt[n]{b^n} = b$$

For example

$$\sqrt{5^2} = 5, \quad \sqrt[3]{(-2)^3} = -2, \quad \sqrt[4]{3^4} = 3$$

and so on.

If b and c are positive, then by a law of exponents and (1.22),

$$(\sqrt{b}\sqrt{c})^2 = (\sqrt{b})^2(\sqrt{c})^2 = bc,$$

which means that

$$\sqrt{bc} = \sqrt{b}\sqrt{c}.$$

Similarly, for all b and c,

$$(\sqrt[3]{b}\sqrt[3]{c})^3 = (\sqrt[3]{b})^3(\sqrt[3]{c})^3 = bc$$

and hence

$$\sqrt[3]{bc} = \sqrt[3]{b}\sqrt[3]{c}.$$

Generally, if n is any positive integer, then

(1.24)
$$\sqrt[n]{bc} = \sqrt[n]{b}\sqrt[n]{c}$$

Similarly

(1.25)
$$\sqrt[n]{\frac{b}{c}} = \frac{\sqrt[n]{b}}{\sqrt[n]{c}}$$

Example 4 Show that (a) $\sqrt{50} = 5\sqrt{2}$; (b) $\sqrt[3]{-108} = -3\sqrt[3]{4}$.

Solutions Applying (1.24), we have

(a)
$$\sqrt{50} = \sqrt{25 \cdot 2} = \sqrt{25}\sqrt{2} = 5\sqrt{2}$$

(b)
$$\sqrt[3]{-108} = \sqrt[3]{(-27)(4)} = \sqrt[3]{-27}\sqrt[3]{4} = -3\sqrt[3]{4}$$

An important type of simplification involving radicals is that of **rationalizing a denominator**. The process involves beginning with a quotient which contains a radical in the denominator and then multiplying numerator and denominator by some expression so that the resulting denominator contains no radicals. The next example illustrates this technique.

Example 5   Rationalize the denominator in each of the following.

(a) $\dfrac{1}{\sqrt{5}}$ (b) $\sqrt{\dfrac{2}{3}}$ (c) $\sqrt[3]{\dfrac{x}{y}}$

Solutions   We may proceed as follows (supply reasons):

(a)
$$\frac{1}{\sqrt{5}} = \frac{1}{\sqrt{5}} \cdot \frac{\sqrt{5}}{\sqrt{5}} = \frac{\sqrt{5}}{(\sqrt{5})^2} = \frac{\sqrt{5}}{5}$$

(b)
$$\sqrt{\frac{2}{3}} = \sqrt{\frac{2}{3} \cdot \frac{3}{3}} = \sqrt{\frac{6}{3^2}} = \frac{\sqrt{6}}{\sqrt{3^2}} = \frac{\sqrt{6}}{3}$$

(c)
$$\sqrt[3]{\frac{x}{y}} = \sqrt[3]{\frac{x}{y} \cdot \frac{y^2}{y^2}} = \frac{\sqrt[3]{xy^2}}{\sqrt[3]{y^3}} = \frac{\sqrt[3]{xy^2}}{y}$$

The proof of the following rule for radicals is left as an exercise. (See Exercise 50.)

(1.26)
$$\boxed{\sqrt[m]{\sqrt[n]{b}} = \sqrt[mn]{b}}$$

where m and n are positive integers.

Example 6   Express each of the following in terms of a single radical.

(a) $\sqrt[3]{\sqrt[4]{b}}$ (b) $\sqrt[3]{\sqrt{64}}$

Solutions   (a)   Using (1.26), with $m = 3$ and $n = 4$,

$$\sqrt[3]{\sqrt[4]{b}} = \sqrt[12]{b}.$$

(b)   Applying (1.26) with $m = 3$ and $n = 2$,

$$\sqrt[3]{\sqrt{64}} = \sqrt[6]{64} = 2.$$

As a check we have

$$\sqrt[3]{\sqrt{64}} = \sqrt[3]{8} = 2.$$

Formulas (1.22)–(1.26) are called the **Laws of Radicals**. We may generalize (1.24) to $\sqrt[n]{xyz} = \sqrt[n]{x}\,\sqrt[n]{y}\,\sqrt[n]{z}$, and so on. In the future we shall also refer to this generalized version as (1.24).

The law stated in the next theorem is a special case of (1.23).

(1.27)   Theorem

> If x is any real number, then $\sqrt{x^2} = |x|$.

Proof.   The theorem is clearly true if $x = 0$. If $x > 0$, then $\sqrt{x^2} = x = |x|$. Finally, if $x < 0$, then $-x > 0$, and since $\sqrt{b^2} = b$ for all $b > 0$, we have $\sqrt{(-x)^2} = -x$. However, $|x| = -x$ if $x < 0$. This gives us $\sqrt{x^2} = \sqrt{(-x)^2} = -x = |x|$, and the theorem is proved.

As an illustration of (1.27), we have

$$\sqrt{(-4)^2} = |-4| = 4.$$

Of course, this equality could also be obtained by writing $\sqrt{(-4)^2} = \sqrt{16} = 4$. Incidentally, we have shown that it is not always true that $\sqrt{x^2} = x$. Indeed, if x is negative, then $\sqrt{x^2} = -x$.

If c is a real number and c^n occurs as a factor in a radical of index n, then c can be removed from the radicand provided the sign of c is taken into account. For example, by (1.24) and (1.23)

$$\sqrt[n]{c^n d} = \sqrt[n]{c^n}\,\sqrt[n]{d} = c\sqrt[n]{d},$$

where we have assumed that the sign of c is such that (1.23) is valid. If, in the preceding equalities, c is negative and $n = 2$, then by (1.27)

$$\sqrt{c^2 d} = \sqrt{c^2}\,\sqrt{d} = |c|\sqrt{d}.$$

A similar situation exists if n is *any* even integer. In order to avoid considering positive and negative cases separately in the examples and exercises to follow, *we shall assume that all letters represent positive real numbers*.

The technique discussed above is particularly useful in simplifying radicals of the form $\sqrt[n]{b}$ where b is an integer. In this case we first obtain the prime factorization of b. If a positive prime power p^n appears as a factor, then p may be taken out from under the radical sign. Several illustrations of this technique were given in Examples 4 and 5. Some additional examples are

$$\sqrt[3]{320} = \sqrt[3]{4^3 \cdot 5} = \sqrt[3]{4^3}\,\sqrt[3]{5} = 4\sqrt[3]{5},$$
$$\sqrt{1400} = \sqrt{2^3 5^2 7} = \sqrt{(2\cdot 5)^2 \cdot 2 \cdot 7} = \sqrt{(2\cdot 5)^2}\,\sqrt{14} = 10\sqrt{14}.$$

A similar method may be used if the radicand contains symbols for unspecified real numbers. For example,

$$\sqrt{x^7} = \sqrt{x^6x} = \sqrt{(x^3)^2x} = \sqrt{(x^3)^2}\sqrt{x} = x^3\sqrt{x}$$
$$\sqrt[4]{x^9y^6} = \sqrt[4]{(x^8y^4)(xy^2)} = \sqrt[4]{(x^2y)^4(xy^2)} = \sqrt[4]{(x^2y)^4}\sqrt[4]{xy^2} = x^2y\sqrt[4]{xy^2}.$$

If we use the term "simplify" when referring to a radical, we mean to proceed as above until the radicand contains no factors with exponent greater than or equal to the index of the radical. Moreover, no fractions should appear under the final radical sign and denominators should be rationalized. The index n should also be as low as possible.

Example 7   Simplify the following.

(a) $\sqrt[3]{16x^3y^8z^4}$ (b) $\sqrt{3a^2b^3}\sqrt{6a^5b}$ (c) $\sqrt{\dfrac{27x^3}{8y^5}}$

Solutions   We may proceed as follows (supply reasons):

(a)
$$\sqrt[3]{16x^3y^8z^4} = \sqrt[3]{(2^3x^3y^6z^3)(2y^2z)}$$
$$= \sqrt[3]{(2xy^2z)^3(2y^2z)}$$
$$= \sqrt[3]{(2xy^2z)^3}\sqrt[3]{2y^2z}$$
$$= 2xy^2z\sqrt[3]{2y^2z}$$

(b)
$$\sqrt{3a^2b^3}\sqrt{6a^5b} = \sqrt{18a^7b^4}$$
$$= \sqrt{(9a^6b^4)(2a)}$$
$$= \sqrt{(3a^3b^2)^2(2a)}$$
$$= \sqrt{(3a^3b^2)^2}\sqrt{2a}$$
$$= 3a^3b^2\sqrt{2a}$$

(c)
$$\sqrt{\frac{27x^3}{8y^5}} = \sqrt{\frac{27x^3}{8y^5}\cdot\frac{2y}{2y}}$$
$$= \sqrt{\frac{54x^3y}{16y^6}}$$
$$= \sqrt{\frac{(9x^2)(6xy)}{(4y^3)^2}}$$
$$= \sqrt{\left(\frac{3x}{4y^3}\right)^2(6xy)}$$
$$= \sqrt{\left(\frac{3x}{4y^3}\right)^2}\sqrt{6xy}$$
$$= \frac{3x}{4y^3}\sqrt{6xy}$$

Radicals with the same index can sometimes be combined by using the Distributive Properties, as illustrated in the next example.

Example 8 Simplify $\sqrt{12} - \sqrt{27} + \sqrt{4x^2y} + \sqrt{y^3}$.

Solution $\sqrt{12} - \sqrt{27} + \sqrt{4x^2y} + \sqrt{y^3} = \sqrt{2^2(3)} - \sqrt{3^3} + \sqrt{(2x)^2(y)} + \sqrt{(y^2)(y)}$

$$= 2\sqrt{3} - 3\sqrt{3} + 2x\sqrt{y} + y\sqrt{y}$$

$$= -\sqrt{3} + (2x + y)\sqrt{y}.$$

EXERCISES

Simplify each of the expressions in Exercises 1–48 (all letters denote positive real numbers).

1 $\sqrt{81}$

2 $\sqrt[3]{-125}$

3 $\sqrt[5]{-64}$

4 $\sqrt[4]{256}$

5 $\dfrac{1}{\sqrt[3]{2}}$

6 $\sqrt{\dfrac{1}{7}}$

7 $\sqrt{192}$

8 $\sqrt[3]{135}$

9 $5\sqrt{20} - \sqrt{45} + 2\sqrt{80}$

10 $\sqrt[4]{162} + \sqrt[4]{32} - \sqrt[4]{2}$

11 $\sqrt{9x^{-4}y^6}$

12 $\sqrt{16a^8b^{-2}}$

13 $\sqrt[3]{8a^6b^{-3}}$

14 $\sqrt[4]{81r^5s^8}$

15 $\sqrt[3]{\dfrac{54a^7}{b^2}}$

16 $\sqrt[5]{\dfrac{-96x^7}{y^3}}$

17 $\sqrt{\dfrac{1}{3u^3v}}$

18 $\sqrt[3]{\dfrac{1}{4x^5y^2}}$

19 $\sqrt[4]{(3x^5y^{-2})^4}$

20 $\sqrt[6]{(2u^{-3}v^4)^6}$

21 $\sqrt[5]{\dfrac{8x^3}{y^4}}\sqrt[5]{\dfrac{4x^4}{y^2}}$

22 $\sqrt{5xy^7}\sqrt{10x^3y^3}$

23 $\sqrt[3]{3t^4v^2}\sqrt[3]{-9t^{-1}v^4}$

24 $\sqrt[3]{(2r-s)^3}$

25 $\sqrt[6]{32p^4y^5z^3}\sqrt[6]{2p^2yz^3}$

26 $(\sqrt{27a^3b^2c})^2$

27 $\sqrt[3]{\dfrac{1}{3x^2}}$

28 $\sqrt[4]{125d\sqrt{25d^6}}$

29 $\sqrt{\dfrac{12x^5}{y^3}}$

30 $\dfrac{\sqrt{6a^5b^3}}{\sqrt{12a^6b}}$

31 $\sqrt{(a-b)^2}$

32 $(\sqrt{8x^5y})^3$

33 $\sqrt[5]{(x^2-5y^3z)^0}$

34 $\sqrt[3]{\dfrac{(3w^{-2}z)^4}{(2wz^{-2})^3}}$

35 $\sqrt[4]{\left(\dfrac{4r^{-2}}{s^8}\right)^{-3}}$

36 $\sqrt{3ab^3c}\,\sqrt{2a^2bc^4}\,\sqrt{6a^3b^4c^3}$

37 $\sqrt{\sqrt{81a^4b^2}}$

38 $\sqrt{\sqrt[3]{64a^7}}$

39 $\left(\sqrt[7]{\dfrac{x^8y^9}{z^2}}\right)^2$

40 $\left(\sqrt[4]{\dfrac{16a^7}{b^3}}\right)^4$

41 $\sqrt{10a^3b}\,\sqrt{2ab^2}$

42 $\sqrt[3]{9xy}\,\sqrt[3]{6x^2y^4}$

43 $\sqrt{\sqrt[3]{a^{12}b}}$

44 $\sqrt[3]{\sqrt[4]{(x^7y^5)^2}}$

45 $\dfrac{1}{\sqrt{a}}\,\dfrac{1}{\sqrt{b}}$

46 $\dfrac{1}{\sqrt{x^2+1}}$

47 $\sqrt{9x^3}+\sqrt{25x^5}-\sqrt{x}$

48 $\sqrt[3]{27a^4b^5}+\sqrt[3]{-8ab^8}$

49 (a) Prove that for all real numbers a and b, $|ab|=|a||b|$. (*Hint:* Write $|ab|=\sqrt{(ab)^2}=\sqrt{a^2b^2}=\sqrt{a^2}\,\sqrt{b^2}$ and use (1.27).)

  (b) Prove that $|a/b|=|a|/|b|$ if $b\neq 0$.

50 Prove (1.26).

6   RATIONAL EXPONENTS

Radicals can be used to introduce rational exponents. Let us begin by defining $b^{1/n}$, where n is a positive integer and $b>0$. This must be done in a manner consistent with our previous work. In particular, if (ii) of (1.18) is to be true, then

$$(b^{1/n})^n = b^{(1/n)n} = b.$$

According to (1.21), with $a=b^{1/n}$, this implies that $b^{1/n}=\sqrt[n]{b}$. We are led, therefore, to the following definition.

(1.28)   Definition

> If b is a real number, n a positive integer, and $\sqrt[n]{b}$ exists, then $b^{1/n}=\sqrt[n]{b}$.

Whenever we use a symbol of the form $b^{1/n}$ in the future, it will be assumed that b and n are chosen so that $\sqrt[n]{b}$ exists.

Let us now consider $b^{m/n}$, where m and n are integers with $n>0$. It is always possible to express a rational number m/n in this way since, for example, we can write $-(m/n)$ as $(-m)/n$. Let us prove that $(b^{1/n})^m=(b^m)^{1/n}$. If $m>0$, then

$$(b^{1/n})^m = (b^{1/n})(b^{1/n})\cdots\cdots(b^{1/n})$$

where there are m factors $b^{1/n}$ on the right. Hence

$$(b^{1/n})^m = \sqrt[n]{b}\,\sqrt[n]{b}\cdot\cdots\cdot\sqrt[n]{b} \qquad (1.28)$$
$$= \sqrt[n]{b\cdot b\cdot\cdots\cdot b} \qquad \text{(ii) of (1.24)}$$
$$= \sqrt[n]{b^m} \qquad (1.17)$$
$$= (b^m)^{1/n} \qquad (1.28)$$

The cases $m \le 0$ are left to the reader. Since $(\sqrt[n]{b})^m = \sqrt[n]{b^m}$, the following definition of rational exponents is meaningful.

(1.29)   Definition

If m/n is a rational number, where n is a positive integer, and if b is a real number such that $\sqrt[n]{b}$ exists, then

$$\boxed{b^{m/n} = (\sqrt[n]{b})^m = \sqrt[n]{b^m}}$$

We may also write (1.29) as follows:

$$b^{m/n} = (b^{1/n})^m = (b^m)^{1/n}.$$

It can be shown that the Laws of Exponents are true for all rational exponents, and henceforth we shall use (1.18) for *rational* as well as integral exponents.

Example 1   Simplify:

(a) $(-27)^{2/3}(4)^{-5/2}$ (b) $(4a^{1/3})(2a^{1/2})$

(c) $(r^2 s^6)^{1/3}$ (d) $\left(\dfrac{2x^{2/3}}{y^{1/2}}\right)^2\left(\dfrac{3x^{-5/6}}{y^{1/3}}\right)$

Solutions   The reader should supply reasons for the following:

(a) $(-27)^{2/3}(4)^{-5/2} = (\sqrt[3]{-27})^2(\sqrt{4})^{-5} = (-3)^2(2)^{-5} = \frac{9}{32}$

(b) $(4a^{1/3})(2a^{1/2}) = 8a^{1/3+1/2} = 8a^{5/6}$

(c) $(r^2 s^6)^{1/3} = (r^2)^{1/3}(s^6)^{1/3} = r^{2/3}s^2$

(d) $\left(\dfrac{2x^{2/3}}{y^{1/2}}\right)^2\left(\dfrac{3x^{-5/6}}{y^{1/3}}\right) = \left(\dfrac{4x^{4/3}}{y}\right)\left(\dfrac{3x^{-5/6}}{y^{1/3}}\right) = \dfrac{12x^{1/2}}{y^{4/3}}$

Rational exponents are often useful for simplifying expressions which involve radicals. The technique is to use (1.28) to transform radicals into expressions with rational exponents, then simplify, and finally change back to radical form. We shall illustrate the procedure in the following example. The reader should supply the reason for each equality.

Example 2   Simplify:

Solutions   (a) $\sqrt[4]{b^2}$ (b) $\dfrac{\sqrt[3]{4}}{\sqrt{2}}$ (c) $\dfrac{\sqrt[4]{a^3}}{\sqrt[3]{a^2}}$ (d) $\sqrt{xy}\sqrt[3]{x^2 y}$

(a)
$$\sqrt[4]{b^2} = (b^2)^{1/4} = b^{1/2} = \sqrt{b}$$

(b)
$$\frac{\sqrt[3]{4}}{\sqrt{2}} = \frac{4^{1/3}}{2^{1/2}} = \frac{4^{2/6}}{2^{3/6}} = \left(\frac{4^2}{2^3}\right)^{1/6} = \left(\frac{16}{8}\right)^{1/6} = 2^{1/6} = \sqrt[6]{2}$$

(c)
$$\frac{\sqrt[4]{a^3}}{\sqrt[3]{a^2}} = \frac{a^{3/4}}{a^{2/3}} = a^{3/4 - 2/3} = a^{1/12} = \sqrt[12]{a}$$

(d)
$$\sqrt{xy}\,\sqrt[3]{x^2 y} = (xy)^{1/2}(x^2 y)^{1/3}$$
$$= x^{1/2}y^{1/2}x^{2/3}y^{1/3}$$
$$= x^{7/6}y^{5/6}$$
$$= xx^{1/6}y^{5/6}$$
$$= x(xy^5)^{1/6}$$
$$= x\sqrt[6]{xy^5}$$

In Chapter Four we shall extend our work to irrational exponents such as $3^{\sqrt{2}}$, 5^{π}, etc.

In scientific areas it is not unusual to work with numbers that are very large or very small. To simplify matters it is customary to write such numbers in the form $a \cdot 10^n$, where a is expressed as a decimal between 1 and 10 and n is an integer. This is referred to as the **scientific form** for real numbers. In applications it is customary to use $\times$ for the multiplication symbol and write $a \times 10^n$ instead of $a \cdot 10^n$. As an illustration, the distance a ray of light travels in one year is approximately 5,900,000,000,000 miles. This may be written in scientific form as 5.9×10^{12}. The positive exponent 12 indicates that the decimal point should be moved 12 places to the *right*. The notation works equally well for small numbers. To illustrate, it is estimated that the weight of an oxygen molecule is 0.000000000000000000000053 grams or, in scientific form, 5.3×10^{-23} grams. The negative exponent indicates that the decimal point should be moved 23 places to the *left*. Some other illustrations of scientific form are

$$513 = 5.13 \times 10^2, \quad 20{,}700 = 2.07 \times 10^4, \quad 92{,}000{,}000 = 9.2 \times 10^7$$
$$0.000648 = 6.48 \times 10^{-4}, \quad 0.00000000043 = 4.3 \times 10^{-10}$$

There are a number of advantages to this notation. It is compact, and it enables the reader to see quickly (without counting zeros) the relative magnitudes of large or small quantities. It may also be used to simplify calculations which involve quantities such as those illustrated in the next two examples.

Example 3   Calculate $(0.000000004)(320,000,000,000)$.

Solution   Introducing scientific form, the given product may be found as follows:

$$(4 \times 10^{-9})(3.2 \times 10^{11}) = (4)(3.2) \times 10^{-9+11}$$
$$= 12.8 \times 10^{2}$$
$$= 1.28 \times 10^{3}$$

Example 4   Calculate $\dfrac{(1,100,000)^2 \sqrt{0.00000004}}{(8,000,000,000)^{2/3}}$.

Solution   Writing each number in scientific form and using the Laws of Exponents, we have

$$\frac{(1.1 \times 10^6)^2 (4 \times 10^{-8})^{1/2}}{(8 \times 10^9)^{2/3}} = \frac{[(1.1)^2 \times 10^{12}][4^{1/2} \times 10^{-4}]}{8^{2/3} \times 10^6}$$
$$= \frac{[1.21 \times 10^{12}][2 \times 10^{-4}]}{4 \times 10^6}$$
$$= \frac{(1.21)(2)}{4} \times 10^{12-4-6}$$
$$= (0.605) \times 10^2$$
$$= 6.05 \times 10$$

Hand-held calculators often employ scientific notation in their display panels. In this case the number 10 used in the scientific notation $a \times 10^n$ is suppressed and only the exponent is shown. For example if we wanted to find $(4,500,000)^2$ on a typical calculator, we would enter the integer 4,500,000 and press the x^2 (or squaring) key . The display panel would show

which is translated as 2.025×10^{13}. Thus,

$$(4,500,000)^2 = 20,250,000,000,000.$$

Many calculators also allow the entry of a number in scientific form. The owner of a calculator should consult the user's manual for details.

EXERCISES

Rewrite the expressions in Exercises 1–6 in terms of fractional exponents.

1 $\sqrt[4]{x^3}$ 　　　　　　　　　　　**2** $\sqrt[3]{x^5}$

3 $\sqrt[3]{(a+b)^2}$ 　　　　　　　　**4** $\sqrt{a+\sqrt{b}}$

5 $\sqrt{x^2+y^2}$ 　　　　　　　　　**6** $\sqrt[3]{r^3-s^3}$

Rewrite the expressions in Exercises 7 and 8 in terms of radicals.

7 (a) $4x^{3/2}$ 　　(b) $(4x)^{3/2}$ 　　(c) $4+x^{3/2}$ 　　(d) $(4+x)^{3/2}$

8 (a) $8y^{1/3}$ 　　(b) $(8y)^{1/3}$ 　　(c) $8-y^{1/3}$ 　　(d) $(8-y)^{1/3}$

Simplify the expressions in Exercises 9–38.

9 $(-64)^{2/3}$ 　　　　　　　　　　**10** $16^{3/2}$

11 $(0.0009)^{3/2}$ 　　　　　　　　**12** $(0.027)^{-1/3}$

13 $(256)^{-3/4}$ 　　　　　　　　　**14** $(-243)^{-2/5}$

15 $(3x^{2/3})(5x^{4/3})$ 　　　　　　**16** $(2u^{5/2})(6u^{1/2})$

17 $(2a^{1/3})(9a^{1/2})$ 　　　　　　**18** $y^{1/6}(8y^{1/3})$

19 $(16t^4)^{-3/2}$ 　　　　　　　　**20** $(-8w^6)^{2/3}$

21 $(3c^{2/3})(4c)^{3/2}$ 　　　　　　**22** $(6x^{1/3}y^{3/2})^2$

23 $\left(\dfrac{16a^8}{p^{-4}}\right)^{3/4}$ 　　　　　　　**24** $\left(\dfrac{w^{-1/3}}{w^{3/2}}\right)^6$

25 $(p^{-3}s^9r^{-6})^{-2/3}$ 　　　　　**26** $\dfrac{(x^{-4}y)^{-1/2}}{(x^2y^3)^{-1/3}}$

27 $\left(\dfrac{r^{1/2}}{s^2}\right)^4\left(\dfrac{s^{-1/3}}{r^{2/3}}\right)^3$ 　　　**28** $\left(\dfrac{1}{4x}\right)^{1/2}$

29 $u^{1/3}u^{1/4}u^{1/6}$ 　　　　　　**30** $a^{1/2}a^{1/3}a^{1/6}$

31 $\left(\dfrac{x^{-2/3}z^{4/3}}{x^{4/5}z^{-1/4}}\right)^0$ 　　　**32** $\dfrac{(a^{-1/5}b^{5/2})^{10}}{(a^{2/3}b^{-3/5})^{15}}$

33 $(4a^2b^{2/3}c^{-1})^{1/2}(3a^{-2}b^{1/3}c^3)^2$ 　　**34** $\left(\dfrac{p^{-2/3}q^{5/6}}{p^{3/4}q^{7/2}}\right)^{12}$

35 $(5x^{3/2}(xy^{-2})^{1/2})^4$ 　　　　**36** $((x^4y^{-3})^{1/5})^{-10}$

37 $a^{1/2}(a^{1/2}+a^{-1/2})$ 　　　　**38** $s^{1/3}(s^{2/3}-s^{5/3})$

Write the expressions in Exercises 39–52 as a radical with least possible index.

39 $\sqrt[6]{x^3}$ 　　　　　　　　　　**40** $\sqrt[4]{u^2}$

41 $\sqrt[4]{c^2d^2}$ 　　　　　　　　**42** $\sqrt[9]{c^3}$

43 $\sqrt[8]{a^4b^2}$ 　　　　　　　　**44** $\sqrt[6]{r^3s^9}$

45 $\sqrt[3]{4}\sqrt{2}$ 　　　　　　　　**46** $\sqrt{5}\sqrt[3]{2}$

47 $\sqrt[3]{x^2 y}\sqrt{xy}$ **48** $\sqrt[3]{xy^2}\sqrt[4]{x^3 y}$

49 $\dfrac{\sqrt[3]{u^2 v}}{\sqrt[6]{u^5 v^4}}$ **50** $\dfrac{\sqrt{ab^3}}{\sqrt[3]{a^2 b}}$

51 $\sqrt[3]{b^2 \sqrt{b}}$ **52** $\sqrt{z\sqrt[3]{z}}$

53 The mass of a hydrogen atom is approximately 0.00000000000000000000000017 grams. Express this number in scientific form.

54 The mass of an electron is approximately 9.1×10^{-31} kilograms. Express this number in decimal form.

Express the numbers in Exercises 55 and 56 in scientific form.

55 (a) 427,000   (b) 0.000000098   (c) 810,000,000

56 (a) 85,200   (b) 0.0000055   (c) 24,900,000

Express the numbers in Exercises 57 and 58 in decimal form.

57 (a) 8.3×10^5 (b) 2.9×10^{-12} (c) 5.63×10^8

58 (a) 2.3×10^7 (b) 7.01×10^{-9} (c) 1.23×10^{10}

Use scientific form to find the numbers in Exercises 59–64.

59 $(2,100,000,000)(0.0000000033)$ **60** $(840,000,000)(0.00000021)$

61 $\dfrac{\sqrt{81,000,000}}{(0.0000002)^3}$ **62** $\dfrac{(300,000)^4}{(8,000,000)^{2/3}}$

63 $\dfrac{\sqrt[5]{(0.00243)^2 (20,000)^3}}{\sqrt[3]{8,000,000}}$ **64** $\dfrac{(160,000)^{3/2}(12,100,000)}{(0.0000011)^2}$

65 In astronomy distances to stars are measured in light years, where 1 light year is the distance a ray of light travels in one year. If the speed of light is 186,000 miles per second, approximate 1 light year.

66 (a) It is estimated that the Milky Way galaxy contains 100 billion stars. Express this number in scientific form.

    (b) The diameter d of the Milky Way galaxy is estimated as 100,000 light years. Express d in miles (refer to Exercise 65).

7   POLYNOMIALS AND ALGEBRAIC EXPRESSIONS

We frequently make use of symbols to denote arbitrary elements of a set. For example, we may use x to denote a real number, although no *particular* real number is specified. A letter which is used to represent any element of a given set is sometimes called a **variable**. Throughout this text, unless otherwise specified, variables will represent real numbers. In some cases it is necessary to restrict the numbers which are represented by a variable. We shall use the following terminology.

> The **domain of a variable** is the set of real numbers represented by the variable.

The domain of a variable x is often referred to as the set of "permissible" or "allowable" values for x. To illustrate, given the expression $\sqrt{x}$, we note that in order to obtain a real number we must have $x \geq 0$, and hence in this case the domain of x is assumed to be the set of nonnegative real numbers. Similarly, when working with the expression $1/(x - 2)$ we must exclude $x = 2$ (Why?) and consequently we take the domain of x as the set of all real numbers different from 2.

If we begin with any collection of variables and real numbers, then an **algebraic expression** is the result obtained by applying additions, subtractions, multiplications, divisions, or the taking of roots. The following are examples of algebraic expressions:

$$x^3 - 2x + \frac{3^{1/9}}{\sqrt{2x}}, \quad \frac{2xy + 3x}{y - 1}, \quad \frac{4yz^{-2} + \left(\dfrac{-7}{x + w}\right)^5}{\sqrt[3]{y^2} + \sqrt{5z}}$$

where x, y, z, and w are variables. If specific numbers are substituted for the variables in an algebraic expression, the resulting real number is called the **value** of the expression for these numbers. For example, the value of the second expression above when $x = -2$ and $y = 3$ is

$$\frac{2(-2)(3) + 3(-2)}{3 - 1} = \frac{-12 - 6}{2} = -9.$$

When we work with algebraic expressions, it will be assumed that the domains are chosen so that variables do not represent numbers which make the expressions meaningless. Thus it is assumed that denominators do not vanish, roots always exist, and so on. For example, if we wish to restrict ourselves to the real number system when working with the first expression above, then it is necessary to have $x > 0$. In the second expression we do not allow $y = 1$. To simplify our work, we shall not always state the domains of variables. If meaningless expressions occur when certain numbers are substituted for the variables, then these numbers are *not* in the domains of the variables.

Certain algebraic expressions are given special names. If x is a variable, then a **monomial** in x is an expression of the form ax^n, where a is a real number and n is a nonnegative integer. The number a is called the **coefficient** of x^n. A **polynomial** in x is any finite sum of monomials in x. Some examples are

$$3x^4 + 5x^3 + (-7)x + 2, \quad x^8 + 7x^2 + (-8)x, \quad 4x^2 + 1.$$

The **general form** of a polynomial in x is

(1.30)

$$a_n x^n + a_{n-1} x^{n-1} + \cdots + a_1 x + a_0$$

where each coefficient a_i is a real number. For each k, $a_k x^k$ is called a **term** of the polynomial. The coefficient a_n of the highest power of x in (1.30) is the **leading coefficient** of the polynomial and, if $a_n \neq 0$, we say that the polynomial has **degree** n. The degrees of the three examples listed above are 4, 8, and 2, respectively. As we have indicated, if a coefficient a_i is zero, we often abbreviate the expression (1.30) by deleting the term $a_i x^i$. If *all* the coefficients of a polynomial are zero, it is called the **zero polynomial** and is denoted by 0. It is customary not to assign a degree to the zero polynomial.

If some of the coefficients are negative, then for convenience we usually use minus signs between appropriate terms. To illustrate, instead of $3x^2 + (-5)x + (-7)$, we write $3x^2 - 5x - 7$ for this polynomial of degree 2. Polynomials in other variables may also be considered. For example, $\frac{2}{3}z^2 - 3z^7 + 8 - \sqrt{5}z^4$ is a polynomial in z of degree 7. We ordinarily arrange the terms in order of decreasing powers of the variable and write $-3z^7 - \sqrt{5}z^4 + \frac{2}{3}z^2 + 8$.

A polynomial in x may be thought of as an algebraic expression obtained by employing only additions, subtractions, and multiplications involving x. In particular, the expressions

$$\frac{1}{x} + 3x, \quad \frac{x-5}{x^2+2}, \quad 3x^2 + \sqrt{x} - 2$$

are not polynomials since they also involve divisions by variables, or roots involving variables.

Since polynomials, and the monomials which make up polynomials, are symbols representing real numbers, all of the rules given in Section 2 can be applied. The principal difference is that symbols such as a, b, and c used in Section 2 will now be replaced by polynomials and other algebraic expressions. If additions, multiplications, and subtractions are carried out with polynomials, we ordinarily simplify the result by using various properties of real numbers.

Example 1   Find the sum of the polynomials $x^3 + 2x^2 - 5x + 7$ and $4x^3 - 5x^2 + 3$.

Solution   Rearranging terms and using the Distributive Properties gives us

$$(x^3 + 2x^2 - 5x + 7) + (4x^3 - 5x^2 + 3)$$
$$= x^3 + 4x^3 + 2x^2 - 5x^2 - 5x + 3 + 7$$
$$= (1 + 4)x^3 + (2 - 5)x^2 + (-5)x + (3 + 7)$$
$$= 5x^3 - 3x^2 - 5x + 10$$

Example 1 illustrates the fact that the sum of any two polynomials in x can be obtained by adding coefficients of like powers of x. In order to keep track of the coefficients it is sometimes convenient to use the following scheme for finding the sum.

$$x^3 + 2x^2 - 5x + 7$$
$$4x^3 - 5x^2 \qquad + 3$$
$$\overline{5x^3 - 3x^2 - 5x + 10}$$

The difference of two polynomials is found by subtracting coefficients of like powers, as indicated by the next example.

Example 2   Subtract $4x^3 - 5x^2 + 3$ from $x^3 + 2x^2 - 5x + 7$.

Solution   We could use the scheme shown previously and subtract, or proceed as follows:

$$(x^3 + 2x^2 - 5x + 7) - (4x^3 - 5x^2 + 3)$$
$$= x^3 + 2x^2 - 5x + 7 - 4x^3 + 5x^2 - 3$$
$$= x^3 - 4x^3 + 2x^2 + 5x^2 - 5x + 7 - 3$$
$$= (1 - 4)x^3 + (2 + 5)x^2 - 5x + (7 - 3)$$
$$= -3x^3 + 7x^2 - 5x + 4$$

The intermediate steps in the previous solution were used for completeness. After the student becomes proficient with such manipulations, these may be omitted. In order to multiply two polynomials, we merely use the distributive law together with laws of exponents and combine like terms.

Example 3   Find the product of $x^2 + 5x - 4$ and $2x^3 + 3x - 1$.

Solution
$$(x^2 + 5x - 4)(2x^3 + 3x - 1)$$
$$= x^2(2x^3 + 3x - 1) + 5x(2x^3 + 3x - 1) - 4(2x^3 + 3x - 1)$$
$$= 2x^5 + 3x^3 - x^2 + 10x^4 + 15x^2 - 5x - 8x^3 - 12x + 4$$
$$= 2x^5 + 10x^4 + (3 - 8)x^3 + (-1 + 15)x^2 + (-5 - 12)x + 4$$
$$= 2x^5 + 10x^4 - 5x^3 + 14x^2 - 17x + 4$$

For convenience, the work above may be arranged as follows.

$$2x^3 + 3x - 1$$
$$\underline{x^2 + 5x - 4}$$

$2x^5$	$+ 3x^3 - x^2$	$=$	$x^2(2x^3 + 3x - 1)$
$10x^4$	$+ 15x^2 - 5x$	$=$	$5x(2x^3 + 3x - 1)$
	$-8x^3 \qquad - 12x + 4 =$		$-4(2x^3 + 3x - 1)$
$2x^5 + 10x^4 - 5x^3 + 14x^2 - 17x + 4 =$			sum of above

We may also consider polynomials in more than one variable. For example, a polynomial in two variables x and y is a sum of terms, each of the form $ax^m y^k$ for some real number a and nonnegative integers m and k. An example is

$$3x^4y + 2x^3y^5 + 7x^2 - 4xy + 8y - 5.$$

In like manner, we may consider polynomials in three variables x, y, z or, for that matter, in *any* number of variables. Addition, subtraction, and multiplication are performed using properties of real numbers. A simple illustration is given in the next example.

Example 4   Find the product of $x^2 + xy + y^2$ and $x - y$.

Solution

$$(x^2 + xy + y^2)(x - y) = (x^2 + xy + y^2)x - (x^2 + xy + y^2)y$$
$$= x^3 + x^2y + xy^2 - x^2y - xy^2 - y^3$$
$$= x^3 - y^3$$

We could also arrange our work as follows:

$$
\begin{array}{l}
x^2 + xy + y^2 \\
\underline{x \;\; - \;\; y} \\
x^3 + x^2y + xy^2 \qquad\quad = x(x^2 + xy + y^2) \\
\quad\;\; - x^2y - xy^2 - y^3 \;\; = (-y)(x^2 + xy + y^2) \\
\hline
x^3 \qquad\qquad\qquad - y^3 \;= \;\text{sum of the above}
\end{array}
$$

Division by a monomial is relatively easy, as seen in the next example.

Example 5   Divide $6x^2y^3 + 4x^3y^2 - 10xy$ by $2xy$.

Solution   Using properties of quotients and Laws of Exponents we obtain

$$\frac{6x^2y^3 + 4x^3y^2 - 10xy}{2xy} = \frac{6x^2y^3}{2xy} + \frac{4x^3y^2}{2xy} - \frac{10xy}{2xy}$$
$$= 3xy^2 + 2x^2y - 5.$$

Certain products occur so frequently in algebra that they deserve special attention. We list some of these in (1.31), where the letters represent real numbers. The reader should check the validity of each formula by actually carrying out the multiplications.

(1.31)   **Product Formulas**

(i)	$(x + y)(x - y) = x^2 - y^2$
(ii)	$(ax + b)(cx + d) = acx^2 + (ad + bc)x + bd$
(iii)	$(x + y)^2 = x^2 + 2xy + y^2$
(iv)	$(x - y)^2 = x^2 - 2xy + y^2$
(v)	$(x + y)^3 = x^3 + 3x^2y + 3xy^2 + y^3$
(vi)	$(x - y)^3 = x^3 - 3x^2y + 3xy^2 - y^3$

Since the symbols x and y used in (1.31) represent real numbers, they may be replaced by algebraic expressions, as illustrated in the next example.

Example 6   Find the products:

(a) $(2r^2 - \sqrt{s})(2r^2 + \sqrt{s})$ (b) $\left(\sqrt{c} + \dfrac{1}{\sqrt{c}}\right)^2$ (c) $(2a - 5b)^3$

Solutions   (a)   Using (i) of (1.31) with $x = 2r^2$ and $y = \sqrt{s}$,

$$(2r^2 - \sqrt{s})(2r^2 + \sqrt{s}) = (2r^2)^2 - (\sqrt{s})^2$$
$$= 4r^4 - s.$$

(b)   Using (iii) of (1.31) with $x = \sqrt{c}$ and $y = 1/\sqrt{c}$,

$$\left(\sqrt{c} + \frac{1}{\sqrt{c}}\right)^2 = (\sqrt{c})^2 + 2\sqrt{c} \cdot \frac{1}{\sqrt{c}} + \left(\frac{1}{\sqrt{c}}\right)^2$$

$$= c + 2 + \frac{1}{c}.$$

(c)   Applying (vi) of (1.31) with $x = 2a$ and $y = 5b$,

$$(2a - 5b)^3 = (2a)^3 - 3(2a)^2(5b) + 3(2a)(5b)^2 - (5b)^3$$
$$= 8a^3 - 60a^2b + 150ab^2 - 125b^3.$$

EXERCISES

In Exercises 1–22 perform the indicated operations.

1   $(4x^3 + 2x^2 - x + 5) + (x^3 - 3x^2 - 5x + 1)$

2   $(x^4 - 3x^2 + 7x + 4) + (x^3 + 3x^2 - 4x - 3)$

3   $(5x^4 - 6x^2 + 9x) - (2x^3 + 3x^2 - 8x + 4)$

4   $(2x^2 - 11x + 13) - (3x^4 + 9x^2 - 10)$

5   $(5y^3 - 6y^2 + y - 7) - (5y^3 + 6y^2 + y + 2)$

6   $(4z^2 - 4z + 1) - (2z + 1)^2$

7   $(2a^4 - 3a^2 + 5) + a(a^3 + 3a - 4)$

8   $(3u + 1)(2u - 3) + 6u(u + 5)$

9   $(3x - 4)(2x^2 + x - 5)$

10   $(4x^3 - x^2 - 7)(3x + 2)$

11   $(r^2 + 2r + 3)(3r^2 - 2r + 4)$

12   $(s + t)(s^2 - st + t^2)$

13 $(6x^3 - 3x^2 - x + 7)(2x^2 + 4x + 5)$ 14 $(2x^2 - xy + y^2)(3x - y)$

15 $(r - t)(r^2 + rt + t^2)$ 16 $(x + y)(x^3 - x^2y + xy^2 - y^3)$

17 $(3x + 1)(2x^2 - x + 2)(x^2 + 4)$ 18 $(3c + 1)(2c^2 + 5)(c^3 + 4)$

19 $\dfrac{8x^2y^3 - 10x^3y}{2x^2y}$ 20 $\dfrac{6a^3b^3 - 9a^2b^2 + 3ab^4}{3ab^2}$

21 $\dfrac{3u^3v^4 - 2u^5v^2 + (u^2v^2)^2}{u^3v^2}$ 22 $\dfrac{6x^2yz^3 - xy^2z}{xyz}$

In Exercises 23–50, use (1.31) to find the indicated products.

23 $(x - 3)(2x + 1)$ 24 $(3x + 2)(3x - 5)$

25 $(2s - 7t)(4s - 5t)$ 26 $(8n - 6p)(7n - 10p)$

27 $(5x^2 + 2y)(3x^2 - 7y)$ 28 $(x + 9y^2)(3x - 4y^2)$

29 $(6t - 5v)(6t + 5v)$ 30 $(8u + 3)^2$

31 $(3r + 10s)^2$ 32 $(4v - 3w)(4v + 3w)$

33 $(4x^2 - 5y^2)^2$ 34 $(10p^2 + 7q^2)^2$

35 $(x - x^{-1})^2$ 36 $(b^3 - b^{-3})^2$

37 $(\sqrt{a} + \sqrt{b})(\sqrt{a} - \sqrt{b})$ 38 $(3\sqrt{s} + \sqrt{t})^2$

39 $(x - 2y)^3$ 40 $(4x - y)^3$

41 $(3r + 4s)^3$ 42 $(2a + 5b)^3$

43 $(x^2 + y^2)^3$ 44 $(u^2 - 3v)^3$

45 $(a^{1/3} - b^{1/3})^3$ 46 $(a + b)^2(a - b)^2$

47 $(x + y + z)(x + y - z)$ 48 $(2a - b + 3c)(2a - b - 3c)$

49 $(3x + 2y + z)^2$ 50 $(x^2 + y^2 + z^2)^2$

8 FACTORING

If a polynomial is written as a product of other polynomials, then each polynomial in the product is called a **factor** of the original polynomial. The process of expressing a polynomial as a product is called **factoring**. For example, since $x^2 - 9 = (x + 3)(x - 3)$, we see that $x + 3$ and $x - 3$ are factors of $x^2 - 9$.

Factoring plays a major role in numerous mathematical applications, since it may be used to reduce the study of a complicated expression to the study of several simpler expressions. For example, important properties of the polynomial $x^2 - 9$ can be determined by examining the factors $x + 3$ and $x - 3$. As another illustration, it can be shown that

$$6x^3 + 37x^2 + 47x - 20 = (2x + 5)(3x - 1)(x + 4).$$

Various properties of the indicated third-degree polynomial can then be found by studying the factors $2x + 5$, $3x - 1$, and $x + 4$. In future chapters you will encounter many specific problems which can be simplified by factoring. This will be especially true when you are finding solutions of equations and inequalities, or studying the behavior of functions.

Any polynomial has, as a factor, *every* nonzero real number c. As an illustration, given $3x^2 + 5x + 2$ and any nonzero real number c, we can write

$$3x^2 + 5x + 2 = c\left(\frac{3}{c}x^2 + \frac{5}{c}x + \frac{2}{c}\right).$$

A factor c of this type is called a **trivial factor**. We shall be interested primarily in **nontrivial factors** of polynomials; that is, factors which contain polynomials of degree greater than zero. An exception to this rule is that if the coefficients are restricted to *integers*, then it is customary to remove a common integral factor from each term of the polynomial. This can be done by means of the distributive law, as in the factorization

$$4x^2y + 8z^3 = 4(x^2y + 2z^3).$$

Before carrying out factorizations of polynomials it is necessary to specify the system from which the coefficients of the factors are to be chosen. In this chapter we shall use the rule that *if a polynomial with integer coefficients is given, then the factors should be polynomials with integer coefficients*. For example,

$$x^2 + x - 6 = (x + 3)(x - 2),$$
$$4x^2 - 9y^2 = (2x - 3y)(2x + 3y).$$

An integer $a > 1$ is prime if it cannot be written as a product of two positive integers greater than 1. A polynomial is said to be **prime** or **irreducible** if it cannot be written as a product of two polynomials of positive degree. When factoring a polynomial the objective is to express it as a product of prime polynomials or powers of prime polynomials. According to the rule given above, $x^2 - 2$ is prime since it cannot be factored as a product of two polynomials of positive degree which have *integral* coefficients. If we allow the factors to have *real* coefficients, then $x^2 - 2$ is not prime, since

$$x^2 - 2 = (x + \sqrt{2})(x - \sqrt{2}).$$

However, as we have said, we shall not allow irrational coefficients in the factorizations made in this chapter. Later, in our work with solutions of equations, this rule will be rescinded.

In general, it is very difficult to factor polynomials of degree greater than 2. In simple cases some of the Product Formulas given in Section 7 are useful. One of the most important formulas is (i) of (1.31) for **the difference of two squares** which we restate as follows:

(1.32)

$$\boxed{a^2 - b^2 = (a + b)(a - b)}$$

The next example illustrates several applications of this formula.

Example 1   Factor each of the following.

(a) $25r^2 - 49s^2$ (b) $81x^4 - y^4$ (c) $16x^4 - (y - 2z)^2$

Solutions   (a)   Applying (1.32) with $a = 5r$ and $b = 7s$ gives us

$$25r^2 - 49s^2 = (5r)^2 - (7s)^2 = (5r + 7s)(5r - 7s).$$

(b)   We make two applications of (1.32) as follows:

$$81x^4 - y^4 = (9x^2)^2 - (y^2)^2$$
$$= (9x^2 + y^2)(9x^2 - v^2)$$
$$= (9x^2 + y^2)(3x + y)(3x - y)$$

(c)   If we write $16x^4 = (4x^2)^2$, then we may use (1.32) with $a = 4x^2$ and $b = y - 2z$ as follows:

$$16x^4 - (y - 2z)^2 = (4x^2)^2 - (y - 2z)^2$$
$$= [(4x^2) + (y - 2z)][(4x^2) - (y - 2z)]$$
$$= (4x^2 + y - 2z)(4x^2 - y + 2z)$$

A factorization of a second-degree polynomial $px^2 + qx + r$ where p, q, and r are integers, must be of the form $(ax + b)(cx + d)$ where a, b, c, and d are integers. It follows that $ac = p$, $bd = r$, and $ad + bc = q$. Evidently there are only a limited number of choices for a, b, c, and d which satisfy these conditions. If none of the choices work, then $px^2 + qx + r$ is prime. This method is also applicable to polynomials of the form $px^2 + qxy + ry^2$.

Example 2   Factor the following:

(a) $6x^2 - 7x - 3$ (b) $4x^2 - 12xy + 9y^2$ (c) $4x^4y - 11x^3y^2 + 6x^2y^3$

Solutions   (a)   If we write

$$6x^2 - 7x - 3 = (ax + b)(cx + d),$$

then the product of a and c is 6; the product of b and d is -3; and $ad + bc = -7$. Trying various possibilities, we arrive at the factorization

$$6x^2 - 7x - 3 = (2x - 3)(3x + 1).$$

(b)   If a factorization as a product of two first-degree polynomials exists, then it must be of the form

$$4x^2 - 12xy + 9y^2 = (ax + by)(cx + dy).$$

By trial we obtain

$$4x^2 - 12xy + 9y^2 = (2x - 3y)(2x - 3y) = (2x - 3y)^2.$$

(c)   Since each term has x^2y as a factor, we begin by writing

$$4x^4y - 11x^3y^2 + 6x^2y^3 = x^2y(4x^2 - 11xy + 6y^2).$$

Then by trial we obtain the following factorization:

$$4x^4y - 11x^3y^2 + 6x^2y^3 = x^2y(4x - 3y)(x - 2y).$$

In some cases if terms in a sum are grouped in suitable fashion, then a factorization can be found by means of the Distributive Properties (1.5), as illustrated in the next example.

Example 3   Factor the following:

(a) $a(x^2 - y) + 2b(x^2 - y)$ (b) $4ac + 2bc - 2ad - bd$

(c) $3x^3 + 2x^2 - 12x - 8$

Solutions   (a)   The given expression is of the form $ac + 2bc$ with $c = x^2 - y^2$. Since $ac + 2bc = (a + 2b)c$, we have

$$a(x^2 - y) + 2b(x^2 - y) = (a + 2b)(x^2 - y).$$

(b)   We group the first two terms and the last two terms, and then use the Distributive Properties as follows:

$$\begin{aligned} 4ac + 2bc - 2ad - bd &= (4ac + 2bc) - (2ad + bd) \\ &= 2c(2a + b) - d(2a + b) \\ &= (2c - d)(2a + b) \end{aligned}$$

(c)   We again begin by grouping the first two terms and the last two terms of the given expression as follows:

$$3x^3 + 2x^2 - 12x - 8 = x^2(3x + 2) - 4(3x + 2).$$

The right-hand side has the form $x^2c - 4c$ where $c = 3x + 2$. Since $x^2c - 4c = (x^2 - 4)c$, we have

$$3x^3 + 2x^2 - 12x - 8 = (x^2 - 4)(3x + 2).$$

Finally, using (1.32) we obtain the factorization

$$3x^3 + 2x^2 - 12x - 8 = (x + 2)(x - 2)(3x + 2).$$

The technique illustrated in Example 3 is called **factorization by grouping**. Each of the next two formulas may be verified by multiplying the two factors on the right-hand side of the equation.

$$x^3 + y^3 = (x + y)(x^2 - xy + y^2)$$
$$x^3 - y^3 = (x - y)(x^2 + xy + y^2)$$

(1.33)

The expressions on the left in (1.33) are referred to as the **sum of two cubes** and the **difference of two cubes** respectively. The student should pay careful attention to where the plus or minus signs occur.

Example 4   Factor the following:

(a) $a^3 + 64b^3$ (b) $8x^6 - 27y^9$

Solutions   (a)   By (1.33), with $x = a$ and $y = 4b$,

$$a^3 + 64b^3 = a^3 + (4b)^3$$
$$= (a + 4b)[a^2 - a(4b) + (4b)^2]$$
$$= (a + 4b)(a^2 - 4ab + 16b^2).$$

(b)   By (1.33),

$$8x^6 - 27y^9 = (2x^2)^3 - (3y^3)^3$$
$$= (2x^2 - 3y^3)[(2x^2)^2 + (2x^2)(3y^3) + (3y^3)^2]$$
$$= (2x^2 - 3y^3)(4x^4 + 6x^2y^3 + 9y^6).$$

EXERCISES

Factor each of the following.

1   $rs + 4st$

2   $4u^2 - 2uv$

3   $3a^2b^2 - 6a^2b$

4   $10xy + 15xy^2$

5   $9x^2y^2 + 15xy^4$

6   $-8p^4qr^2 - 4p^3q^3r^2$

7   $4x^2 + 5x - 6$

8   $21x^2 + 29x - 10$

9   $6c^2 + 25c + 24$

10   $20y^2 - 41y + 20$

11   $4r^2 - 25t^2$

12   $36a^2 - 49b^2$

13   $50x^2 + 45xy - 18y^2$

14   $45x^2 + 38xy + 8y^2$

15   $16w^4 - 9s^2$

16   $25p^2 - 16v^4$

17   $36z^2 + 60z + 25$

18   $64y^2 + 112y + 49$

19   $27x^3 - y^3$

20   $x^3 + 8y^3$

21   $27a^3 + 64b^3$

22   $8r^3 - 27s^3$

23   $8x^6 - 125$

24   $216 - y^6$

25   $6ax - 3ay + 2bx - by$

26   $5ru + 10vr + 2ut + 4vt$

27   $40zw + 8x^2w - 35z - 7x^2$

28   $18ck + 4dk + 9cj + 2dj$

29   $a^3 - a^2b + ab^2 - b^3$

30   $6w^8 + 17w^4 + 12$

31   $a^6 - b^6$

32   $x^8 - 16$

33   $x^4 + 25$

34   $a^2 + a + 1$

35   $6x^2 + 42x + 60$

36   $16x^2 + 40x - 24$

37   $4x^2 - 24x + 36$

38   $60x^2 - 85x + 30$

39   $75x^2 + 120x + 48$

40   $64x^2 - 16$

41   $36 - 9x^2$

42   $18x^2 - 50$

43   $2x^2y + xy - 2xz - z$

44   $3ac - 6bd + 3ad - 6bc$

45   $12x^2z + 8y^2z - 15x^2w - 10y^2w$

46   $4x^3 + 6x^2y - 4y^2x - 6y^3$

47   $y^6 + 7y^3 - 8$

48   $8c^6 + 19c^3 - 27$

49   $(x + y)^3 - 27$

50   $(a + b)^4 - 1$

51   $x^2 + 2x + 5$

52   $4x^3 + 4x^2 + x$

53   $x^{16} - 1$

54   $x^{16} + 1$

9   FRACTIONAL EXPRESSIONS

Quotients of algebraic expressions are called **fractional expressions**. As a special case, a quotient of two polynomials is called a **rational expression**. Some examples are

$$\frac{x^2 - 5x + 1}{x^3 + 7}, \quad \frac{z^2x^4 - 3yz}{5z}, \quad \text{and} \quad \frac{1}{4xy}.$$

Many problems in mathematics involve combining rational expressions and then simplifying the result. Since rational expressions are quotients containing symbols which represent real numbers, the rules for quotients (1.11) may be used. Of course, the letters a, b, c, and d will now be replaced by polynomials. Of particular importance in simplification problems is the formula

$$\boxed{\frac{ad}{bd} = \frac{a}{b}}$$

which is obtained by multiplying the numerator and denominator of the given fraction by $1/d$. This rule is sometimes phrased "a common factor in the numerator and denominator may be *canceled* from the quotient." To use this technique in problems, we factor both the numerator and denominator of the given rational expression into prime factors and then cancel common factors which occur in the numerator and denominator. We refer to the resulting expression as being *simplified*, or *reduced to lowest terms*.

Example 1   Simplify $\dfrac{3x^2 - 5x - 2}{x^2 - 4}$.

Solution   Factoring the numerator and denominator and canceling common factors gives us

$$\frac{3x^2 - 5x - 2}{x^2 - 4} = \frac{(3x + 1)(x - 2)}{(x + 2)(x - 2)} = \frac{3x + 1}{x + 2}.$$

In the preceding example we canceled the common factor $x - 2$; that is, we divided numerator and denominator by $x - 2$. This simplification is valid only if $x - 2 \neq 0$, that is, $x \neq 2$. However, 2 is not in the domain of x since it leads to a zero denominator when substituted in the original expression. Hence our manipulations are valid. We shall always assume such restrictions when simplifying rational expressions.

Example 2   Simplify $\dfrac{2 - x - 3x^2}{6x^2 - x - 2}$.

Solution
$$\frac{2 - x - 3x^2}{6x^2 - x - 2} = \frac{(1 + x)(2 - 3x)}{(2x + 1)(3x - 2)} = \frac{-(1 + x)}{2x + 1}$$

The fact that $(2 - 3x) = -(3x - 2)$ accounts for the minus sign in the final answer. Another method of attack is to change the form of the numerator as follows:

$$\frac{2 - x - 3x^2}{6x^2 - x - 2} = \frac{-(3x^2 + x - 2)}{6x^2 - x - 2}$$

$$= -\frac{(3x - 2)(x + 1)}{(3x - 2)(2x + 1)} = -\frac{x + 1}{2x + 1}.$$

Multiplication and division are performed using rules for quotients and then simplifying, as illustrated in the next example.

Example 3   Perform the indicated operations and simplify:

(a) $\dfrac{x^2 - 6x + 9}{x^2 - 1} \cdot \dfrac{2x - 2}{x - 3}$ (b) $\dfrac{x + 2}{2x - 3} \div \dfrac{x^2 - 4}{2x^2 - 3x}$

Solutions   (a)
$$\frac{x^2 - 6x + 9}{x^2 - 1} \cdot \frac{2x - 2}{x - 3} = \frac{(x - 3)^2}{(x + 1)(x - 1)} \cdot \frac{2(x - 1)}{x - 3}$$

$$= \frac{2(x - 3)^2(x - 1)}{(x + 1)(x - 1)(x - 3)}$$

$$= \frac{2(x - 3)}{x + 1}$$

(b)
$$\frac{x+2}{2x-3} \div \frac{x^2-4}{2x^2-3x} = \frac{x+2}{2x-3} \cdot \frac{2x^2-3x}{x^2-4}$$

$$= \frac{(x+2)x(2x-3)}{(2x-3)(x+2)(x-2)}$$

$$= \frac{x}{x-2}$$

When adding or subtracting two rational expressions, it is customary to find a common denominator and use the rules

$$\frac{a}{d} + \frac{c}{d} = \frac{a+c}{d}; \quad \frac{a}{d} - \frac{c}{d} = \frac{a-c}{d}.$$

If different denominators occur, a common denominator may be introduced by multiplying numerator and denominator of each of the given fractions by suitable polynomials. It is usually desirable to use the **least common denominator (l.c.d)** of the two fractions. The l.c.d. can be found by obtaining the prime factorization for each denominator and then forming the product of the different prime factors, using the *highest* exponent which appears with each prime factor. Let us begin with a numerical example of this technique.

Example 4   Express as a rational number in lowest terms:

$$\frac{7}{24} + \frac{5}{18}$$

Solution   The prime factorizations of the denominators are $24 = (2^3)(3)$ and $18 = (2)(3^2)$. To find the l.c.d. we form the product of the different prime factors, using the highest exponent associated with each factor. This gives us $(2^3)(3^2)$, or 72. We now change each fraction to an equal fraction with denominator 72 and add as follows:

$$\frac{7}{24} + \frac{5}{18} = \frac{7}{24} \cdot \frac{3}{3} + \frac{5}{18} \cdot \frac{4}{4}$$

$$= \frac{21}{72} + \frac{20}{72} = \frac{41}{72}$$

If we use the addition formula for rational numbers in (1.11) we obtain

$$\frac{7}{24} + \frac{5}{18} = \frac{(7)(18) + (24)(5)}{(24)(18)}$$

$$= \frac{126 + 120}{432} = \frac{246}{432}$$

which reduces to 41/72.

The method for finding the l.c.d. for rational expressions is analogous to that illustrated in the solution of Example 4. The only difference is that we use factorizations of polynomials instead of integers.

Example 5   Change the following to a rational expression in lowest terms.

$$\frac{6}{x(3x-2)} + \frac{5}{3x-2} - \frac{2}{x^2}$$

Solution   The denominators are already in factored form. Evidently the l.c.d. is $x^2(3x-2)$. In order to obtain three fractions having that denominator we multiply numerator and denominator of the first fraction by x, of the second by x^2, and of the third by $3x-2$. This gives us

$$\frac{6}{x(3x-2)} + \frac{5}{3x-2} - \frac{2}{x^2} = \frac{6x}{x^2(3x-2)} + \frac{5x^2}{x^2(3x-2)} - \frac{2(3x-2)}{x^2(3x-2)}$$

$$= \frac{6x + 5x^2 - (6x-4)}{x^2(3x-2)}$$

$$= \frac{5x^2 + 4}{x^2(3x-2)}$$

Example 6   Simplify:

$$\frac{2x+5}{x^2+6x+9} + \frac{x}{x^2-9} + \frac{1}{x-3}$$

Solution   We begin by factoring denominators as follows:

$$\frac{2x+5}{x^2+6+9} + \frac{x}{x^2-9} + \frac{1}{x-3} = \frac{2x+5}{(x+3)^2} + \frac{x}{(x+3)(x-3)} + \frac{1}{x-3}$$

Since the l.c.d. is $(x+3)^2(x-3)$ we multiply numerator and denominator of the first fraction by $x-3$, of the second by $x+3$, and of the third by $(x+3)^2$ and add as follows:

$$\frac{(2x+5)(x-3)}{(x+3)^2(x-3)} + \frac{x(x+3)}{(x+3)^2(x-3)} + \frac{(x+3)^2}{(x+3)^2(x-3)}$$

$$= \frac{(2x^2-x-15) + (x^2+3x) + (x^2+6x+9)}{(x+3)^2(x-3)}$$

$$= \frac{4x^2+8x-6}{(x+3)^2(x-3)} = \frac{2(2x^2+4x-3)}{(x+3)^2(x-3)}$$

It is sometimes necessary to simplify quotients in which the numerator and denominator are not polynomials, as illustrated in the next two examples.

Example 7 Simplify $\dfrac{1 - \dfrac{2}{x+1}}{x - \dfrac{1}{x}}$.

Solution

$$\dfrac{1 - \dfrac{2}{x+1}}{x - \dfrac{1}{x}} = \dfrac{\dfrac{(x+1)-2}{x+1}}{\dfrac{x^2-1}{x}}$$

$$= \dfrac{\dfrac{x-1}{x+1}}{\dfrac{x^2-1}{x}}$$

$$= \dfrac{x-1}{x+1} \cdot \dfrac{x}{x^2-1}$$

$$= \dfrac{(x-1)x}{(x+1)(x+1)(x-1)}$$

$$= \dfrac{x}{(x+1)^2}$$

Example 8 Simplify $\dfrac{x^{-2}+y^{-2}}{(xy)^{-1}}$.

Solution

$$\dfrac{x^{-2}+y^{-2}}{(xy)^{-1}} = \dfrac{\dfrac{1}{x^2}+\dfrac{1}{y^2}}{\dfrac{1}{xy}}$$

$$= \dfrac{\dfrac{y^2+x^2}{x^2y^2}}{\dfrac{1}{xy}}$$

$$= \dfrac{y^2+x^2}{x^2y^2} \cdot \dfrac{xy}{1}$$

$$= \dfrac{y^2+x^2}{xy}$$

Another method of attacking this problem is to multiply numerator and denominator of the original expression by x^2y^2 and simplify.

The denominators of certain fractional expressions contain sums or differences involving radicals. In some cases the denominators can be rationalized, as illustrated in the next example.

Example 9   Simplify $\dfrac{1}{\sqrt{x} - \sqrt{y}}$.

Solution   Multiplying numerator and denominator by $\sqrt{x} + \sqrt{y}$ we obtain

$$\frac{1}{\sqrt{x} - \sqrt{y}} = \frac{1}{\sqrt{x} - \sqrt{y}} \cdot \frac{\sqrt{x} + \sqrt{y}}{\sqrt{x} + \sqrt{y}}$$

$$= \frac{\sqrt{x} + \sqrt{y}}{(\sqrt{x})^2 - \sqrt{y}\sqrt{x} + \sqrt{x}\sqrt{y} - (\sqrt{y})^2}$$

$$= \frac{\sqrt{x} + \sqrt{y}}{x - y}.$$

EXERCISES

Simplify each of the following.

1 $\dfrac{6x^2 + 7x - 10}{6x^2 + 13x - 15}$

2 $\dfrac{10x^2 + 29x - 21}{5x^2 - 23x + 12}$

3 $\dfrac{12y^2 + 3y}{20y^2 + 9y + 1}$

4 $\dfrac{4z^2 + 12z + 9}{2z^2 + 3z}$

5 $\dfrac{6 - 7a - 5a^2}{10a^2 - a - 3}$

6 $\dfrac{6y - 5y^2}{25y^2 - 36}$

7 $\dfrac{4x^3 - 9x}{10x^4 + 11x^3 - 6x^2}$

8 $\dfrac{16x^4 + 8x^3 + x^2}{4x^3 + 25x^2 + 6x}$

9 $\dfrac{6r}{3r - 1} - \dfrac{4r}{2r + 5}$

10 $\dfrac{3s}{s^2 + 1} - \dfrac{6}{2s - 1}$

11 $\dfrac{2x + 1}{2x - 1} - \dfrac{x - 1}{x + 1}$

12 $\dfrac{3u + 2}{u - 4} + \dfrac{4u + 1}{5u + 2}$

13 $\dfrac{9t - 6}{8t^3 - 27} \cdot \dfrac{4t^2 - 9}{12t^2 + 10t - 12}$

14 $\dfrac{a^2 + 4a + 3}{3a^2 + a - 2} \cdot \dfrac{3a^2 - 2a}{2a^2 + 13a + 21}$

15 $\dfrac{5a^2 + 12a + 4}{a^4 - 16} \div \dfrac{25a^2 + 20a + 4}{a^2 - 2a}$

16 $\dfrac{x^3 - 8}{x^2 - 4} \div \dfrac{x}{x^3 + 8}$

17 $\dfrac{2}{3x + 1} - \dfrac{9}{(3x + 1)^2}$

18 $\dfrac{4}{(5x - 2)^2} + \dfrac{x}{5x - 2}$

19 $\dfrac{1}{c} - \dfrac{c + 2}{c^2} + \dfrac{3}{c^3}$

20 $\dfrac{6}{3t} + \dfrac{t + 5}{t^3} + \dfrac{1 - 2t^2}{t^4}$

21 $\dfrac{5}{x-1}+\dfrac{8}{(x-1)^2}-\dfrac{3}{(x-1)^3}$

22 $\dfrac{8}{x}+\dfrac{3}{2x-4}+\dfrac{7x}{x^2-4}$

23 $\dfrac{2}{x}+\dfrac{7}{x^2}+\dfrac{5}{2x-3}+\dfrac{1}{(2x-3)^2}$

24 $\dfrac{4}{x}+\dfrac{3x^2+5}{x^3}-\dfrac{6}{2x+1}$

25 $\dfrac{p^4+3p^3-8p-24}{p^3-2p^2-9p+18}$

26 $\dfrac{2ac+bc-6ad-3bd}{6ac+2ad+3bc+bd}$

27 $\left(\dfrac{1}{2x+2h+3}-\dfrac{1}{2x+3}\right)\div h$

28 $\left(\dfrac{7}{5x+5h-2}-\dfrac{7}{5x-2}\right)\div h$

29 $\dfrac{(x+h)^{-3}-x^{-3}}{h}$

30 $\dfrac{(x+h)^{-2}-x^{-2}}{h}$

31 $\dfrac{5}{7x-3}-\dfrac{2}{2x+1}+\dfrac{4x}{14x^2+x-3}$

32 $2+\dfrac{3}{x}+\dfrac{7x}{3x+10}$

33 $\dfrac{\dfrac{a}{b}-\dfrac{b}{a}}{\dfrac{1}{a}+\dfrac{1}{b}}$

34 $\dfrac{\dfrac{1}{x+1}-5}{\dfrac{1}{x}-x}$

35 $\dfrac{\dfrac{x}{y^2}-\dfrac{y}{x^2}}{\dfrac{1}{y^2}-\dfrac{1}{x^2}}$

36 $\dfrac{(r/s)+(s/r)}{(r^2/s^2)-(s^2/r^2)}$

37 $\dfrac{\dfrac{5}{x+1}+\dfrac{2x}{x+3}}{\dfrac{x}{x+1}+\dfrac{7}{x+3}}$

38 $\dfrac{\dfrac{3}{w}-\dfrac{6}{2w+1}}{\dfrac{5}{w}+\dfrac{8}{2w+1}}$

39 $\dfrac{a^2-b^3a^{-1}}{b^2a^{-1}+a+b}$

40 $\dfrac{(c+d)^{-1}-(c-d)^{-1}}{(c+d)^{-1}+(c-d)^{-1}}$

41 $\dfrac{1}{\sqrt{a}+\sqrt{b}}$

42 $\dfrac{\sqrt{a}+\sqrt{b}}{\sqrt{a}-\sqrt{b}}$

43 $\dfrac{1}{\sqrt{x}}+\dfrac{1}{\sqrt{y}}$

44 $\dfrac{r}{\sqrt{s}}-\dfrac{s}{\sqrt{r}}$

45 $\dfrac{c}{1-\sqrt{c}}$

46 $\dfrac{3}{\sqrt{t}+4}$

47 $\dfrac{1}{a+b+\sqrt{c}}$

48 $\dfrac{1}{x+\sqrt{y+z}}$

49 $\dfrac{1}{h}\left(\dfrac{1}{x+h}-\dfrac{1}{x}\right)$

50 $\left(\dfrac{1}{\sqrt{x}+\sqrt{y}+\sqrt{z}}\right)^0$

10 REVIEW

Concepts

Define or explain each of the following.

1 The Commutative Properties of real numbers
2 The Associative Properties
3 The Distributive Properties
4 Rational and irrational numbers
5 The integers
6 Prime number
7 Coordinate line
8 A number a is greater than a number b.
9 A number a is less than a number b.
10 The absolute value of a real number
11 The distance between points on a coordinate line
12 Integer exponents
13 The Laws of Exponents
14 Principal nth root
15 The radical notation
16 Rationalizing a denominator
17 Rational exponents
18 Scientific form for real numbers
19 Variable
20 Domain of a variable
21 Algebraic expression
22 Monomial
23 Polynomial
24 Degree of a polynomial
25 Prime polynomial
26 Irreducible polynomial
27 Factoring
28 Rational expression

Exercises

1 Express each of the following as a rational number with least positive numerator.

(a) $\left(\dfrac{2}{3}\right)\left(-\dfrac{5}{8}\right)$ (b) $\dfrac{3}{4}+\dfrac{6}{5}$ (c) $\dfrac{5}{8}-\dfrac{6}{7}$ (d) $\dfrac{3}{4}\div\dfrac{6}{5}$

2 Replace the □ with either $<$, $>$ or $=$.

(a) $-0.1\ \square\ -0.01$ (b) $\sqrt{9}\ \square\ -3$ (c) $1/6\ \square\ 0.166$

3 Express in terms of inequalities:

(a) x is negative.

(b) a is between 1/2 and 1/3.

(c) The absolute value of x is not greater than 4.

4 Rewrite without using the absolute value symbol:

(a) $|-7|$ (b) $|-5|/(-5)$ (c) $|3^{-1} - 2^{-1}|$

5 If points $A, B,$ and C on a coordinate line have coordinates $-8, 4,$ and $-3,$ respectively, find the following:

(a) $d(A, C)$ (b) $d(C, A)$ (c) $d(B, C)$

6 Prove or disprove each of the following:

(a) $(x + y)^2 = x^2 + y^2$

(b) $\dfrac{1}{\sqrt{x+y}} = \dfrac{1}{\sqrt{x}} + \dfrac{1}{\sqrt{y}}$

(c) $\dfrac{1}{\sqrt{c} - \sqrt{d}} = \dfrac{\sqrt{c} + \sqrt{d}}{c - d}$

Simplify the expressions in Exercises 7–32.

7 $(3a^2b)^2(2ab^3)$

8 $\dfrac{6r^3y^2}{2r^5y}$

9 $\dfrac{(3x^2y^{-3})^{-2}}{x^{-5}y}$

10 $\left(\dfrac{a^{2/3}b^{3/2}}{a^2b}\right)^6$

11 $(-2p^2q)^3(p/4q^2)^2$

12 $c^{-4/3}c^{3/2}c^{1/6}$

13 $\left(\dfrac{xy^{-1}}{\sqrt{z}}\right)^4 \div \left(\dfrac{x^{1/3}y^2}{z}\right)^3$

14 $\left(\dfrac{-64x^3}{z^6y^9}\right)^{2/3}$

15 $((a^{2/3}b^{-2})^3)^{-1}$

16 $(4x^3y^2z)^{-0}$

17 $\dfrac{r^{-1} + s^{-1}}{(rs)^{-1}}$

18 $(u + v)^3(u + v)^{-2}$

19 $\sqrt[3]{(x^4y^{-1})^6}$

20 $\sqrt[3]{8x^5y^3z^4}$

21 $\dfrac{1}{\sqrt[3]{4}}$

22 $\sqrt{\dfrac{a^2b^3}{c}}$

23 $\sqrt[3]{4x^2y}\sqrt[3]{2x^5y^2}$

24 $\sqrt[4]{(-4a^3b^2c)^2}$

25 $\dfrac{1}{\sqrt{t}}\left(\dfrac{1}{\sqrt{t}} - 1\right)$

26 $\sqrt{\sqrt[3]{(c^3d^6)^4}}$

27 $\dfrac{\sqrt{12x^4y}}{\sqrt{3x^2y^5}}$

28 $\sqrt[3]{(a + 2b)^3}$

29 $\sqrt[3]{x^2y}\sqrt[4]{x^3y^2}$

30 $\sqrt{x} - \sqrt{9x^3}$

31 $\dfrac{1}{\sqrt{a} - \sqrt{b}}$

32 $(\sqrt[3]{m^3 + n^3})^3$

33   Express in scientific form.

   (a)   93,700,000,000   (b)   0.00000402

34   Express as a decimal.

   (a) 6.8×10^7 (b) 7.3×10^{-4}

In each of Exercises 35–42, perform the indicated operations.

35 $(3x^3 - 4x^2 + x - 7) + (x^4 - 2x^3 + 3x^2 + 5)$

36 $(4z^4 - 3z^2 + 1) - z(z^3 + 4z^2 - 4)$

37 $(x + 4)(x + 3) - (2x - 1)(x - 5)$

38 $(4x - 5)(2x^2 + 3x - 7)$

39 $(3y^3 - 2y^2 + y + 4)(y^2 - 3)$

40 $(3x + 2)(x - 5)(5x + 4)$

41 $(a - b)(a^3 + a^2b + ab^2 + b^3)$

42 $\dfrac{9p^4q^3 - 6p^2q^4 + 5p^3q^2}{3p^2q^2}$

Find the products in Exercises 43–50.

43 $(3a - 5b)(2a + 7b)$ **44** $(4r^2 - 3s)^2$

45 $(13a^2 + 4b)(13a^2 - 4b)$ **46** $(a^3 - a^{-3})^2$

47 $(2a + b)^3$ **48** $(c^2 - d^2)^3$

49 $(3x + 2y)^2(3x - 2y)^2$ **50** $(a + b + c + d)^2$

Factor the expressions in Exercises 51–62.

51 $60xw + 70w$ **52** $2r^4s^3 - 8r^2s^5$

53 $28x^2 + 4x - 9$ **54** $16a^4 + 24a^2b^2 + 9b^4$

55 $2wy + 3yx - 8wz - 12zx$ **56** $2c^3 - 12c^2 + 3c - 18$

57 $8x^3 + 64y^3$ **58** $u^3v^4 - u^6v$

59 $p^8 - q^8$ **60** $x^4 - 8x^3 + 16x^2$

61 $w^6 + 1$ **62** $3x + 6$

Simplify the expressions in Exercises 63–70.

63 $\dfrac{6x^2 - 7x - 5}{4x^2 + 4x + 1}$ **64** $\dfrac{r^3 - t^3}{r^2 - t^2}$

65 $\dfrac{6x^2 - 5x - 6}{x^2 - 4} \div \dfrac{2x^2 - 3x}{x + 2}$ **66** $\dfrac{2}{4x - 5} - \dfrac{5}{10x + 1}$

67 $\dfrac{7}{x + 2} + \dfrac{3x}{(x + 2)^2} - \dfrac{5}{x}$ **68** $\dfrac{x + x^{-2}}{1 + x^{-2}}$

69 $\dfrac{1}{x} - \dfrac{2}{x^2 + x} - \dfrac{3}{x + 3}$ **70** $(a^{-1} + b^{-1})^{-1}$

CHAPTER 2

Equations
and
Inequalities

For hundreds of years one of the main concerns in algebra has been the solutions of equations. More recently the study of inequalities has reached the same level of importance. Both of these topics are used extensively in applications of mathematics. In this chapter we shall discuss the rudiments of solving equations and inequalities. All variables will represent real numbers. In a later chapter we shall study equations that involve complex numbers.

1 LINEAR EQUATIONS

If x is a variable, then expressions such as

$$x + 3 = 0, \quad x^2 - 5 = 4x, \quad \text{or} \quad (x^2 - 9)\sqrt[3]{x + 1} = 0$$

are called **equations** in x. If certain numbers are substituted for x in these equations true statements are obtained, whereas other numbers produce false statements. For example, the equation $x + 3 = 0$ leads to a false statement for every value of x except -3. If 2 is substituted for x in the equation $x^2 - 5 = 4x$ we obtain $4 - 5 = 8$, or $-1 = 8$, a false statement. However, if we let $x = 5$, then we obtain $(5)^2 - 5 = 4 \cdot 5$, or $20 = 20$, which is true. In general, given any equation in x, if a true statement is obtained when x is replaced by some real number a from the domain of x, then a is called a **solution** or a **root** of the equation. We also say that a **satisfies** the equation. To **solve** an equation means to find all the solutions.

Sometimes every number in the domain of the variable is a solution of an equation. In this case the equation is called an **identity**. For example,

$$x^2 - 4 = (x + 2)(x - 2)$$

is an identity since it is true for every number in the domain of x. If there are numbers in the domain of x which are not solutions, then the equation is called a **conditional equation**.

The solutions of an equation depend on the system of numbers under consideration. For example, if we demand that solutions be rational numbers, then the equation $x^2 = 2$ has no solutions since there is no rational number whose square is 2. However, if we allow *real* numbers, then the solutions are $-\sqrt{2}$ and $\sqrt{2}$. Similarly, the equation $x^2 = -1$ has no real solutions; however, we shall see later that this equation has solutions if *complex* numbers are allowed.

Two equations are said to be **equivalent** if they have exactly the same solutions. For example, the equations

$$x - 1 = 2, \quad x = 3, \quad 5x = 15, \quad \text{and} \quad 2x + 1 = 7$$

are all equivalent.

One method of solving an equation is to replace it by a chain of equivalent equations, each in some sense simpler than the preceding one and terminating in an equation for which the solutions are obvious. This is often accomplished by using various properties of real numbers. For example, we may add the same expression to both sides of an equation without changing the solutions. Similarly, we may subtract the same expression from both sides of an equation. We can also multiply or divide both sides of an equation by an expression which represents a nonzero real number. The following example illustrates these remarks.

Example 1   Solve the equation $2x - 5 = 3$.

Solution   The following is a chain of equations, each of which is equivalent to the preceding one. (Why?)

$$2x - 5 = 3$$
$$(2x - 5) + 5 = 3 + 5$$
$$2x = 8$$
$$\tfrac{1}{2}(2x) = \tfrac{1}{2}(8)$$
$$x = 4$$

Since the last equation has only one solution, 4, and since the last equation is equivalent to the first, it follows that 4 is the only solution of $2x - 5 = 3$.

In order to guard against errors in manipulations and simplifications, answers should be checked by substitution in the original equation. If we apply a check in Example 1 we obtain $2(4) - 5 = 3$, which reduces to $3 = 3$, a true statement.

We sometimes say that an equation such as that given in Example 1 *has the solution* $x = 4$, in the sense that substitution of 4 for x produces a true statement. As another illustration, the equation $x^2 = 9$ is said to have solutions $x = 3$ and $x = -3$.

If we inadvertently multiply both sides of an equation by an expression which equals zero for some value of x, then an equivalent equation is not always obtained. The following example illustrates a manner in which this may happen.

Example 2   Solve the equation

$$\frac{3x}{x - 2} = 1 + \frac{6}{x - 2}.$$

Solution   Multiplying both sides by $x - 2$ and simplifying leads to

$$\left(\frac{3x}{x-2}\right)(x-2) = 1(x-2) + \left(\frac{6}{x-2}\right)(x-2)$$
$$3x = (x-2) + 6$$
$$2x = 4$$
$$x = 2.$$

Let us check to see whether 2 is a solution of the given equation. Substituting 2 for x we obtain

$$\frac{3(2)}{2-2} = 1 + \frac{6}{2-2}, \quad \text{or} \quad \frac{6}{0} = 1 + \frac{6}{0}.$$

Since division by 0 is not permissible, 2 is not a solution. Actually, the given equation has no solutions, for we have shown that if the equation is true for some value of x, then that value must be 2. However, as we have seen, 2 is not a solution.

The preceding example indicates that *it is essential to check answers obtained after multiplying both sides of an equation by an expression containing variables.*

In this section we shall restrict our efforts to equations which are equivalent to equations of the form

$$ax + b = 0$$

where a and b are real numbers and $a \neq 0$. An equation of this type is called a **linear equation** in x. To solve the equation we first add $-b$ to both sides and then multiply by $1/a$ as follows:

$$(ax + b) + (-b) = 0 + (-b)$$
$$ax = -b$$
$$\frac{1}{a}(ax) = \frac{1}{a}(-b)$$
$$x = -\frac{b}{a}$$

This shows that *if* $ax + b = 0$ has a solution, then it must be $-b/a$. To verify that $-b/a$ is actually a solution, we substitute it for x in $ax + b = 0$, obtaining

$$a\left(-\frac{b}{a}\right) + b = 0 \quad \text{or} \quad (-b) + b = 0$$

which is a true statement. We have established the following fact.

If $a \neq 0$, then the equation $ax + b = 0$ has precisely one solution, $x = -b/a$.

In Section 3 we shall consider equations having more than one solution.

Example 3   Solve $(8x - 2)(3x + 4) = (4x + 3)(6x - 1)$.

Solution   The following equations are all equivalent (supply reasons).

$$(8x - 2)(3x + 4) = (4x + 3)(6x - 1)$$
$$24x^2 + 26x - 8 = 24x^2 + 14x - 3$$
$$26x - 14x = -3 + 8$$
$$12x = 5$$
$$x = \frac{5}{12}$$

Hence the solution of the given equation is 5/12.

Example 4   Solve $\dfrac{3}{2x - 4} - \dfrac{5}{x + 3} = \dfrac{2}{x - 2}$.

Solution   If we rewrite the equation as

$$\frac{3}{2(x - 2)} - \frac{5}{x + 3} = \frac{2}{x - 2}$$

we see that the l.c.d. of the three fractions is $2(x - 2)(x + 3)$. Multiplying both sides by this l.c.d. leads to

$$3[(x + 3)] - 5[2(x - 2)] = 2[2(x + 3)].$$

Simplifying, we obtain

$$3x + 9 - 10x + 20 = 4x + 12$$
$$3x - 10x - 4x = 12 - 9 - 20$$
$$-11x = -17$$
$$x = \frac{17}{11}.$$

Since we multiplied by an expression involving x we must check this result in the given equation. Substituting for x we have

$$\frac{3}{2(17/11) - 4} - \frac{5}{(17/11) + 3} = \frac{2}{(17/11) - 2}.$$

We leave it to the reader to show that this reduces to $-22/5 = -22/5$, a true statement. Hence the given equation has the solution 17/11.

Students sometimes have difficulty in determining whether an equation is conditional or an identity. An identity will often be indicated when, after applying properties of real numbers, an equation of the form $p = p$ is obtained, where p is some expression. To illustrate, if we multiply both sides of the equation

$$\frac{x}{x^2 - 4} = \frac{x}{(x + 2)(x - 2)}$$

by $x^2 - 4$ we obtain $x = x$. This alerts us to the fact that we may have an identity on our hands; however, it does not prove anything. A standard method for verifying that an equation is an identity is to show, using properties of real numbers, that the expression which appears on one side of the given equation can be transformed into the expression which appears on the other side of the given equation. That is easy to do in the above illustration, since we know that $x^2 - 4 = (x + 2)(x - 2)$. Of course, to show that an equation is not an identity, we need only find one real number in the domain of the variable that fails to satisfy the original equation.

EXERCISES

Solve the equations in Exercises 1–38.

1 $4x + 7 = 0$

2 $3x + 16 = 0$

3 $\sqrt{2}x - 5 = 0$

4 $\sqrt{3}x - 2 = 0$

5 $5x + 3 = 7x - 2$

6 $8x - 5 = 6x + 4$

7 $3(7y - 2) = 2(4y + 1)$

8 $2(9z + 2) - 5(z - 8) = 0$

9 $\frac{2}{3}t + 4 = 2 - \frac{5}{2}t$

10 $\frac{3}{4}u - 1 = 2 + \frac{1}{3}u$

11 $0.2(4 - 3x) + 0.3x = 2.6$

12 $0.7x - 1.2 = 0.3(2x - 1)$

13 $\frac{12 - 7w}{6} = \frac{2w + 1}{9}$

14 $\frac{3r + 2}{8} = 1 - \frac{r}{12}$

15 $\frac{18 - 5p}{3p + 2} = \frac{7}{3}$

16 $\frac{6}{5v - 2} = \frac{9}{7v + 3}$

17 $4 - \frac{3}{x} = 6 - \frac{5}{x}$

18 $\frac{5}{q} + \frac{2}{q} - \frac{10}{q} = 12$

19 $(6x - 5)^2 = (4x + 3)(9x - 2)$

20 $(x - 7)^2 - 4 = (x + 1)^2$

21 $(4x - 3)(2x + 3) - 8x(x - 4) = 0$

22 $(2x + 3)(3x - 2) = 6x^2 + 1$

23 $\frac{6s + 7}{4s - 1} = \frac{3s + 8}{2s - 4}$

24 $\frac{8t + 5}{10t - 7} = \frac{4t - 3}{5t + 7}$

25 $\frac{1}{3} + \frac{2}{6x + 3} = \frac{3}{2x + 1}$

26 $\frac{-3}{4x - 2} + \frac{2}{2x - 1} = \frac{7}{2}$

27 $\frac{5}{4a - 2} - \frac{1}{6a - 3} = \frac{4}{5}$

28 $\frac{8}{3b + 6} - \frac{1}{2b + 4} = \frac{3}{4}$

29 $\frac{1}{2x - 1} = \frac{3}{4x - 2}$

30 $\frac{6}{2x + 11} + 5 = 5$

31 $2 - \dfrac{5}{3x - 7} = 2$

32 $\dfrac{4}{5x + 2} - \dfrac{7}{15x + 6} = 0$

33 $\dfrac{7}{y^2 - 4} - \dfrac{4}{y + 2} = \dfrac{5}{y - 2}$

34 $\dfrac{4}{2u - 3} + \dfrac{10}{4u^2 - 9} = \dfrac{1}{2u + 3}$

35 $(x + 3)^3 - (3x - 1)^2 = x^3 + 4$

36 $(x - 1)^3 = (x + 1)^3 - 6x^2$

37 $\dfrac{9x}{3x - 1} = 2 + \dfrac{3}{3x - 1}$

38 $\dfrac{2x}{2x + 3} + \dfrac{6}{4x + 6} = 5$

Prove that the equations in Exercises 39–44 are identities.

39 $(3x + 1)^2 - 9x^2 = 6x + 1$

40 $(4x - 5)(3x + 2) + 7x = 12x^2 - 10$

41 $\dfrac{x^2 - 1}{x - 1} = x + 1$

42 $\dfrac{3x - 2}{x} = 3 - \dfrac{2}{x}$

43 $\dfrac{5x^2 + 2}{x} = \dfrac{2}{x} + 5x$

44 $\dfrac{9x^2 - 16}{3x + 4} = 3x - 4$

45 For what value of c is -3 a solution of the equation $3x + 1 - 5c = 2c + x - 10$?

46 For what value of b is 8 a solution of the equation $4x + 3b = 7$?

47 Determine values for a and b such that $5/3$ is a solution of the equation $ax + b = 0$. Are these the only possible values for a and b? Explain.

48 In each of the following determine whether the two given equations are equivalent.
(a) $x^2 = 4, \quad x = 2$
(b) $x = \sqrt{4}, \quad x = 2$
(c) $2x = 4, \quad x = 2$

49 (a) Find an equation in the following chain which is not equivalent to the preceding equation.

$$x^2 - x - 2 = x^2 - 4$$
$$(x + 1)(x - 2) = (x + 2)(x - 2)$$
$$x + 1 = x + 2$$
$$1 = 2$$

(b) Find the solutions of the first equation in part (a).

50 Find an equation in the following chain which is not equivalent to the preceding equation.

$$x + 3 = 0$$
$$5x - 4x = -3$$
$$5x + 6 = 4x + 3$$
$$x^2 + 5x + 6 = x^2 + 4x + 3$$
$$(x + 3)(x + 1) = (x + 2)(x + 3)$$
$$x + 1 = x + 2$$
$$0 = 1$$

2   APPLICATIONS

Formulas or equations involving variables are used in all fields which deal with numbers. For certain applications it is necessary to solve for a particular variable in terms of the remaining variables which appear in the formula. This is done by treating the equation as if the desired variable were the only one present and transforming the original equation into an equivalent equation in which that variable is isolated on one side, as illustrated in the next three examples.

Example 1   If a sum of money P (the principal) is invested at a simple interest rate of r per cent per year, then the interest I at the end of t years is given by $I = Prt$. Solve for r in terms of the remaining variables.

Solution   We begin by writing

$$Prt = I.$$

In order to isolate r we multiply both sides by $1/Pt$, obtaining

$$\frac{1}{Pt} \cdot Prt = \frac{1}{Pt} \cdot I.$$

It follows that

$$r = \frac{I}{Pt}.$$

Example 2   The relationship between the temperature F on the Fahrenheit scale and the temperature C on the Celsius scale is given by

$$C = \tfrac{5}{9}(F - 32).$$

Solve for F in terms of C.

Solution   We may proceed as follows (supply reasons):

$$C = \tfrac{5}{9}(F - 32)$$
$$\tfrac{9}{5}C = F - 32$$
$$\tfrac{9}{5}C + 32 = F$$
$$F = \tfrac{9}{5}C + 32.$$

Example 3   The formula $R = \dfrac{R_1 R_2}{R_1 + R_2}$ is used in electrical theory, where R_1, R_2, and R are positive. Solve for R_1 in terms of R and R_2.

Solution The following equations are equivalent to the given equation.

$$(R_1 + R_2)R = (R_1 + R_2)\left(\frac{R_1 R_2}{R_1 + R_2}\right)$$

$$R_1 R + R_2 R = R_1 R_2$$

$$R_1 R - R_1 R_2 = -R_2 R$$

$$R_1(R - R_2) = -R_2 R$$

$$R_1 = \frac{-R_2 R}{R - R_2}$$

$$R_1 = \frac{R_2 R}{R_2 - R}$$

Problems often occur in everyday life which can be solved by means of equations or other mathematical tools. Some problems are described orally, from one person to another. Others are stated using written words, as is the case in textbooks. For this reason they are often called "word problems" by students and teachers of mathematics. They may also be referred to as "practical problems." We shall use the terminology "applied problem" for any problem which involves an application of mathematics to some other field.

Due to the unlimited variety of applied problems it is difficult to state specific rules for finding solutions. However, it is possible to develop a general strategy for attacking such problems. Below are listed some guidelines which may be helpful, provided the problem can be formulated in terms of an equation in one variable. In a later chapter we shall introduce techniques which involve several variables.

Guidelines for Solving Applied Problems

---

1. If the problem is stated in written words, read it carefully several times and think about the given facts, together with the unknown quantity that is to be found.

2. Introduce a letter to denote the unknown quantity. This is one of the most crucial steps in the solution! Phrases containing words such as "what," "find," "how much," "how far," or "when" should alert you to the unknown quantity.

3. If possible, draw a picture and label it appropriately.

4. Make a list of known facts together with any relationships involving the unknown quantity. A relationship may often be described by means of an equation in which written statements, instead of letters or numbers, appear on either one or both sides of the equals sign.

5. After analyzing the list in step 4, and perhaps rereading the problem several more times, formulate an equation which describes precisely what is stated in words.

6. Solve the equation formulated in step 5.

7.   Check the solutions obtained in step 6 by referring to the original statement of the problem. Carefully note whether the solution agrees with the stated conditions.

8.   Don't become discouraged if you are unable to solve a given problem. It takes a great deal of effort and practice to become proficient in solving applied problems. Keep trying!

Example 4   A student has test scores of 64 and 78. What score on a third test will give the student an average of 80?

Solution   We shall follow the Guidelines which precede this example. Reading the problem carefully, as suggested in step 1, we note that the unknown quantity is the score on the third test. Accordingly, as in step 2, we introduce a letter as follows:

$$x = \text{score on the third test.}$$

Drawing a picture, as mentioned in step 3, is inappropriate for this problem, so we go on to step 4 and look for relationships involving x. Since the average of the three scores is found by adding them and dividing by 3, we may write

$$\frac{64 + 78 + x}{3} = \text{average of the three scores 64, 78, } x.$$

From the statement of the problem we obtain

$$80 = \text{average desired.}$$

Consequently, x must satisfy the equation

$$\frac{64 + 78 + x}{3} = 80.$$

The last equation is the one referred to in step 5 of the Guidelines. We next solve the equation (see Guideline 6) as follows:

$$64 + 78 + x = 240$$
$$142 + x = 240$$
$$x = 240 - 142$$
$$x = 98$$

Guideline 7 tells us to check by referring to the original statement. If three test scores are 64, 78 and 98, then the average is

$$\frac{64 + 78 + 98}{3} = \frac{240}{3} = 80.$$

Hence a score of 98 on the third test will give the student an average of 80. Happily, we can ignore Guideline 8.

In the remaining examples we shall not point out the explicit Guidelines which are used in the solutions. The reader should be able to determine that without being told.

Example 5 A store holding a clearance sale advertises that all prices have been discounted 20%. If a certain article is on sale for $28, what was its price before the sale?

Solution We begin by noting that the unknown quantity is the presale price. It is convenient to arrange our work as follows, where the quantities are measured in dollars.

$$x = \text{presale price}$$
$$0.20x = \text{discount}$$
$$28 = \text{sale price}$$

The sale price is determined as follows:

$$(\text{presale price}) - (\text{discount}) = (\text{sale price}).$$

This leads to the equation

$$x - 0.20x = 28$$

which we solve as follows:

$$x - \tfrac{1}{5}x = 28$$
$$\tfrac{4}{5}x = 28$$
$$x = (\tfrac{5}{4})28 = 35$$

Hence the price before the sale was $35.

To check this answer we note that if a $35 article is discounted 20%, then the discount (in dollars) is $(0.20)(35) = 7$, and the selling price is $35 - 7$, or $28.

Example 6 A man has $15,000 to invest. He plans to deposit part of it in a savings account paying 5% simple interest and the remainder in an investment fund yielding 8% simple interest. How much should he invest in each to obtain a 7% return on his money after one year?

Solution The simple interest formula $I = Prt$ was given in Example 1. In the present example, $t = 1$ and hence the interest is given by $I = Pr$. If we let x denote the amount deposited in the savings account, then the remainder, $15,000 - x$, will be put into the investment fund. This leads to the following equalities:

$$x = \text{amount invested at } 5\%$$
$$15,000 - x = \text{amount invested at } 8\%$$
$$0.05x = \text{interest on } x \text{ dollars at } 5\%$$
$$0.08(15,000 - x) = \text{interest on } 15,000 - x \text{ dollars at } 8\%$$
$$0.07(15,000) = \text{total interest desired.}$$

Since the total interest must equal the combined interest from the two investments, we see that

$$0.05x + 0.08(15000 - x) = 0.07(15000)$$
$$0.05x + 1200 - 0.08x = 1050$$
$$-0.03x = -150$$
$$x = \frac{-150}{-0.03} = 5000.$$

Consequently, $5,000 should be deposited in the savings account and $10,000 in the investment fund.

Checking, we see that if $5,000 is placed in the savings account, then the interest obtained is $(0.05)(\$5,000) = \$250$. If $10,000 is placed in the investment fund then the interest is $(0.08)(\$10,000) = \$800$. Hence the total interest is $1,050, which is 7% of $15,000.

An important type of problem which may be solved using linear equations involves mixing two substances to obtain a prescribed mixture. For such problems it is often helpful to draw a picture, as illustrated in the next two examples.

Example 7   A chemist has 10 ml of a solution which contains a 30% concentration of acid. How many ml of pure acid must be added in order to increase the concentration to 50%?

Solution   Since we wish to find the amount of pure acid to add, we let

$$x = \text{ml of acid to be added.}$$

The picture in Figure 2.1 is self-explanatory. This leads to the equation

$$3 + x = 0.5(10 + x)$$

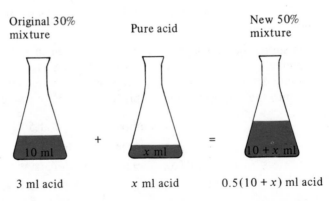

Original 30% mixture   Pure acid   New 50% mixture

10 ml   +   x ml   =   10 + x ml

3 ml acid   x ml acid   0.5(10 + x) ml acid

Figure 2.1

or equivalently,

$$3 + x = 5 + 0.5x$$
$$0.5x = 2$$
$$\tfrac{1}{2}x = 2$$
$$x = 4$$

Hence 4 ml of the acid should be added to the original solution.

To check, we note that if 4 ml of acid is added to the given solution, then the new solution contains 14 ml, 7 of which are acid. This is the desired 50% concentration.

Example 8 A radiator contains 8 quarts of a mixture of water and antifreeze. If 40% of the mixture is antifreeze, how much of the mixture should be drained and replaced by pure antifreeze in order that the resultant mixture will contain 60% antifreeze?

Solution Let

$$x = \text{the number of qts to be drained.}$$

Since there were 8 quarts in the original 40% mixture, we may picture the problem as shown in Figure 2.2. This gives us the equation

$$0.4(8 - x) + x = 0.6(8)$$

or equivalently

$$3.2 - 0.4x + x = 4.8$$
$$-0.4x + x = 4.8 - 3.2$$
$$0.6x = 1.6$$
$$x = \frac{1.6}{0.6} = \frac{8}{3}$$

Thus 8/3 quarts should be drained from the original mixture.

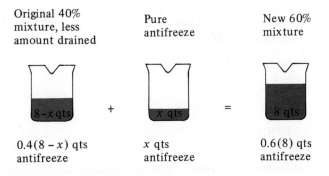

Original 40%
mixture, less
amount drained

Pure
antifreeze

New 60%
mixture

8−x qts   +   x qts   =   8 qts

0.4(8 − x) qts
antifreeze

x qts
antifreeze

0.6(8) qts
antifreeze

Figure 2.2

To check, let us first note that the amount of antifreeze in the original 8 quart mixture was 0.4(8), or 3.2 quarts. In draining 8/3 quarts of the original 40 % mixture, we lose 0.4(8/3) quarts of antifreeze and hence there remain $(3.2) - 0.4(8/3)$ quarts of antifreeze. If we then add 8/3 quarts of pure antifreeze, the amount of antifreeze in the final mixture is $(3.2) - 0.4(8/3) + 8/3$ quarts. This reduces to 4.8, which is 60 % of 8.

Many applied problems have to do with objects that move at a constant, or uniform, rate. If an object travels at a uniform (or average) rate r, then the distance d traversed in time t is given by $d = rt$. Of course we assume that the units are properly chosen; that is, if r is in feet per second, then t is in seconds, and so on.

Example 9   Two cities A and B are connected by means of a highway 150 miles long. An automobile leaves A at 1:00 P.M. and travels at a uniform rate of 40 miles per hour toward B. Thirty minutes later, another automobile leaves A and travels toward B at a uniform rate of 55 miles per hour. At what time will the second car overtake the first car?

Solution   Let t denote the time, in hours, *after* 1:30 P.M. At 1:30 P.M. the first automobile has already traveled 20 miles. Hence at time t *after* 1:30 P.M. the distance it has traveled is $20 + 40t$ miles. Since the second automobile starts the trip at 1:30 P.M., the distance it has traveled at time t is $55t$ miles. We wish to find the time t at which the distances traveled by the two automobiles are equal. This will be true when

$$55t = 20 + 40t.$$

Solving for t, we obtain

$$55t - 40t = 20$$
$$15t = 20$$
$$t = \tfrac{4}{3}.$$

Consequently, $t = 1\tfrac{1}{3}$ hours, or equivalently, 1 hour and 20 minutes. Since this is the amount of time after 1:30 P.M., it follows that the second car overtakes the first at 2:50 P.M.

To check our answer we note that at 2:50 P.M. the first car has traveled for $1\tfrac{5}{6}$ hours and its distance from A is $40(11/6) = 220/3$ miles. At 2:50 P.M. the second car has traveled for $1\tfrac{1}{3}$ hours and is $55(4/3) = 220/3$ miles from A. Hence they are together at 2:50 P.M.

EXERCISES

The formulas in Exercises 1–20 occur in mathematics and its applications. Solve each for the indicated variable in terms of the remaining variables.

1 $A = \frac{1}{2}bh$ for h

2 $C = 2\pi r$ for r

3 $V = \frac{1}{3}\pi r^2 h$ for h

4 $F = g\dfrac{m_1 m_2}{d^2}$ for m_1

5 $\dfrac{1}{R} = \dfrac{1}{R_1} + \dfrac{1}{R_2} + \dfrac{1}{R_3}$ for R_2

6 $\dfrac{x}{a} + \dfrac{y}{b} = 1$ for y

7 $S = P + Prt$ for P

8 $F = \frac{9}{5}C + 32$ for C

9 $V = \frac{1}{3}\pi h^2 (3r - h)$ for r

10 $s = \frac{1}{2}gt^2 + v_0 t$ for v_0

11 $S = \dfrac{a - rl}{1 - r}$ for r

12 $S = a + (n - 1)d$ for n

13 $Ft = mv_1 - mv_2$ for m

14 $\dfrac{1}{f} = \dfrac{1}{f_1} + \dfrac{1}{f_2}$ for f_1

15 $A = \frac{1}{2}(b_1 + b_2)h$ for b_1

16 $A = 2\pi r(r + h)$ for h

17 $a = \dfrac{v_2 - v_1}{t}$ for v_1

18 $l = l_0(1 + ct)$ for c

19 $R = \dfrac{nE - rI}{nI}$ for n

20 $E = \dfrac{T_1 - T_2}{T_1}$ for T_1

21 A newspaper boy collects $13.45 in dimes and quarters. If there are 70 coins in all, how many quarters does he have?

22 A girl has 125 coins consisting of nickels and pennies. If the total amount is $4.25, how many coins of each type does she have?

23 Find four consecutive integers whose sum is 550.

24 Find two consecutive integers such that the difference of their squares is 133.

25 The relationship between the temperature F on the Fahrenheit scale and the temperature C on the Celsius scale is given by

$$C = \tfrac{5}{9}(F - 32).$$

Find the temperature at which the reading is the same on both scales.

26 Refer to Exercise 25. When will the Celsius reading be twice the Fahrenheit reading?

27 A student in an algebra course has test scores of 75, 82, 71, and 84. What score on the next test will raise the student's average to 80?

28 Going into the final exam a student has test scores of 72, 80, 65, 78, and 60. If the final exam counts as 1/3 of the final grade, what score must the student receive in order to end up with an average of 76?

29 A businesswoman wishes to invest $30,000 in two different funds which yield annual profits of 6% and $8\frac{1}{2}$%, respectively. How much should she invest in each in order to realize a profit of $2,100 after one year?

30 A banker plans to lend part of $24,000 at a simple interest rate of 7% and the remainder at $8\frac{1}{2}$%. How should he allot the loans in order to obtain a return of $7\frac{1}{2}$% after one year?

31 A college student has $3,000 in two different savings accounts, paying interest at the rates of $4\frac{1}{2}$% and 5%, respectively. If the total yearly interest is $144.10 how much is deposited in each account?

32 A man has $4,000 more invested at $7\frac{1}{2}\%$ simple interest than he has at 6%. If his total yearly interest is $467.40, how much is invested at each rate?

33 Two boys who are 224 meters apart start walking toward each other at the same time. If they walk at rates of 1.5 and 2 meters per second, respectively, when will they meet? How far will each have walked?

34 A jogger starts from a certain point and runs at a constant rate of 6 mph. Five minutes later a second jogger begins at the same point, running at a rate of 8 mph and following the same course. How long will it take the second jogger to catch up with the first?

35 If the radius of a circle is increased by 2 cm its area increases by $16\pi\,\text{cm}^2$. What is the original radius of the circle?

36 A rectangle is twice as long as it is wide. If the length and width are decreased by 2 cm and 3 cm, respectively, the area is decreased by $30\,\text{cm}^2$. Find the original dimensions.

37 After playing 100 games, a major league baseball team has a record of 0.650. If it wins only 50% of its games for the remainder of the season, when will its record be 0.600?

38 A projectile is fired horizontally at a target and the sound of its impact is heard 1.5 seconds later. If the speed of the projectile is 3300 ft/sec and the speed of sound is 1100 ft/sec, how far away is the target?

39 Twenty liters of a solution contains 20% of a certain chemical. How much water should be added so that the resulting solution contains 15% of the chemical?

40 How many grams of an alloy containing 25% silver should be melted with 50 grams of an alloy containing 60% silver in order to obtain an alloy containing 50% silver?

41 How much water should be added to 1 liter of pure acid in order to obtain a solution which is 25% acid?

42 An automobile 20 feet long overtakes a truck that is 40 feet long which is traveling at 50 mph. At what constant speed must the automobile travel in order to pass the truck in 5 seconds?

43 A bus traveled from one city to another at an average rate of 50 mph. On the return trip the average rate was 45 mph and the elapsed time was 15 minutes longer. What was the total distance traveled?

44 At a concert certain tickets sold for $5.50 and others sold for $3.75. If a total of 520 tickets were sold for $2,448.75, how many of each kind were sold?

45 It takes a boy 90 minutes to mow his father's yard, but his sister can do it in 60 minutes. How long would it take them to mow the lawn if they worked together, using two lawnmowers?

46 Using water from one outlet, a swimming pool can be filled in 8 hours. A second, larger, outlet used alone can fill the pool in 5 hours. How long would it take to fill the pool if both outlets are used simultaneously?

47 How much water must be evaporated from 500 grams of a 10% salt solution in order to obtain a solution containing 15% salt?

48 A chemist has two acid solutions, the first containing 20% acid and the second 35% acid. How many ml of each should be mixed to obtain 50 ml of a solution containing 30% acid?

49 Generalize Exercise 47 as follows: Let a, b, and d be positive real numbers with $a < b < 100$. Given a solution of d grams of salt water that is $a\%$ salt, how much water must be evaporated in order that the resulting mixture will be $b\%$ salt? Express the answer in terms of a, b, and d.

50 Let a, b, c, and d be positive real numbers where $a < c < b < 100$. Generalize Exercise 48 to the case where the given solutions contain $a\%$ and $b\%$ acid respectively, and where d ml of a solution containing $c\%$ acid is desired. Express the answer in terms of a, b, c, and d.

3   QUADRATIC EQUATIONS

A **quadratic equation** is an equation of the form

(2.1)
$$ax^2 + bx + c = 0$$

where a, b, and c are real numbers with $a \neq 0$. An important technique for solving quadratic equations makes use of the fact that if p and q are real numbers such that $pq = 0$, then either $p = 0$ or $q = 0$ (see (1.7)). It follows that if the left side of (2.1) can be expressed as a product of two first-degree polynomials, then the solutions can be found by setting each factor equal to 0, as illustrated in the next example.

Example 1   Solve the equation $3x^2 + x - 10 = 0$.

Solution   The equation may be written in the form

$$(3x - 5)(x + 2) = 0$$

and hence either

$$3x - 5 = 0 \quad \text{or} \quad x + 2 = 0.$$

The solutions of these linear equations are

$$x = \tfrac{5}{3} \quad \text{and} \quad x = -2.$$

This shows that if a real number satisfies the quadratic equation then it must equal either 5/3 or -2. The fact that both numbers actually are roots may be seen by direct substitution in the given equation.

The method used for solving the equation in Example 1 is called the **method of factoring**.

Example 2   Solve $x^2 - 8x + 16 = 0$.

Solution   Factoring we see that the equation is equivalent to

$$(x - 4)^2 = 0 \quad \text{or} \quad (x - 4)(x - 4) = 0.$$

Setting each factor equal to zero, we obtain $x - 4 = 0$ and $x - 4 = 0$. Hence the given equation has one solution, $x = 4$.

Since $x - 4$ appears as a factor twice in the previous solution, the real number 4 is called a **double root**, or **root of multiplicity two**, of the equation $x^2 - 8x + 16 = 0$.

Given a quadratic equation of the form $x^2 - d = 0$, where $d \geq 0$, we may write $x^2 - d = 0$, or equivalently,

$$(x + \sqrt{d})(x - \sqrt{d}) = 0.$$

Setting each factor equal to zero gives us the solutions $-\sqrt{d}$ and $\sqrt{d}$. We frequently use $\pm\sqrt{d}$ as an abbreviation for solutions of this type. Thus for $d \geq 0$ we have shown that

$$\boxed{\text{If}\quad x^2 = d, \quad \text{then} \quad x = \pm\sqrt{d}.}$$

The process of solving $x^2 = d$ as indicated in the box may be referred to by the phrase "take the square root of both sides of the equation." Note that in so doing we introduce a positive and a negative square root, and not only the principal square root defined in Section 5 of Chapter One.

Example 3 Solve the equation $x^2 = 5$.

Solution Taking the square root of both sides gives us $x = \pm\sqrt{5}$. Thus the solutions are $\sqrt{5}$ and $-\sqrt{5}$.

We shall now derive a general formula for solving any equation of the form

$$ax^2 + bx + c = 0$$

where $a \neq 0$. We begin by dividing both sides by a, obtaining

$$x^2 + \frac{b}{a}x + \frac{c}{a} = 0$$

or equivalently

$$x^2 + \frac{b}{a}x = -\frac{c}{a}.$$

Next we shall add a certain number to both sides so that the resulting expression on the left may be written in the form $(x + d)^2$ for some real number d. This can be accomplished by adding the square of half the coefficient of x as follows:

$$x^2 + \frac{b}{a}x + \left(\frac{b}{2a}\right)^2 = \left(\frac{b}{2a}\right)^2 - \frac{c}{a}.$$

We may now write

$$\left(x + \frac{b}{2a}\right)^2 = \frac{b^2 - 4ac}{4a^2}$$

$$x + \frac{b}{2a} = \pm\sqrt{\frac{b^2 - 4ac}{4a^2}}$$

$$x = -\frac{b}{2a} \pm \sqrt{\frac{b^2 - 4ac}{4a^2}}.$$

Note that since $4a^2 > 0$, the radicand $(b^2 - 4ac)/4a^2$ will be nonnegative if and only if $b^2 - 4ac \geq 0$. We may change the radical on the right-hand side of the last equation as follows:

$$\pm\sqrt{\frac{b^2 - 4ac}{4a^2}} = \pm\frac{\sqrt{b^2 - 4ac}}{\sqrt{(2a)^2}} = \pm\frac{\sqrt{b^2 - 4ac}}{|2a|}.$$

If $a > 0$, then $|2a| = 2a$ and the expression equals

$$\pm\frac{\sqrt{b^2 - 4ac}}{2a}.$$

If $a < 0$, then $|2a| = -2a$ and we obtain

$$\pm\frac{\sqrt{b^2 - 4ac}}{2a}.$$

Consequently, whether a is positive or negative,

$$x = -\frac{b}{2a} \pm \frac{\sqrt{b^2 - 4ac}}{2a}.$$

We have shown that if the quadratic equation (2.1) has roots, then they are given by the two numbers in the last formula. Moreover, it can be shown by direct substitution that the two numbers do satisfy (2.1). This gives us the following very important fact.

(2.2)   Quadratic Formula

If $a \neq 0$, then the roots of the equation $ax^2 + bx + c = 0$ are given by

$$x = \frac{-b \pm \sqrt{b^2 - 4ac}}{2a}.$$

The number $b^2 - 4ac$ which appears under the radical sign in the quadratic formula is called the **discriminant** of the quadratic equation. It can be used to determine the nature of the roots of the equation as follows, where a, b, and c are real numbers.

(i)   If $b^2 - 4ac = 0$, the equation has a double root.
(ii)   If $b^2 - 4ac > 0$, the equation has two real and unequal solutions.
(iii) If $b^2 - 4ac < 0$, the equation has no real solutions.

Example 4   Solve $4x^2 + x - 3 = 0$.

Solution   Letting $a = 4$, $b = 1$, and $c = -3$ in (2.2), we obtain

$$x = \frac{-1 \pm \sqrt{1 - 4(4)(-3)}}{2(4)}$$

$$= \frac{-1 \pm \sqrt{49}}{8}$$

$$= \frac{-1 \pm 7}{8}.$$

Hence the solutions are

$$x = \frac{-1 + 7}{8} = \frac{3}{4} \quad \text{and} \quad x = \frac{-1 - 7}{8} = -1.$$

Example 4 could also have been solved by factoring. If we write $(4x - 3)(x + 1) = 0$ and set each factor equal to zero, we obtain the solutions $x = 3/4$ and $x = -1$.

Example 5   Solve $2x^2 - 6x + 3 = 0$.

Solution   Letting $a = 2$, $b = -6$, and $c = 3$ in the quadratic formula (2.2) we obtain

$$x = \frac{6 \pm \sqrt{(-6)^2 - 4(2)(3)}}{2(2)}$$

$$= \frac{6 \pm \sqrt{12}}{4}$$

$$= \frac{6 \pm 2\sqrt{3}}{4}$$

$$= \frac{3 \pm \sqrt{3}}{2}.$$

Hence the solutions are $x = (3 \pm \sqrt{3})/2$.

The following example illustrates the case of a double root.

Example 6   Solve the equation $9x^2 - 30x + 25 = 0$.

Solution   Letting $a = 9$, $b = -30$, and $c = 25$ in the quadratic formula gives us

$$x = \frac{30 \pm \sqrt{(-30)^2 - 4(9)(25)}}{2(9)}$$

$$= \frac{30 \pm \sqrt{900 - 900}}{18}$$

$$= \frac{30 \pm 0}{18} = \frac{5}{3}.$$

Consequently there is only one solution, 5/3.

There are many applied problems which lead to quadratic equations. One is illustrated in the following example.

Example 7   A box with a square base and no top is to be made from a square piece of tin by cutting out 3-inch squares from each corner and folding up the sides. If the box is to hold 48 cubic inches, what size piece of tin should be used?

Solution   If we let x denote the length of the side of the piece of tin, then the length of the base of the box is $x - 6$ (see Figure 2.3). Since the area of the base is $(x - 6)^2$ and the height is 3, the volume of the box is $3(x - 6)^2$. Moreover, since the box is to hold 48 cubic inches,

$$3(x - 6)^2 = 48.$$

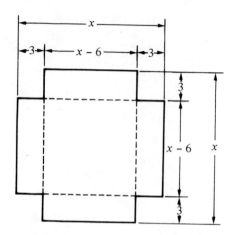
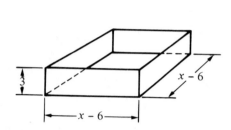

Figure 2.3

Solving for x we obtain

$$(x - 6)^2 = 16$$
$$x - 6 = \pm 4$$
$$x = 6 \pm 4.$$

Consequently either $x = 10$ or $x = 2$.

Let us now check each of these numbers. Referring to Figure 2.3 we see that 2 is unacceptable since no box is possible in this case. (Why?) However, if we begin with a 10-inch square of tin, cut out 3-inch corners and fold, we obtain a box having dimensions 4 inches, 4 inches, and 3 inches. The box has the desired volume of 48 cubic inches. Thus 10 inches is the answer to the problem.

As illustrated in Example 7, even though an equation is formulated correctly, it is possible, owing to the physical nature of a given problem, to arrive at meaningless solutions. These solutions should be discarded. For example, we would not accept the answer -7 years for the age of an individual nor $\sqrt{50}$ for the number of automobiles in a parking lot.

EXERCISES

Solve the equations in Exercises 1–10 by factoring.

1   $6x^2 + 11x - 10 = 0$

2   $15x^2 + x - 6 = 0$

3   $20x^2 - 33x + 7 = 0$

4   $4x^2 + 16x + 15 = 0$

5   $4y^2 + 29y + 30 = 0$

6   $8z^2 + 19z - 27 = 0$

7   $16t^2 - 24t + 9 = 0$

8   $9u^2 - 30u + 25 = 0$

9   $54r^2 - 9r - 30 = 0$

10   $60s^2 - 85s - 35 = 0$

Use the quadratic formula to solve the equations in Exercises 11–24.

11 $2x^2 - x - 3 = 0$

12 $3x^2 - 2x - 8 = 0$

13 $u^2 + 2u - 6 = 0$

14 $v^2 + 3v - 5 = 0$

15 $2x^2 - 4x - 5 = 0$

16 $3x^2 - 6x + 2 = 0$

17 $4y^2 - 20y + 25 = 0$

18 $9t^2 + 6t + 1 = 0$

19 $5s^2 + 9s + 3 = 0$

20 $3z^2 - 8z - 2 = 0$

21 $\dfrac{x + 1}{3x + 2} = \dfrac{x - 2}{2x - 3}$

22 $\dfrac{5}{w^2} - \dfrac{10}{w} + 2 = 0$

23 $9x^2 + 4x = 0$

24 $16x^2 - 9 = 0$

The formulas in Exercises 25–32 occur in mathematics and its applications. Solve for the indicated variable in terms of the remaining variables (all letters denote positive numbers).

25 $V = \frac{1}{3}\pi r^2 h$ for r

26 $K = \frac{1}{2}mv^2$ for v

27 $F = g\dfrac{m_1 m_2}{d^2}$ for d

28 $V = \frac{4}{3}\pi a^2 b$ for a

29 $s = \frac{1}{2}gt^2 + v_0 t$ for t

30 $A = 2\pi r(r + h)$ for r

31 $\dfrac{x^2}{a^2} - \dfrac{y^2}{b^2} = 1$ for y

32 $s = \sqrt{(r_1 - r_2)^2 + h^2}$ for r_2

33 Find two consecutive odd integers whose product is 255.

34 Find two consecutive even integers, the sum of whose squares is 1060.

35 The diameter of a circle is 10 cm. What change in the radius will decrease the area by 16π cm^2?

36 The hypotenuse of a right triangle is 5 cm long. Find the lengths of the two legs if their sum is 6 cm.

37 A rectangular plot of ground having dimensions 120 ft by 160 ft is surrounded by a walk of uniform width. If the area of the walk is 1425 ft^2, what is its width?

38 It can be shown by mathematical induction (see Chapter 10), that the sum of the first n positive integers 1, 2, 3, ..., n equals $\frac{1}{2}n(n + 1)$. For what value of n will the sum equal 276?

39 An airplane flying north at 320 mph passed over a point on the ground at 2:00 P.M. Another airplane at the same altitude passed over the point at 2:15 P.M., flying east at 300 mph. At what time were the airplanes 500 miles apart?

40 A box with an open top is to be constructed by cutting out 3-inch squares from a rectangular sheet of tin whose length is twice its width. What size sheet will produce a box having a volume of 60 in^3?

41 A piece of wire 100 inches long is cut into two pieces and then each piece is bent into the shape of a square. If the sum of the enclosed areas is 397 in^2, find the lengths of each piece of wire.

42 Generalize Exercise 41 to the case where the wire is l inches long and the area is A in^2. Express the lengths of the two pieces in terms of l and A.

43   A projectile is fired straight upward with an initial speed of 800 ft/sec. The number of feet s above the ground after t seconds is given by $s = -16t^2 + 800t$.
(a)   When will the projectile be 3200 feet above the ground?
(b)   When will it hit the ground?
(c)   What is its maximum height?

44   If a projectile is fired upward from a height of s_0 feet above the ground with an initial speed of v_0 feet per second, then its height above the ground at time t is given by $s = -16t^2 + v_0 t + s_0$. Solve for t in terms of s, v_0, and s_0. When will the projectile hit the ground? Express your answer in terms of v_0 and s_0.

45   If one solution of the equation $kx^2 + 7x + 3k = 0$ is -5, find the other solution.

46   If a, b, and c are rational numbers, can the equation $ax^2 + bx + c = 0$ have one rational and one irrational root? Explain.

47   If r_1 and r_2 are the two roots of $ax^2 + bx + c = 0$, show that $r_1 + r_2 = -b/a$ and $r_1 r_2 = c/a$.

48   Prove that the roots of $ax^2 + bx + c = 0$ are numerically equal but opposite in sign to the roots of $ax^2 - bx + c = 0$.

49   Given the equation $4x^2 - 4xy + 1 - y^2 = 0$ use the quadratic formula to solve for (a) x in terms of y, (b) y in terms of x.

50   Given the equation $2x^2 - xy = 3y^2 + 1$, use the quadratic formula to solve for (a) x in terms of y, (b) y in terms of x.

4   MISCELLANEOUS EQUATIONS

The method of factoring used in the previous section can also be applied to certain nonquadratic equations. Specifically, if an equation can be expressed in factored form, *with zero on one side*, then solutions may often be obtained by setting each factor equal to zero, as illustrated in the next two examples.

Example 1   Solve $x^3 + 2x^2 - x - 2 = 0$.

Solution   The left side may be factored by grouping, as follows:

$$x^3 + 2x^2 - x - 2 = 0$$
$$x^2(x + 2) - (x + 2) = 0$$
$$(x^2 - 1)(x + 2) = 0$$
$$(x + 1)(x - 1)(x + 2) = 0$$

Setting each factor equal to 0 gives us

$$x + 1 = 0, \quad x - 1 = 0, \quad x + 2 = 0$$

or equivalently,

$$x = -1, \quad x = 1, \quad x = -2.$$

That these three numbers are solutions may be checked by substitution in the original equation.

Example 2 Solve $x^{3/2} = x^{1/2}$.

Solution We may proceed as follows:

$$x^{3/2} - x^{1/2} = 0$$
$$x^{1/2}(x - 1) = 0$$

Hence $x^{1/2} = 0$ or $x - 1 = 0$, from which we obtain the solutions $x = 0$ and $x = 1$.

A common error, when an equation such as $x^{3/2} = x^{1/2}$ is given (see Example 2), is to divide both sides by $x^{1/2}$, obtaining $x = 1$. Observe that this leads to the loss of the solution $x = 0$.

Certain equations that are not linear or quadratic can be transformed into one of those forms by a suitable manipulation. If equations involve radicals or fractional exponents, the method of raising both sides to a positive integral power is often used. When this is done, the solutions of the new equation always contain the solutions of the original equation. For example, the solutions of

$$2x - 3 = \sqrt{x + 6}$$

are also solutions of

$$(2x - 3)^2 = (\sqrt{x + 6})^2.$$

In some cases the new equation has *more* solutions than the original equation. To illustrate, if we start with the equation $x = 3$ and square both sides we obtain $x^2 = 9$. Note that the given equation has only one solution, 3, whereas the new equation has two solutions, 3 and -3. Any solution of the new equation which is not a solution of the original equation is called an **extraneous solution**. Since extraneous solutions may arise, it is *absolutely essential* to check all solutions obtained after raising both sides of an equation to some power.

Example 3 Solve $\sqrt[3]{x^2 - 1} = 2$.

Solution If we cube both sides, then the solutions of the given equation are included among the solutions of the following:

$$(\sqrt[3]{x^2 - 1})^3 = 2^3$$
$$x^2 - 1 = 8$$
$$x^2 = 9.$$

Hence the only possible solutions of $\sqrt[3]{x^2 - 1} = 2$ are 3 or -3.

We next check each of these numbers by substitution in $\sqrt[3]{x^2 - 1} = 2$. Substituting 3 for x in the equation we obtain $\sqrt[3]{3^2 - 1} = 2$, or $\sqrt[3]{8} = 2$, which is a true statement. Thus 3 is a solution. Similarly, -3 is a solution. Hence the solutions of the given equation are 3 and -3.

Example 4   Solve $3 + \sqrt{3x + 1} = x$.

Solution   We begin by isolating the radical on one side as follows:

$$\sqrt{3x + 1} = x - 3.$$

Next we square both sides and simplify, obtaining

$$(\sqrt{3x + 1})^2 = (x - 3)^2$$
$$3x + 1 = x^2 - 6x + 9$$
$$x^2 - 9x + 8 = 0$$
$$(x - 1)(x - 8) = 0$$

Since the last equation has solutions 1 and 8, it follows that 1 and 8 are the only possible solutions of the original equation.

We now check each of these by substitution in $3 + \sqrt{3x + 1} = x$. Letting $x = 1$ gives us

$$3 + \sqrt{4} = 1 \qquad \text{or} \qquad 5 = 1$$

which is false. Consequently 1 is not a solution. Letting $x = 8$ in the given equation we obtain

$$3 + \sqrt{25} = 8 \qquad \text{or} \qquad 3 + 5 = 8$$

which is true. Hence the equation $3 + \sqrt{3x + 1} = x$ has only one solution, $x = 8$.

For certain equations involving radicals it is necessary to use the process of raising sides to powers several times, as illustrated in the next example.

Example 5   Solve $\sqrt{2x - 3} - \sqrt{x + 7} + 2 = 0$.

Solution   Let us begin by writing

$$\sqrt{2x - 3} = \sqrt{x + 7} - 2.$$

Squaring both sides we obtain

$$2x - 3 = (x + 7) - 4\sqrt{x + 7} + 4$$

which simplifies to

$$x - 14 = -4\sqrt{x + 7}.$$

Squaring both sides of the last equation and simplifying gives us

$$x^2 - 28x + 196 = 16(x + 7)$$
$$x^2 - 28x + 196 = 16x + 112$$
$$x^2 - 44x + 84 = 0$$
$$(x - 42)(x - 2) = 0.$$

Hence the only possible solutions of the given equation are 42 and 2.

We next check each of these by substitution in the original equation. Substituting $x = 42$ gives us

$$\sqrt{84 - 3} - \sqrt{42 + 7} + 2 = 0$$

or

$$9 - 7 + 2 = 0$$

which is false. Hence 42 is not a solution. If we substitute $x = 2$ we obtain

$$\sqrt{4 - 3} - \sqrt{2 + 7} + 2 = 0$$

or

$$1 - 3 + 2 = 0$$

which is true. Hence the given equation has one solution, $x = 2$.

An equation in a variable x is said to be of **quadratic type** if it can be written in the form

$$au^2 + bu + c = 0$$

where $a \neq 0$ and u is an expression in some variable. If we find the solutions in terms of u, then the solutions of the original equation can be obtained by referring to the specific form of u. The technique is illustrated in the following example.

Example 6   Solve $x^{2/3} + x^{1/3} - 6 = 0$.

Solution   If we let $u = x^{1/3}$, then the equation can be written

$$u^2 + u - 6 = 0 \quad \text{or} \quad (u + 3)(u - 2) = 0$$

which has solutions

$$u = -3 \quad \text{and} \quad u = 2.$$

Since $u = x^{1/3}$ we have

$$x^{1/3} = -3 \quad \text{or} \quad x^{1/3} = 2.$$

Cubing gives us

$$x = -27 \quad \text{or} \quad x = 8.$$

We next check each of these by substitution. Letting $x = -27$ in the given equation we obtain

$$[(-27)^{1/3}]^2 + (-27)^{1/3} - 6 = 9 - 3 - 6 = 0.$$

Thus -27 is a solution. Similarly it can be shown that 8 is a solution. Hence the solutions of the given equations are -27 and 8.

Example 7   Solve $x^4 - 3x^2 + 1 = 0$.

Solution   Letting $u = x^2$ gives us

$$u^2 - 3u + 1 = 0.$$

Using the quadratic formula, we obtain

$$u = \frac{3 \pm \sqrt{9-4}}{2} = \frac{3 \pm \sqrt{5}}{2}.$$

Since $u = x^2$ we have

$$x^2 = \frac{3 \pm \sqrt{5}}{2} \quad \text{or} \quad x = \pm \sqrt{\frac{3 \pm \sqrt{5}}{2}}.$$

Thus there are four possible solutions:

$$\sqrt{\frac{3+\sqrt{5}}{2}}, \quad -\sqrt{\frac{3+\sqrt{5}}{2}}, \quad \sqrt{\frac{3-\sqrt{5}}{2}}, \quad -\sqrt{\frac{3-\sqrt{5}}{2}}.$$

By checking it can be shown that each is a solution of the original equation.

EXERCISES

Solve the equations in Exercises 1–36.

1   $4x^3 + 12x^2 - 9x - 27 = 0$

2   $2x^3 + 5x^2 - 8x - 20 = 0$

3   $6x^5 + 10x^4 = 3x^3 + 5x^2$

4   $25z^4 + 5z = 125z^3 + z^2$

5   $y^{3/2} = 4y$

6   $2x^3 = 5x^2$

7   $\dfrac{1}{x} = \dfrac{1}{x^2}$

8   $\sqrt{x}\sqrt[3]{x} - 3\sqrt[3]{x} - 2\sqrt{x} + 6 = 0$

9   $\sqrt{7 - 5x} = 8$

10   $\sqrt{2x - 9} = 3^{-1}$

11   $2 + \sqrt[3]{1 - 5t} = 0$

12   $\sqrt[3]{6 - s^2} + 5 = 0$

13   $\sqrt[5]{2x^2 + 1} - 2 = 0$

14   $\sqrt[4]{2x^2 - 1} = x$

15   $3\sqrt{2x - 3} + 2\sqrt{7 - x} = 11$

16   $\sqrt{2x + 15} - 2 = \sqrt{6x + 1}$

17   $\sqrt{7 - 2x} - \sqrt{5 + x} = \sqrt{4 + 3x}$

18   $4\sqrt{1 + 3x} + \sqrt{6x + 3} = \sqrt{-6x - 1}$

19 $\sqrt{11 + 8x} + 1 = \sqrt{9 + 4x}$

20 $2\sqrt{x} - \sqrt{x - 3} = \sqrt{5 + x}$

21 $\sqrt{2\sqrt{x + 1}} = \sqrt{3x - 5}$

22 $\sqrt{5\sqrt{x}} = \sqrt{2x - 3}$

23 $\sqrt{1 + 4\sqrt{x}} = \sqrt{x} + 1$

24 $\sqrt{x + 1} = \sqrt{x - 1}$

25 $4x^4 - 37x^2 + 9 = 0$

26 $2x^4 - 9x^2 + 4 = 0$

27 $3z^4 - 5z^2 + 1 = 0$

28 $2y^4 + y^2 - 5 = 0$

29 $3x^{2/3} + 8x^{1/3} - 3 = 0$

30 $2t^{1/3} - 5t^{1/6} + 2 = 0$

31 $3w - 19\sqrt{w} + 20 = 0$

32 $2x^{-2/3} + 7x^{-1/3} - 4 = 0$

33 $\dfrac{2}{(x - 1)^2} + \dfrac{3}{x - 1} - 2 = 0$

34 $\dfrac{4}{(x^2 - 1)^2} - \dfrac{5}{x^2 - 1} + 1 = 0$

35 $\left(\dfrac{t}{t + 1}\right)^2 + \dfrac{2t}{t + 1} - 15 = 0$

36 $6u^{-1/2} - 17u^{-1/4} + 5 = 0$

The formulas in Exercises 37–42 occur in mathematics and its applications. Solve for the indicated variable in terms of the remaining variables (all letters denote positive real numbers).

37 $S = \pi r \sqrt{r^2 + h^2}$ for h

38 $d = \frac{1}{2}\sqrt{4R^2 - C^2}$ for C

39 $y = \dfrac{b}{a}\sqrt{a^2 - x^2}$ for x

40 $T = 2\pi\sqrt{\dfrac{l}{g}}$ for l

41 $x^{2/3} + y^{2/3} = a^{2/3}$ for y

42 $y = (\sqrt[3]{a} - \sqrt[3]{x})^3$ for x

5 INEQUALITIES

If a and b are real numbers, then either of the inequalities $a > b$ or $b < a$ means that $a - b$ is positive (see (1.12)). In particular, we may write

$$7 > 3, \quad 2 > -5, \quad -6 < -4, \quad \text{or} \quad \sqrt{2} < \pi.$$

If we refer to a coordinate line, then $a > b$ (or $b < a$) implies that the point with coordinate a lies to the right of the point with coordinate b. In addition, if $b > c$, then we have a situation similar to that shown in Figure 2.4, and evidently $a > c$. This provides a geometric demonstration of the following general rule for inequalities:

$$\text{If } a > b \text{ and } b > c, \quad \text{then} \quad a > c.$$

Figure 2.4

When working with an inequality we often employ techniques which are similar to those used for solving equations. As we shall see, adding the same real number to both sides leaves the inequality sign invariant. To illustrate, if we add 2 to both sides of $7 > 3$ we obtain

$$7 + 2 > 3 + 2 \quad \text{or} \quad 9 > 5.$$

This fact is not surprising, since adding 2 to both sides of $7 > 3$ amounts to shifting the corresponding points on a coordinate line 2 units to the right, as illustrated in Figure 2.5. Similarly, adding -2 (or equivalently subtracting 2) would shift the points to the left. These are special cases of the following general rule:

If $a > b$, then $a + c > b + c.$

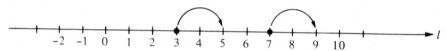

Figure 2.5

As indicated in (iii) of (2.3) which follows, we can also multiply both sides of an inequality by the same positive real number. For example, multiplying both sides of $7 > 3$ by 2 gives us

$$7 \cdot 2 > 3 \cdot 2 \quad \text{or} \quad 14 > 6.$$

A major difference from our work with equations is that if we multiply by a negative real number, then as in (iv) of (2.3), the inequality sign is reversed. To illustrate, multiplying both sides of $7 > 3$ by -2 gives us

$$7(-2) < 3(-2) \quad \text{or} \quad -14 < -6.$$

The four fundamental properties of inequalities referred to above may be stated as follows, where a, b, and c denote real numbers.

(2.3)

(i)	If $a > b$ and $b > c$, then $a > c$.
(ii)	If $a > b$, then $a + c > b + c$.
(iii)	If $a > b$ and $c > 0$, then $ac > bc$.
(iv)	If $a > b$ and $c < 0$, then $ac < bc$.

We can supply algebraic proofs for all four properties by making use of the fact that both the sum and product of any two positive real numbers are positive. Thus to prove rule (i) we first note that if $a > b$ and $b > c$, then $a - b$ and $b - c$ are both positive. Consequently, the sum $(a - b) + (b - c)$ is positive. Since the sum reduces to $a - c$, we see that $a - c$ is positive, which means that $a > c$.

To establish rule (ii) we again note that if $a > b$, then $a - b$ is positive. Since $(a + c) - (b + c) = a - b$, it follows that $(a + c) - (b + c)$ is positive; that is, $a + c > b + c$.

For rule (iii), if $a > b$ and $c > 0$, then $a - b$ and c are both positive and hence so is the product $(a - b)c$. Consequently $ac - bc$ is positive; that is, $ac > bc$.

Finally, to establish rule (iv) we first note that if $c < 0$, then $0 - c$, or $-c$, is positive. In addition, if $a > b$, then $a - b$ is positive and hence the product $(a - b)(-c)$ is positive. However, $(a - b)(-c) = -ac + bc$ and hence $bc - ac$, is positive. This means that $bc > ac$, or $ac < bc$.

Similar results hold for the symbol $<$. Specifically, the following can be proved.

(2.4)

(i)	If $a < b$ and $b < c$, then $a < c$.
(ii)	If $a < b$, then $a + c < b + c$.
(iii)	If $a < b$ and $c > 0$, then $ac < bc$.
(iv)	If $a < b$ and $c < 0$, then $ac > bc$.

We shall next consider inequalities which contain variables. The domains of the variables will be sets of real numbers. As a first illustration, let us consider

$$x^2 - 3 < 2x + 4.$$

If certain numbers such as 4 or 5 are substituted for x, we obtain the false statements $13 < 12$ or $22 < 14$, respectively. Other numbers such as 1 or 2 produce the true statements $-2 < 6$ or $1 < 8$, respectively. In general, if we are given an inequality in x and if a true statement is obtained when x is replaced by a real number a, then a is called a **solution** of the inequality. Thus 1 and 2 are solutions of the inequality $x^2 - 3 < 2x + 4$, whereas 4 and 5 are not solutions. To **solve** an inequality means to find all solutions. We say that two inequalities are **equivalent** if they have exactly the same solutions.

As with equations, a standard method for solving an inequality is to replace it with a chain of equivalent inequalities, terminating in one for which the solutions are obvious. The main tools used in applying this method are the properties listed in (2.3) and (2.4). For example, if x represents a real number, then adding the same expression in x to both sides leads to an equivalent inequality. We may multiply both sides of an inequality by an expression containing x if we are certain that the expression is positive for all values of x under consideration. To illustrate, multiplication by $x^4 + 3x^2 + 5$ would be permissible since this expression is always positive. If we multiply both sides of an inequality by an expression that is always negative, such as $-7 - x^2$, then the inequality sign is reversed.

The reader should supply reasons for the solutions of the following inequalities, referring to (2.3) and (2.4) if necessary.

Example 1   Solve the inequality $-3x + 4 > 11$.

Solution   The following inequalities are all equivalent.

$$-3x + 4 < 11$$
$$(-3x + 4) + (-4) < 11 + (-4)$$
$$-3x < 7$$
$$(-\tfrac{1}{3})(-3x) > (-\tfrac{1}{3})(7)$$
$$x > -\tfrac{7}{3}$$

Since the last inequality is equivalent to the first, the solutions of $-3x + 4 < 11$ consist of all real numbers x such that $x > -7/3$.

Example 2   Solve the inequality $4x - 3 < 2x + 5$.

Solution   The following is a list of equivalent inequalities:

$$4x - 3 < 2x + 5$$
$$(4x - 3) + 3 < (2x + 5) + 3$$
$$4x < 2x + 8$$
$$4x + (-2x) < (2x + 8) + (-2x)$$
$$2x < 8$$
$$\tfrac{1}{2}(2x) < \tfrac{1}{2}(8)$$
$$x < 4$$

Hence the solutions of the given inequality consist of all real numbers x such that $x < 4$.

It is convenient to give graphical interpretations for solutions of inequalities. By the **graph** of a set of real numbers we mean the collection of all points on a coordinate line l which correspond to the numbers. To **sketch a graph** we darken or color an appropriate portion of l. The graph corresponding to the solutions of Example 2 consists of all points to the left of the point with coordinate 4 and is sketched in Figure 1.5. The parenthesis in the figure indicates that the point corresponding to 4 is not part of the graph. If we wish to include such an **endpoint** of a graph we use a bracket instead of a parenthesis. This will be illustrated in Example 4.

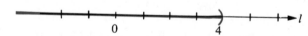

Figure 1.5

If $a < b$, the symbol (a, b) is sometimes used for all real numbers between a and b. This may be expressed in set notation as follows:

$$(a, b) = \{x : a < x < b\}$$

where the notation on the right is translated "the set of all x such that $a < x < b$."
The set (a, b) is called an **open interval**. The graph of (a, b) consists of all points on a
coordinate line which lie between the points corresponding to a and b. In Figure 2.7
we have sketched the graph of a general open interval (a, b) and also the special open
intervals $(-1, 3)$ and $(2, 4)$. The parentheses in the figure indicate that the endpoints
of the intervals are not to be included. For convenience we shall use the terms *open
interval* and *graph of an open interval* interchangeably.

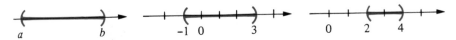

Figure 2.7. Open intervals (a, b), $(-1, 3)$ and $(2, 4)$

The solutions of an inequality may often be described in terms of intervals, as
illustrated in the next example.

Example 3 Solve the inequality $-6 < 2x - 4 < 2$ and represent the solutions graphically.

Solution A real number x is a solution of the given inequality if and only if it is a solution of
both of the inequalities

$$-6 < 2x - 4 \quad \text{and} \quad 2x - 4 < 2.$$

The first inequality is equivalent to each of the following:

$$-6 + 4 < (2x - 4) + 4$$
$$-2 < 2x$$
$$\tfrac{1}{2}(-2) < \tfrac{1}{2}(2x)$$
$$-1 < x.$$

The second inequality is equivalent to each of the following:

$$2x - 4 < 2$$
$$2x < 6$$
$$x < 3.$$

Thus x is a solution of the given inequality if and only if *both*

$$-1 < x \quad \text{and} \quad x < 3,$$

that is,

$$-1 < x < 3.$$

Hence the solutions are all numbers in the open interval $(-1, 3)$. The graph is
sketched in Figure 2.8.

Figure 2.8

An alternate (and shorter) solution may be given by working with both inequalities simultaneously as follows:

$$-6 < 2x - 4 < 2$$
$$-6 + 4 < 2x < 2 + 4$$
$$-2 < 2x < 6$$
$$-1 < x < 3.$$

The last inequality agrees with the first solution.

For convenience we refer to the interval $(-1, 3)$ as the **interval solution** of the inequality in Example 3, meaning that the solutions of the inequality consist of all numbers in that interval. We shall use this terminology in other problems, whenever applicable.

If we wish to include an endpoint of an interval, a bracket is used instead of a parenthesis. **Closed intervals**, denoted by $[a, b]$, and **half-open intervals**, denoted by $[a, b)$ or $(a, b]$, are defined as follows:

$$[a, b] = \{x : a \le x \le b\}$$
$$[a, b) = \{x : a \le x < b\}$$
$$(a, b] = \{x : a < x \le b\}$$

Typical graphs are sketched in Figure 1.7.

Figure 1.7

Example 4   Solve the inequality

$$-5 \le \frac{4 - 3x}{2} < 1$$

and illustrate the solutions graphically.

Solution   A number x is a solution if and only if it satisfies both of the inequalities

$$-5 \le \frac{4 - 3x}{2} \quad \text{and} \quad \frac{4 - 3x}{2} < 1.$$

We could work with each of these separately or proceed as in the alternate solution of Example 3. Let us employ the latter technique as follows:

$$-5 \le \frac{4 - 3x}{2} < 1$$

$$-10 \le 4 - 3x < 2$$

$$-10 - 4 \le -3x < 2 - 4$$

$$-14 \le -3x < -2$$

$$\left(-\frac{1}{3}\right)(-14) \ge \left(-\frac{1}{3}\right)(-3x) > \left(-\frac{1}{3}\right)(-2)$$

$$\frac{14}{3} \ge x > \frac{2}{3}$$

$$\frac{2}{3} < x \le \frac{14}{3}$$

Hence the interval solution of the inequality is $(2/3, 14/3]$. The graph is sketched in Figure 1.8

Figure 1.8

The next example is an elementary illustration of an applied problem involving inequalities.

Example 5   The temperature readings on the Fahrenheit and Celsius scales are related by the equation $C = \frac{5}{9}(F - 32)$. What range of F corresponds to $30 \le C \le 40$?

Solution   The following inequalities are equivalent:

$$30 \le C \le 40$$

$$30 \le \tfrac{5}{9}(F - 32) \le 40$$

$$\tfrac{9}{5}(30) \le F - 32 \le \tfrac{9}{5}(40)$$

$$54 \le F - 32 \le 72$$

$$86 \le F \le 104$$

Thus a Celsius temperature range from $30°$ C to $40°$ C is the same as a Fahrenheit range from $86°$ F to $104°$ F.

In order to describe solutions of inequalities such as $x < 4, x > -2, x \leq 7$, or $x \geq 3$, it is convenient to employ **infinite intervals**. In particular, if a is a real number we define

$$(-\infty, a) = \{x : x < a\}$$

that is, $(-\infty, a)$ denotes the set of all real numbers less than a. The symbol ∞ (**infinity**) is merely a notational device and is not to be interpreted as representing a real number. To illustrate, the interval solution in Example 1 is the infinite interval $(-\infty, -7/3)$ and that in Example 2 is $(-\infty, 4)$. The graph of $(-\infty, 4)$ is sketched in Figure 2.6. If we wish to include the point corresponding to a we write

$$(-\infty, a] = \{x : x \leq a\}.$$

Other types of infinite intervals are defined by

$$(a, \infty) = \{x : x > a\}.$$
$$[a, \infty) = \{x : x \geq a\}.$$

Example 6   Solve $\dfrac{1}{x - 2} > 0$ and sketch the graph corresponding to the solutions.

Solution   Since the numerator is positive, the given fraction is positive if and only if $x - 2 > 0$ or equivalently, $x > 2$. Hence the interval solution is $(2, \infty)$. The graph is sketched in Figure 2.11.

Figure 2.11

EXERCISES

1   Given $-6 < -4$, what inequality is obtained if
   (a)   5 is added to both sides?
   (b)   5 is subtracted from both sides?
   (c)   both sides are multiplied by $1/2$?
   (d)   both sides are multiplied by $-1/2$?

2 Given $3 > -3$, what inequality is obtained if
    (a) 3 is added to both sides?
    (b) -3 is added to both sides?
    (c) both sides are divided by 3?
    (d) both sides are divided by -3?

In each of Exercises 3–12, express the inequality in interval notation and sketch a graph of the interval.

3 $2 < x < 5$ 4 $-1 < x < 2$

5 $-1 < x \le 3$ 6 $-2 \ge x \ge -3$

7 $4 \ge x \ge 1$ 8 $0 < x \le 4$

9 $x > -1$ 10 $x < 1$

11 $x \le 2$ 12 $x \ge -3$

Express each of the intervals in Exercises 13–18 as an inequality in the variable x.

13 $(-1, 7)$ 14 $[5, 12]$

15 $(8, 9]$ 16 $[-5, 0)$

17 $(5, \infty)$ 18 $(-\infty, 7]$

Solve the inequalities in Exercises 19–42 and express the solutions in terms of intervals.

19 $5x - 6 > 11$ 20 $3x - 5 < 10$

21 $2 - 7x \le 16$ 22 $7 - 2x \ge -3$

23 $3x + 1 < 5x - 4$ 24 $6x - 5 > 9x + 1$

25 $4 - \frac{1}{2}x \ge -7 + \frac{1}{4}x$ 26 $\frac{1}{3}x - 4 \le \frac{1}{4}x - 6$

27 $-4 < 3x + 5 < 8$ 28 $6 > 2x - 6 > 4$

29 $3 \ge \dfrac{7 - x}{2} \ge 1$ 30 $-2 \le \dfrac{5 - 3x}{4} \le \dfrac{1}{2}$

31 $0 < 2 - \frac{3}{4}x \le \frac{1}{2}$ 32 $-3 < \frac{1}{2}x - 4 \le 0$

33 $(3x + 1)(2x - 4) < (6x - 2)(x + 5)$

34 $(x + 2)(x - 5) < (x - 1)^2$ 35 $(x + 2)^2 > x(x - 2)$

36 $2x(8x - 1) \ge (4x + 1)(4x - 3)$

37 $\dfrac{5}{3x + 7} > 0$ 38 $\dfrac{1}{6 - 2x} < 0$

39 $(1 - 4x)^{-1} < 0$ 40 $(x + 4)^{-1} > 0$

41 $\dfrac{1}{(x - 1)^2} > 0$ 42 $\dfrac{1}{x^2 + 1} < 0$

43 The relationship between the Fahrenheit and Celsius temperature scales is given by $C = (5/9)(F - 32)$. If $60 \le F \le 80$, express the corresponding range for C in terms of an inequality.

44 In the study of electricity, Ohm's Law states that $R = E/I$ where E is measured in volts, I in amperes, and R in ohms. If $E = 110$, what values of R correspond to $I \leq 10$?

45 According to Hooke's Law, the force F (in pounds) required to stretch a certain spring x inches beyond its natural length is given by $F = (4.5)x$. If $10 \leq F \leq 18$, what is the corresponding range for x?

46 Boyle's Law for a certain gas states that $pv = 200$, where p denotes the pressure (lbs/in^2) and v denotes the volume (in^3). If $25 \leq v \leq 50$, what is the corresponding range for p?

47 If $0 < a < b$, prove that $(1/a) > (1/b)$. Why is the restriction $0 < a$ necessary?

48 If $0 < a < b$, prove that $a^2 < b^2$. Why is the restriction $0 < a$ necessary?

49 If $a < b$ and $c < d$, prove that $a + c < b + d$.

50 If $a < b$ and $c < d$ is it always true that $ac < bd$? Explain.

6   MORE ON INEQUALITIES

Among the most important inequalities which occur in advanced mathematics are those involving absolute values. From the discussion in Section 3 of Chapter One, if a represents a real number, then $|a|$ is the distance on a coordinate line between the origin and the point corresponding to a. It follows that if we are given an inequality of the form $|a| < 5$, then the solutions consist of all real numbers a such that $-5 < a < 5$, or equivalently, the open interval $(-5, 5)$. In general, if b is any positive real number, then it can be shown that

(2.5)

$$\boxed{|a| < b \quad \text{if and only if} \quad -b < a < b}$$

It can also be proved that if $b \geq 0$, then

(2.6)

$$\boxed{|a| > b \quad \text{if and only if} \quad \text{either } a > b \text{ or } a < -b}$$

(2.7)

$$\boxed{|a| = b \quad \text{if and only if} \quad a = b \text{ or } a = -b}$$

Example 1　Solve each of the following.

(a)　$|x| < 4$　　(b)　$|x| > 4$　　(c)　$|x| = 4$

Solution　(a)　Using (2.5) with $a = x$ and $b = 4$, we see that

$$|x| < 4 \quad \text{if and only if} \quad -4 < x < 4.$$

Hence the interval solution is the open interval $(-4, 4)$.

(b) By (2.6), $|x| > 4$ means that either $x > 4$ or $x < -4$. Thus the solutions consist of all real numbers in the two infinite intervals $(-\infty, -4)$ and $(4, \infty)$.

(c) From (2.7),

$$|x| = 4 \quad \text{if and only if} \quad x = 4 \text{ or } x = -4.$$

For solutions of the type obtained in (b) of Example 1 it is sometimes convenient to use the union symbol $\cup$ and write

$$(-\infty, -4) \cup (4, \infty)$$

to denote all real numbers x such that either x is in $(-\infty, -4)$ or x is in $(4, \infty)$.

Example 2 Solve the inequality $|x - 3| < 0.1$.

Solution By (2.5) with $a = x - 3$ and $b = 0.1$, the given inequality is equivalent to

$$-0.1 < x - 3 < 0.1$$

and hence to

$$-0.1 + 3 < (x - 3) + 3 < 0.1 + 3$$

or

$$2.9 < x < 3.1.$$

Consequently the interval solution of the given inequality is $(2.9, 3.1)$.

Example 3 Solve $|2x + 3| > 9$ and sketch the graph corresponding to the solutions.

Solution By (2.6), with $a = 2x + 3$ and $b = 9$, the solutions of the given inequality are the solutions of the *two* inequalities

$$2x + 3 > 9 \quad \text{and} \quad 2x + 3 < -9.$$

The first inequality is equivalent to $2x > 6$, or $x > 3$. This gives us the infinite interval $(3, \infty)$. The second inequality is equivalent to $2x < -12$, or $x < -6$, which leads to the interval $(-\infty, -6)$. Consequently the solutions of $|2x + 3| > 9$ consist of the numbers in the union $(-\infty, -6) \cup (3, \infty)$. The graph is sketched in Figure 2.12.

Figure 2.12

Before proceeding, let us make an important remark about notation. The solutions in Example 3 consist of all real numbers x such that either $x > 3$ or $x < -6$. Some amateurs employ the *incorrect* form $3 < x < -6$ to denote these inequalities. The notation $a < b < c$ should be used if and only if b satisfies *both* of the conditions $a < b$ and $b < c$. Hence the expression $3 < x < -6$ is absurd, since it is impossible for x to satisfy $3 < x$ and $x < -6$ simultaneously.

Most of the inequalities considered up to this time were equivalent to inequalities containing only first-degree polynomials. Let us now investigate inequalities involving polynomials of degree greater than 1.

Example 4   Solve $2x^2 < x + 3$.

Solution   The following inequalities are equivalent:

$$2x^2 < x + 3$$
$$2x^2 - x - 3 < 0$$
$$(2x - 3)(x + 1) < 0$$

Consequently, x is a solution of the given inequality if and only if the product $(2x - 3)(x + 1)$ is negative. In order for the latter to occur, the factors $2x - 3$ and $x + 1$ must have opposite signs. Let us therefore examine the signs of each factor.

The factor $2x - 3$ is positive whenever x is a solution of any of the following inequalities:

$$2x - 3 > 0, \qquad 2x > 3, \qquad x > 3/2.$$

Hence $2x - 3$ is positive if x is in the infinite interval $(3/2, \infty)$. In like manner, $2x - 3$ is negative if x is in the interval $(-\infty, 3/2)$. Similarly, the factor $x + 1$ is positive if

$$x + 1 > 0 \qquad \text{or} \qquad x > -1;$$

that is, if x is in the interval $(-1, \infty)$. However, $x + 1$ is negative whenever x is in the interval $(-\infty, -1)$. It is convenient to use the graphical technique shown in Figure 2.13 to display the intervals in which the factors $2x - 3$ and $x + 1$ are positive or negative. Referring to the figure we see that the factors have opposite signs, and hence the product is negative, whenever $-1 < x < 3/2$. Hence the interval solution of the given inequality is $(-1, 3/2)$.

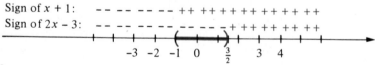

Sign of $x + 1$:
Sign of $2x - 3$:

Figure 2.13

Example 5   Solve the inequality $x^2 - 7x + 10 > 0$ and represent the solutions graphically.

Solution   Since the given inequality may be written

$$(x - 5)(x - 2) > 0$$

it follows that x is a solution if and only if both factors $x - 5$ and $x - 2$ are positive, or both are negative. The diagram in Figure 2.14 indicates the signs of these factors for various real numbers. Evidently, both factors are positive if x is in the interval $(5, \infty)$, and both are negative if x is in $(-\infty, 2)$. Hence the solutions are given by $(-\infty, 2) \cup (5, \infty)$. The graph is sketched in Figure 2.14

Sign of x - 5:
-- -- -- -- -- -- ++ ++ ++ ++
Sign of x - 2:
-- -- -- ++ ++ ++ ++ ++ ++ ++
-1 0 1 2 3 4 5 6

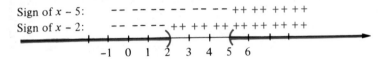

Figure 2.14

Example 6 Solve $\dfrac{x+1}{x+2} \leq 3$ and sketch the graph of the solutions.

Solution The following inequalities are equivalent:

$$\frac{x+1}{x+2} \leq 3$$

$$\frac{x+1}{x+2} - 3 \leq 0$$

$$\frac{x+1-3x-6}{x+2} \leq 0$$

$$\frac{-2x-5}{x+2} \leq 0$$

$$\frac{2x+5}{x+2} \geq 0.$$

The fraction $\dfrac{2x+5}{x+2}$ equals zero if $2x + 5 = 0$, or $x = -5/2$. Hence $-5/2$ is a solution. It remains to determine the solutions of the inequality

$$\frac{2x+5}{x+2} > 0.$$

Since a quotient is positive if and only if the numerator and denominator have the same signs, we must determine where both $2x + 5$ and $x + 2$ are positive or where both are negative. Referring to the diagram in Figure 2.15, we see that both factors are positive if x is in the interval $(-2, \infty)$ and both are negative if x is in the interval $(-\infty, -5/2)$. Since $-5/2$ is also a solution, it follows that the solutions of the given inequality are given by $(-\infty, -5/2] \cup (-2, \infty)$. The graph is sketched in Figure 2.16.

Sign of x + 2
-- -- -- ++ ++ ++ ++ ++ ++
Sign of 2x + 5
-- -- -+ ++ ++ ++ ++ ++ ++
-5/2 -2 -1 0 1

Figure 2.15

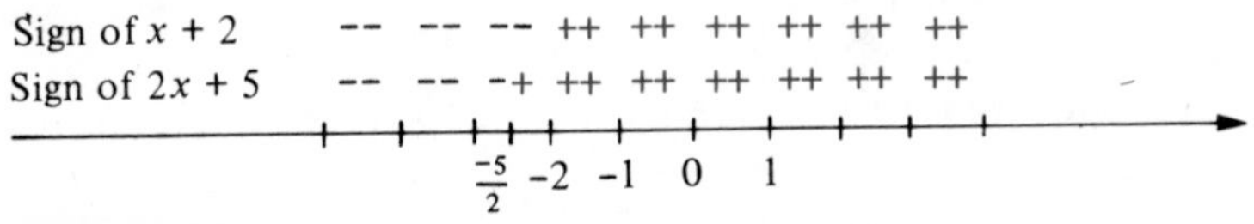

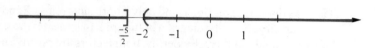

Figure 2.16

Incidentally, it would have been incorrect to begin the solution by multiplying both sides of $(x + 1)/(x + 2) \le 3$ by $(x + 2)$, obtaining $(x + 1) \le 3(x + 2)$, since $x + 2$ is not always positive for every x under consideration. The student may find it instructive to see what answer is obtained by using this incorrect approach.

Example 7 Solve the inequality $(x + 2)(x - 1)(x - 5) > 0$ and represent the solutions graphically.

Solution If a number is substituted for x in $(x + 2)(x - 1)(x - 5)$, the result will be positive provided all three factors are positive or provided two of the factors are negative and one is positive. As in the previous sections we find that $x + 2$ is negative in $(-\infty, -2)$ and positive in $(-2, \infty)$. Similarly, the factor $x - 1$ is negative in $(-\infty, 1)$ and positive in $(1, \infty)$. Finally, $x - 5$ is negative in $(-\infty, 5)$ and positive in $(5, \infty)$. These facts are displayed in Figure 2.17. We see from this diagram that all factors are positive if x is in $(5, \infty)$ and that two factors are negative and one is positive if x is in $(-2, 1)$. Hence the solutions are given by $(-2, 1) \cup (5, \infty)$. The graph is sketched in Figure 2.18.

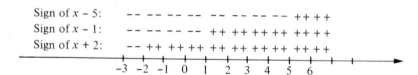

Figure 2.17

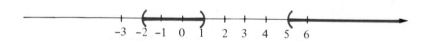

Figure 2.18

EXERCISES

Solve the inequalities in Exercises 1–38 and express the solutions in terms of intervals.

1 $\|x\| < 2$	**2** $\|x\| < 25$	**3** $\|x\| > 6$
4 $\|x\| > 1/2$	**5** $\|-x\| \le 10$	**6** $\|x\| \ge 7$

7 $|x - 10| < 0.05$

8 $|x - 5| < 0.001$

9 $\left|\dfrac{x + 4}{3}\right| \le 2$

10 $\left|\dfrac{x + 1}{5}\right| \le 1$

11 $|5 - 3x| < 7$

12 $|7x + 4| < 10$

13 $|2x + 4| > 8$

14 $|6x - 7| \ge 4$

15 $\dfrac{3}{|5 - 2x|} \le 1$

16 $\left|\dfrac{4 - x}{2}\right| > 9$

17 $x^2 - x - 6 < 0$

18 $x^2 + 6x + 5 < 0$

19 $x(2x + 3) > 5$

20 $x(3x - 1) > 4$

21 $x^2 \le 10x$

22 $4x^2 \ge x$

23 $\dfrac{2x + 1}{10 - 3x} < 0$

24 $\dfrac{9 - 4x}{x + 1} \ge 0$

25 $\dfrac{6}{x^2 - 9} < 0$

26 $\dfrac{10}{25x^2 - 16} > 0$

27 $\dfrac{7x}{x^2 - 16} \ge 0$

28 $\dfrac{x - 2}{x^2 - 9} < 0$

29 $\dfrac{x^2 - 25}{4x^2 - 9} \le 0$

30 $\dfrac{10}{x^4 - 16} > 0$

31 $\dfrac{x + 4}{2x - 1} < 3$

32 $\dfrac{1}{x - 2} > \dfrac{3}{x + 1}$

33 $x^3 < x$

34 $x^4 + 3x^2 > 4$

35 $x^3 - x^2 - 4x + 4 \ge 0$

36 $(x^2 - 4x + 4)(3x - 7) < 0$

37 $\dfrac{x^2 - x - 2}{x^2 - 4x + 3} \ge 0$

38 $\dfrac{x^2 + 2x - 3}{x^2 - 4x} \le 0$

39 If a projectile is fired straight upward from level ground with an initial velocity of 72 ft/sec, its altitude s (in feet) after t seconds is given by $s = -16t^2 + 72t$. During what time interval will the projectile be at least 32 feet above the ground?

40 The period T (sec) of a simple pendulum of length l (cm) is given by $T = 2\pi\sqrt{l/g}$, where g is a physical constant. If, under certain conditions, $g = 980$ and $98 \le l \le 100$, what is the corresponding range for T?

7 REVIEW

Concepts

Define or discuss each of the following.

1 Solution of an equation

2 Root of an equation

3 Identity

4 Conditional equation

5 Equivalent equations

6 Linear equation

7 Quadratic equation

8 The quadratic formula

9 Discriminant of a quadratic equation

10 Extraneous solution

11 Properties of inequalities

12 Solution of an inequality

13 Equivalent inequalities

14 The graph of a set of real numbers

15 Open interval

16 Closed interval

17 Half-open interval

18 Infinite interval

Exercises

Solve the equations and inequalities in Exercises 1–30.

1 $\dfrac{3x + 1}{5x + 7} = \dfrac{6x + 11}{10x - 3}$

2 $2 - \dfrac{1}{x} = 1 + \dfrac{4}{x}$

3 $\dfrac{2}{x + 5} - \dfrac{3}{2x + 1} = \dfrac{5}{6x + 3}$

4 $\dfrac{7}{x - 2} - \dfrac{6}{x^2 - 4} = \dfrac{3}{2x + 4}$

5 $\dfrac{1}{\sqrt{x}} - 2 = \dfrac{1 - 2\sqrt{x}}{\sqrt{x}}$

6 $2x^2 + 5x - 12 = 0$

7 $3x^2 + 4x - 5 = 0$

8 $\dfrac{x}{3x + 1} = \dfrac{x - 1}{2x + 3}$

9 $(x - 2)(x + 1) = 3$

10 $4x^4 - 33x^2 + 50 = 0$

11 $x^{2/3} - 2x^{1/3} - 15 = 0$

12 $20x^3 + 8x^2 - 35x - 14 = 0$

13 $\dfrac{1}{x} + 6 = \dfrac{5}{\sqrt{x}}$

14 $\sqrt[3]{4x - 5} - 2 = 0$

15 $\sqrt{7x + 2} + x = 6$

16 $\sqrt{x + 4} = \sqrt[4]{6x + 19}$

17 $\sqrt{3x + 1} - \sqrt{x + 4} = 1$

18 $10 - 7x < 4 + 2x$

19 $-\dfrac{1}{2} < \dfrac{2x + 3}{5} < \dfrac{3}{2}$

20 $(3x - 1)(10x + 4) \geq (6x - 5)(5x - 7)$

21 $\dfrac{6}{10x + 3} < 0$

22 $|4x + 7| < 21$

23 $|16 - 3x| \geq 5$

24 $2 < |x - 6| < 4$

25 $10x^2 + 11x - 6 > 0$

26 $x^2 - 3x \leq 10$

27 $\dfrac{3}{2x + 3} < \dfrac{1}{x - 2}$

28 $\dfrac{x + 1}{x^2 - 25} \leq 0$

29 $x^3 > x^2$

30 $(x^2 - x)(x^2 - 5x + 6) < 0$

In Exercises 31–36 solve for the indicated variable in terms of the remaining variables.

31 $S = \pi(r + R)s$ for R

32 $S = 2(ab + bc + ac)$ for a

33 $n = \dfrac{\pi P R^4}{8VL}$ for R

34 $V = \frac{4}{3}\pi r^3$ for r

35 $\dfrac{1}{R} = (n - 1)\left(\dfrac{1}{R_1} + \dfrac{1}{R_2}\right)$ for R_2

36 $V = \frac{1}{3}\pi h(R_1^2 + R_2^2 + R_1 R_2)$ for R_1

37   The diagonal of a square is 50 cm long. What change in the length of a side will increase the area by 46 cm²?

38   The surface area S of a sphere of radius r is given by the formula $S = 4\pi r^2$. If $r = 6$ inches, what change in radius will increase the surface area by 36π square inches?

39   An airplane flew with the wind for 30 minutes and returned the same distance in 45 minutes. If the cruising speed of the airplane is 320 mph, find the speed of the wind.

40   A merchant wishes to mix peanuts costing $1.50 per pound with cashews costing $4.00 per pound, obtaining 50 pounds of a mixture costing $2.40 per pound. How many pounds of each should be used?

41   A chemist has 80 ml of a solution containing 25% acid. How many ml should be removed and replaced by pure acid in order to obtain a solution containing 40% acid?

42   A motorist averaged 45 mph driving outside the city limits and 25 mph within the city limits. If an 80-mile trip took the motorist two hours, how much time was spent driving within the city limits?

43   The width of a page in a book is 2 inches smaller than its length. The printed area is 72 square inches, with 1-inch margins at the top and bottom and $\frac{1}{2}$-inch margins at each side. Find the dimensions of the page.

44   A man puts a fence around a rectangular field and then subdivides the field into three smaller rectangular plots by placing two fences parallel to one of the sides. If the area of the field is 31,250 square yards and 1,000 yards of fencing was used, find the dimensions of the field.

CHAPTER 3

Functions

One of the most useful concepts in mathematics is that of function. Indeed, it is safe to say that without the notion of function, little progress could be made in mathematics or in any area of science. In the first section of this chapter we consider coordinate systems in two dimensions. The remainder of the chapter contains a discussion of functions and graphs.

1 COORDINATE SYSTEMS IN TWO DIMENSIONS

In Section 3 of Chapter One we indicated how coordinates may be assigned to points on a straight line. Coordinate systems can also be introduced in planes by means of ordered pairs. The term **ordered pair** refers to two real numbers, of which one is designated as the "first" number and the other as the "second." The symbol (a, b) is used to denote the ordered pair consisting of the real numbers a and b where a is first and b is second. There are many uses for ordered pairs. They were used in Chapter 2 to denote open intervals. In this chapter we shall use ordered pairs to represent points in a plane. Although ordered pairs are employed in different situations, there is little chance for confusion, since it should always be clear from the discussion whether the symbol (a, b) represents an interval, a point, or some other mathematical object. We consider two ordered pairs (a, b) and (c, d) equal, and write

$$(a, b) = (c, d) \quad \text{if and only if} \quad a = c \text{ and } b = d$$

This implies, in particular, that $(a, b) \neq (b, a)$ if $a \neq b$.

A **rectangular**, or **Cartesian,*** **coordinate system** may be introduced in a plane by considering two perpendicular coordinate lines in the plane which intersect in the origin O on each line. Unless specified otherwise, the same unit of length is chosen on each line. Usually one of the lines is horizontal with positive direction to the right, and the other line is vertical with positive direction upward, as indicated by the arrowheads in Figure 3.1. The two lines are called **coordinate**

---

*The term "Cartesian" is used in honor of the French mathematician and philosopher René Descartes (1596–1650), who was one of the first to employ such coordinate systems.

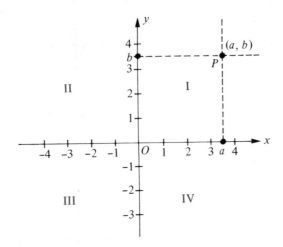

Figure 3.1

axes and the point O is called the **origin**. The horizontal line is often referred to as the **x-axis** and the vertical line as the **y-axis**, and they are labeled x and y, respectively. The plane is then called a **coordinate plane** or, with the preceding notation for coordinate axes, an **xy-plane**. Although the symbols x and y are used to denote lines as well as numbers, there should be no misunderstanding as to what these letters represent when they appear alongside of coordinate lines as in Figure 3.1. In certain applications different labels such as d, t, etc., are used for the coordinate lines. The coordinate axes divide the plane into four parts called the **first, second, third**, and **fourth quadrants** and labeled I, II, III, and IV, respectively, as shown in Figure 3.1.

Each point P in an xy-plane may be assigned a unique ordered pair. If vertical and horizontal lines through P intersect the x- and y-axes at points with coordinates a and b, respectively (see Figure 3.1), then P is assigned the ordered pair (a, b). The number a is called the **x-coordinate** (or **abscissa**), of P, and b is called the **y-coordinate** (or **ordinate**), of P. We sometimes say that P *has coordinates* (a, b). Conversely, every ordered pair (a, b) determines a point P in the xy-plane with coordinates a and b. Specifically, P is the point of intersection of lines perpendicular to the x-axis and y-axis at the points having coordinates a and b, respectively. This establishes a one-to-one correspondence between the set of all points in the xy-plane and the set of all ordered pairs. It is sometimes convenient to refer to the *point* (a, b) meaning the point with abscissa a and ordinate b. The symbol $P(a, b)$ will denote the point P with coordinates (a, b). To **plot a point** $P(a, b)$ means to locate, in a coordinate plane, the point P with coordinates (a, b). This point is represented by a dot in the appropriate position, as illustrated in Figure 3.1.

Note that abscissas are positive for points in quadrants I or IV and negative for points in quadrants II or III. Ordinates are positive for points in quadrants I or II and negative for points in quadrants III or IV. Some typical points in a coordinate plane are plotted in Figure 3.2.

We shall next derive a formula for finding the distance between any two points in a coordinate plane. The distance between two points P and Q will be denoted by $d(P, Q)$. If $P = Q$, then we agree that $d(P, Q) = 0$, whereas if $P \neq Q$ the distance is positive. Let us consider any two points $P_1(x_1, y_1)$ and $P_2(x_2, y_2)$ in the

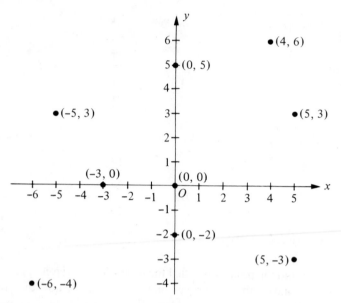

Figure 3.2

plane. If the points lie on the same horizontal line then $y_1 = y_2$, and we may denote the points by $P_1(x_1, y_1)$ and $P_2(x_2, y_1)$. If lines through P_1 and P_2 parallel to the y-axis intersect the x-axis at $A_1(x_1, 0)$ and $A_2(x_2, 0)$, as shown in (i) of Figure 3.3, then we see that $d(P_1, P_2) = d(A_1, A_2)$. However, by (1.19), $d(A_1, A_2) = |x_2 - x_1|$ and hence

$$d(P_1, P_2) = |x_2 - x_1|.$$

Since $|x_2 - x_1| = |x_1 - x_2|$, the last formula is valid whether P_1 lies to the left of P_2 or to the right of P_2. Moreover, the formula is independent of the quadrants in which the points lie.

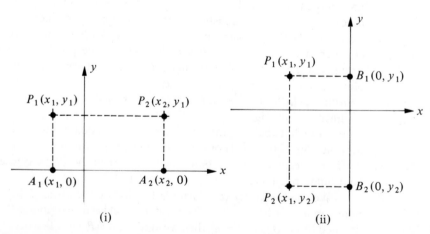

(i)   (ii)

Figure 3.3

In similar fashion, if P_1 and P_2 are on the same vertical line, then $x_1 = x_2$, and we may denote the points by $P_1(x_1, y_1)$ and $P_2(x_1, y_2)$. If we consider the points $B_1(0, y_1)$ and $B_2(0, y_2)$ on the y-axis as shown in (ii) of Figure 3.3 then

$$d(P_1, P_2) = d(B_1, B_2) = |y_2 - y_1|.$$

Finally, let us consider the general case, in which the points $P_1(x_1, y_1)$ and $P_2(x_2, y_2)$ do not lie on the same horizontal or vertical line. The line through $P_1(x_1, y_1)$ parallel to the x-axis and the line through $P_2(x_2, y_2)$ parallel to the y-axis intersect at some point P_3. Since P_3 has the same y-coordinate as P_1 and the same x-coordinate as P_2, we can denote it by $P_3(x_2, y_1)$(see Figure 3.4). From the

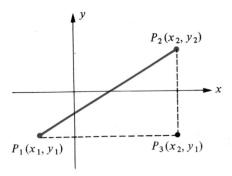

Figure 3.4

previous discussion $d(P_1, P_3) = |x_2 - x_1|$ and $d(P_3, P_2) = |y_2 - y_1|$. Since P_1, P_2, and P_3 form a right triangle with hypotenuse from P_1 to P_2 we have, by the Pythagorean Theorem,

$$[d(P_1, P_2)]^2 = [d(P_1, P_3)]^2 + [d(P_3, P_2)]^2$$

and hence

$$[d(P_1, P_2)]^2 = |x_2 - x_1|^2 + |y_2 - y_1|^2.$$

Using the fact that $d(P_1, P_2)$ is nonnegative and that $|a|^2 = a^2$ for every real number a, we obtain the following formula.

(3.1)   **Distance Formula**

> If $P_1(x_1, y_1)$ and $P_2(x_2, y_2)$ are points in a coordinate plane, then the **distance between P_1 and P_2** is given by
>
> $$d(P_1, P_2) = \sqrt{(x_2 - x_1)^2 + (y_2 - y_1)^2}.$$

Although we referred to the special case indicated in Figure 3.4, the argument used in the proof of the distance formula is independent of the positions of P_1 and P_2.

Example 1   Plot the points $A(-1, -3)$, $B(6, 1)$, $C(2, -5)$, and prove that the triangle with vertices A, B, C is a right triangle.

Solution   The points and the triangle are shown in Figure 3.5. From plane geometry, a triangle is a right triangle if and only if the sum of the squares of two of its sides is equal to the square of the remaining side. Using the distance formula we obtain

$$d(A, B) = \sqrt{(-1 - 6)^2 + (-3 - 1)^2}$$
$$= \sqrt{49 + 16} = \sqrt{65}$$
$$d(B, C) = \sqrt{(6 - 2)^2 + (1 + 5)^2}$$
$$= \sqrt{16 + 36} = \sqrt{52}$$
$$d(A, C) = \sqrt{(-1 - 2)^2 + (-3 + 5)^2}$$
$$= \sqrt{9 + 4} = \sqrt{13}.$$

Since $[d(A, B)]^2 = [d(B, C)]^2 + [d(A, C)]^2$, the triangle is a right triangle with hypotenuse joining A to B.

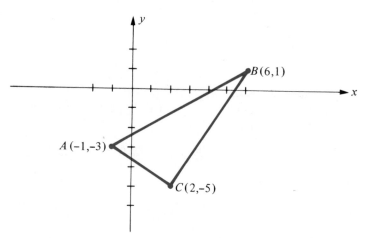

Figure 3.5

It is easy to obtain a formula for the midpoint of a line segment. Let $P_1(x_1, y_1)$ and $P_2(x_2, y_2)$ be two points in a coordinate plane and let M be the midpoint of the segment $P_1 P_2$. The lines through P_1 and P_2 parallel to the y-axis intersect the x-axis at $A_1(x_1, 0)$ and $A_2(x_2, 0)$ and, from plane geometry, the line through M parallel to the y-axis bisects the segment $A_1 A_2$ (see Figure 3.6). If $x_1 < x_2$, then $x_2 - x_1 > 0$, and hence

$$d(A_1, A_2) = x_2 - x_1.$$

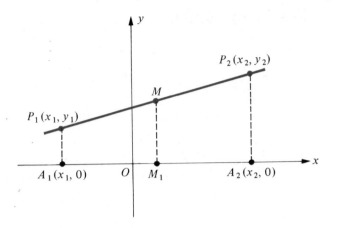

Figure 3.6

Since M_1 is halfway from A_1 to A_2, the abscissa of M_1 is

$$x_1 + \tfrac{1}{2}(x_2 - x_1) = x_1 + \tfrac{1}{2}x_2 - \tfrac{1}{2}x_1$$
$$= \tfrac{1}{2}x_1 + \tfrac{1}{2}x_2$$
$$= \frac{x_1 + x_2}{2}.$$

It follows that the abscissa of M is also $(x_1 + x_2)/2$. It can be shown in similar fashion that the ordinate of M is $(y_1 + y_2)/2$. Moreover, these formulas hold for all positions of P_1 and P_2. This gives us the following result.

(3.2) **Midpoint Formula**

The midpoint of the line segment from $P_1(x_1, y_1)$ to $P_2(x_2, y_2)$ is

$$\left(\frac{x_1 + x_2}{2}, \frac{y_1 + y_2}{2} \right).$$

Example 2   Find the midpoint M of the line segment from $P_1(-2, 3)$ to $P_2(4, -2)$. Plot the points P_1, P_2, M and verify that $d(P_1, M) = d(P_2, M)$.

Solution   Applying (3.2), the coordinates of M are

$$\left(\frac{-2 + 4}{2}, \frac{3 + (-2)}{2} \right) \quad \text{or} \quad \left(1, \frac{1}{2} \right).$$

The three points P_1, P_2, and M are plotted in Figure 3.7. Using the distance formula we obtain

$$d(P_1, M) = \sqrt{(-2 - 1)^2 + (3 - \tfrac{1}{2})^2} = \sqrt{9 + (\tfrac{25}{4})}$$
$$d(P_2, M) = \sqrt{(4 - 1)^2 + (-2 - \tfrac{1}{2})^2} = \sqrt{9 + (\tfrac{25}{4})}$$

Hence $d(P_1, M) = d(P_2, M)$.

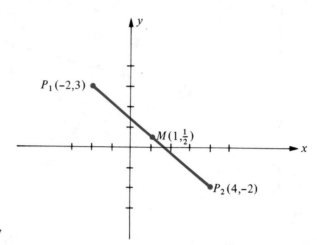

Figure 3.7

EXERCISES

1 Plot the following points on a rectangular coordinate system:
   $A(5, -2)$, $B(-5, -2)$, $C(5, 2)$, $D(-5, 2)$, $E(3, 0)$, $F(0, 3)$

2 Plot the points $A(-3, 1)$, $B(3, 1)$, $C(-2, -3)$, $D(0, 3)$, and $E(2, -3)$ on a rectangular coordinate system and then draw the line segments AB, BC, CD, DE, and EA.

3 Plot $A(0, 0)$, $B(1, 1)$, $C(3, 3)$, $D(-1, -1)$, and $E(-2, -2)$. Describe the set of all points of the form (x, x) where x is a real number.

4 Plot $A(0, 0)$, $B(1, -1)$, $C(2, -2)$, $D(-1, 1)$, and $E(-3, 3)$. Describe the set of all points of the form $(a, -a)$ where a is a real number.

5 Describe the set of all points $P(x, y)$ in a coordinate plane such that
   (a) $x = 3$ (b) $y = -1$ (c) $x \geq 0$ (d) $xy > 0$ (e) $y < 0$

6 Describe the set of all points $P(x, y)$ in a coordinate plane such that
   (a) $y = 0$ (b) $x = -5$ (c) $x/y < 0$ (d) $xy = 0$ (e) $y > 1$

In Exercises 7–12, find (a) the distance $d(A, B)$ between the given points A and B; (b) the midpoint of the segment AB.

7   $A(4, -3), B(6, 2)$ 8 $A(-2, -5), B(4, 6)$

9   $A(-5, 0), B(-2, -2)$ 10 $A(6, 2), B(6, -2)$

11   $A(7, -3), B(3, -3)$ 12 $A(-4, 7), B(0, -8)$

In Exercises 13 and 14 prove that the triangle with the indicated vertices is a right triangle and find its area.

13   $A(8, 5), B(1, -2), C(-3, 2)$

14   $A(-6, 3), B(3, -5), C(-1, 5)$

15 Prove that the following points are vertices of a square:
    $A(-4, 2), B(1, 4), C(3, -1), D(-2, -3)$.

16 Prove that the following points are vertices of a parallelogram: $A(-4, -1)$, $B(0, -2)$, $C(6, 1)$, $D(2, 2)$.

17 Given $A(-3, 8)$, find the coordinates of the point B such that $M(5, -10)$ is the midpoint of AB.

18 Given $A(5, -8)$ and $B(-6, 2)$, find the point on AB that is three-fourths of the way from A to B.

19 Given $A(-4, -3)$ and $B(6, 1)$, prove that $P(5, -11)$ is on the perpendicular bisector of AB.

20 Given $A(-4, -3)$ and $B(6, 1)$, find a formula which expresses the fact that $P(x, y)$ is on the perpendicular bisector of AB.

21 Find a formula which expresses the fact that $P(x, y)$ is a distance 5 from the origin. Describe the totality of all such points.

22 If r is a positive real number, find a formula that states that $P(x, y)$ is a distance r from a fixed point $C(h, k)$. Describe the totality of all such points.

23 Find all points on the y-axis that are 6 units from the point $(5, 3)$.

24 Find all points on the x-axis that are 5 units from $(-2, 4)$.

25 Let S denote the set of points of the form $(2x, x)$ where x is a real number. Find the point in S that is in the third quadrant and is a distance 5 from the point $(1, 3)$.

26 Let S denote the set of points of the form (x, x) where x is a real number. Find all points in S that are a distance 3 from the point $(-2, 1)$.

27 For what values of a is the distance between $(a, 3)$ and $(5, 2a)$ greater than $\sqrt{26}$?

28 Given the points $A(-2, 0)$ and $B(2, 0)$, find a formula not containing radicals that expresses the fact that the sum of the distances from $P(x, y)$ to A and to B, respectively, is 5.

29 Prove that the midpoint of the hypotenuse of any right triangle is equidistant from the vertices. (*Hint:* Label the vertices of the triangle $O(0, 0)$, $A(a, 0)$, and $B(0, b)$.)

30 Prove that the diagonals of any parallelogram bisect each other. (*Hint:* Label three of the vertices of the parallelogram $O(0, 0)$, $A(a, b)$, and $C(0, c)$.)

2 GRAPHS

If W is a set of ordered pairs, then we may speak of the point $P(x, y)$ in a coordinate plane which corresponds to the ordered pair (x, y) in W. The **graph** of W is the set of all points which correspond to the ordered pairs in W. The phrase "sketch the graph of W" means to illustrate the significant features of the graph geometrically on a coordinate plane, as illustrated in the next example.

Example 1 Sketch the graph of $W = \{(x, y): -1 < x \le 4, 2 \le y < 3\}$.

Solution The set notation describing W may be translated "the set of all ordered pairs (x, y) such that $-1 < x \le 4$ and $2 \le y < 3$." Thus a point $P(x, y)$ is on the graph of W if and only if the abscissa x is greater than -1 and less than or equal to 4, while the ordinate y is greater than or equal to 2 and less than 3. Hence the graph of W consists of all points inside the shaded rectangle in Figure 3.8, together with the points on the lower and right-hand sides of the rectangle.

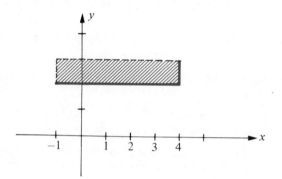

Figure 3.8

Example 2   Sketch the graph of $W = \{(x, y): y = 2x - 1\}$.

Solution   We begin by finding points with coordinates of the form (x, y) where the ordered pair (x, y) is in W. It is convenient to list these coordinates in tabular form as shown below, where for each real number x the corresponding value for y is $2x - 1$.

x	-2	-1	0	1	2	3
y	-5	-3	-1	1	3	5

After plotting, it appears that the points with these coordinates all lie on a line and we sketch the graph accordingly (see Figure 3.9). Ordinarily the few points we have plotted would not be enough to illustrate the graph; however, in this elementary case we can be reasonably sure that the graph is a line. It will be proved in Section 5 that our conjecture is correct.

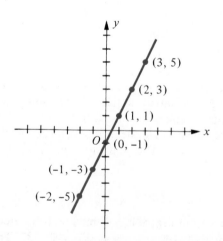

Figure 3.9

It is impossible to sketch the entire graph in Example 2 since x may be assigned values which are numerically as large as desired. Nevertheless, we often call a drawing of the type given in Figure 3.9 *the graph of* W or *a sketch of the graph*, where it is understood that the drawing is only a device for visualizing the actual graph and the line does not terminate as shown in the figure. In general, when sketching a graph you should illustrate enough of the graph so that the remaining parts are evident.

The graph in Example 2 is determined by the equation $y = 2x - 1$ in the sense that for every real number x, the equation can be used to find a number y such that (x, y) is in W. Given an equation in x and y, we say that an ordered pair (a, b) is a **solution** of the equation if equality is obtained when a is substituted for x and b for y. For example, $(2, 3)$ is a solution of $y = 2x - 1$ since substitution of 2 for x and 3 for y leads to $3 = 4 - 1$, or $3 = 3$. Two equations in x and y are said to be **equivalent** if they have exactly the same solutions. The solutions of an equation in x and y determine a set S of ordered pairs, and we define the **graph of the equation** as the graph of S. Notice that the solutions of the equation $y = 2x - 1$ are the pairs (a, b) such that $b = 2a - 1$, and hence the solutions are identical with the set W given in Example 2. Consequently the graph of the equation $y = 2x - 1$ is the same as the graph of W (see Figure 3.9).

For some of the equations we shall encounter in this chapter the technique used for sketching the graph will consist of plotting a sufficient number of points until some pattern emerges, and then sketching the graph accordingly. This is obviously a crude (and often inaccurate) way to arrive at the graph; however, it is a method often employed at the beginning of elementary courses. As we progress through this text, techniques will be introduced which will enable us to sketch a variety of graphs without plotting many points. In order to give accurate descriptions of graphs when complicated expressions are involved, it is usually necessary to employ more advanced mathematical tools of the types introduced in the study of calculus.

Example 3 Sketch the graph of the equation $y = x^2$.

Solution In order to obtain the graph, it is necessary to plot more points than in the previous example. Increasing successive abscissas by $\frac{1}{2}$, we obtain the following table.

x	-3	$-\frac{5}{2}$	-2	$-\frac{3}{2}$	-1	$-\frac{1}{2}$	0	$\frac{1}{2}$	1	$\frac{3}{2}$	2	$\frac{5}{2}$	3
y	9	$\frac{25}{4}$	4	$\frac{9}{4}$	1	$\frac{1}{4}$	0	$\frac{1}{4}$	1	$\frac{9}{4}$	4	$\frac{25}{4}$	9

Larger numerical values of x produce even larger values of y. For example, the points $(3, 9)$, $(4, 16)$, $(5, 25)$, and $(6, 36)$ are on the graph, as are $(-3, 9)$, $(-4, 16)$, $(-5, 25)$, and $(-6, 36)$. Plotting the points given in the table and drawing a smooth curve through these points, we obtain the sketch in Figure 3.10, where we have labeled only the points with integer coordinates.

The graph in Example 3 is called a **parabola**. The lowest point $(0, 0)$ is called the **vertex** of the parabola and we say that the parabola **opens upward**. If the graph were inverted, as would be the case for $y = -x^2$, then the parabola **opens downward**. The y-axis is called the **axis of the parabola**.

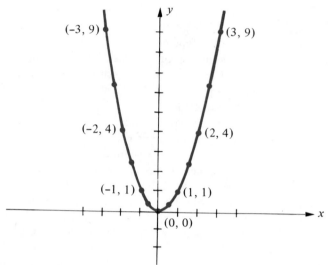

Figure 3.10. $y = x^2$

If the coordinate plane in Figure 3.10 is folded along the y-axis, then the graph which lies in the left half of the plane coincides with that in the right half. We say that **the graph is symmetric with respect to the y-axis**. As in (i) of Figure 3.11, a graph is symmetric with respect to the y-axis provided that the point

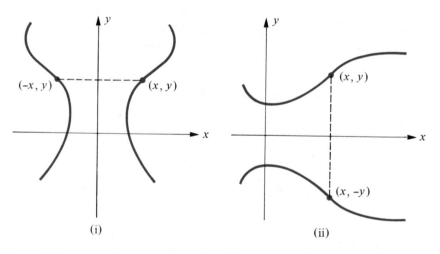

Figure 3.11

$(-x, y)$ is on the graph whenever (x, y) is on the graph. Similarly, as in (ii) of Figure 3.11, **a graph is symmetric with respect to the x-axis** if whenever a point (x, y) is on the graph, then $(x, -y)$ is also on the graph. In the latter case if we fold the coordinate plane along the x-axis, that part of the graph which lies above the x-axis will coincide with the part which lies below. The previous remarks give us the following useful result.

(3.3)   **Tests for Symmetry with Respect to an Axis**

> (i)   The graph of an equation is symmetric with respect to the y-axis if substitution of $-x$ for x does not change the solutions of the equation.
>
> (ii) The graph of an equation is symmetric with respect to the x-axis if substitution of $-y$ for y does not change the solutions of the equation.

If, in the equation of Example 3, we substitute $-x$ for x, we obtain $y = (-x)^2$, which is the same as $y = x^2$. Hence by (i) of (3.3) the graph is symmetric with respect to the y-axis.

If symmetry with respect to an axis exists, then it is sufficient to determine the graph in half of the coordinate plane, since the remainder of the graph is a mirror image, or reflection, of that half.

Example 4   Sketch the graph of $y^2 = x$.

Solution   Since substitution of $-y$ for y does not change the equation, the graph is symmetric with respect to the x-axis. (See (ii) of (3.3).) It is sufficient, therefore, to plot points with nonnegative ordinates and then reflect through the x-axis. Since $y^2 = x$, the ordinates of points above the x-axis are given by $y = \sqrt{x}$. Coordinates of some points on the graph are tabulated below. A portion of the graph is sketched in Figure 3.12. The graph is a parabola with vertex at the origin and which opens to the right. In this case the x-axis is the axis of the parabola.

x	0	1	2	3	4	9
y	0	1	$\sqrt{2}$	$\sqrt{3}$	2	3

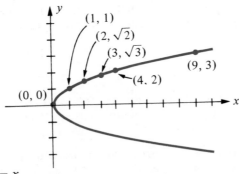

Figure 3.12. $y^2 = x$

Another type of symmetry which certain graphs possess is called **symmetry with respect to the origin**. In this situation, whenever a point (x, y) is on the graph, then $(-x, -y)$ is also on the graph, as illustrated in Figure 3.13. Evidently we have the following result.

(3.4)   Test for Symmetry with Respect to the Origin

> The graph of an equation is symmetric with respect to the origin if the simultaneous substitution of $-x$ for x and $-y$ for y does not change the solutions of the equation.

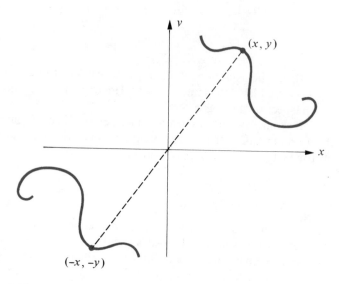

Figure 3.13

Example 5   Sketch the graph of the equation $y = x^3$.

Solution   If we substitute $-x$ for x and $-y$ for y, then

$$-y = (-x)^3 = -x^3.$$

Multiplying both sides by -1, we see that the latter equation has the same solutions as the given equation $y = x^3$. Hence from (3.4) the graph is symmetric with respect to the origin. The following table lists some points on the graph.

x	0	$\frac{1}{4}$	$\frac{1}{2}$	$\frac{3}{4}$	1	$\frac{3}{2}$	2
y	0	$\frac{1}{64}$	$\frac{1}{8}$	$\frac{27}{64}$	1	$\frac{27}{8}$	8

By symmetry (or substitution) we see that the points $(-\frac{1}{4}, -\frac{1}{64}), (-\frac{1}{2}, -\frac{1}{8})$, etc., are on the graph. Plotting points leads to the graph in Figure 3.14.

When a geometric figure in a coordinate plane is given, it is sometimes possible to find its equation in the sense that the figure is the graph of the equation. We shall demonstrate how this can be accomplished for circles.

If $C(h, k)$ is a point in a coordinate plane, then a circle in the plane with center C and radius r may be defined as the collection of all points in the plane that are r units from C. If $P(x, y)$ is an arbitrary point in the plane, then as illustrated in (i) of Figure 3.15, P is on the circle if and only if $d(C, P) = r$.

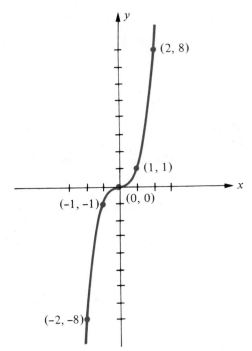

Figure 3.14. $y = x^3$

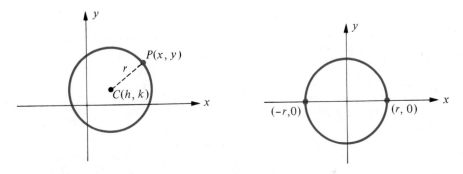

Figure 3.15.   (i) $(x - h)^2 + (y - k)^2 = r^2$ (ii) $x^2 + y^2 = r^2$

Using the distance formula (3.1) gives us the equation

$$\sqrt{(x - h)^2 + (y - k)^2} = r.$$

The equivalent equation

(3.5)

$$(x - h)^2 + (y - k)^2 = r^2, \qquad r > 0$$

is called the **standard equation of a circle of radius r and center (h, k).**

If $h = 0$ and $k = 0$, then (3.5) reduces to

(3.6)
$$x^2 + y^2 = r^2$$

which is an equation of a circle of radius r with center at the origin (see (ii) of Figure 3.15). If $r = 1$, the graph of (3.6) is a **unit circle** with center at the origin. A point $P(x, y)$ is on this unit circle if and only if $x^2 + y^2 = 1$.

Example 6   Find an equation of the circle having center $C(-2, 3)$ and containing the point $D(4, 5)$.

Solution   The circle is illustrated in Figure 3.16. Since D is on the circle, the radius r is

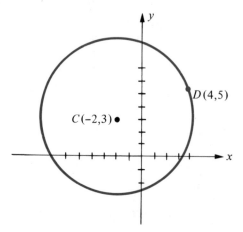

Figure 3.16

$d(C, D)$. By the distance formula (3.1) we have

$$r = d(C, D) = \sqrt{(-2 - 4)^2 + (3 - 5)^2}$$
$$= \sqrt{36 + 4} = \sqrt{40}.$$

Applying (3.5) with $h = -2$ and $k = 3$ we obtain

$$(x + 2)^2 + (y - 3)^2 = 40.$$

EXERCISES

In each of Exercises 1–8 sketch the graph of the given set W of ordered pairs.

1   $W = \{(x, y): x = 4\}$

2   $W = \{(x, y): y = -3\}$

3   $W = \{(x, y): xy < 0\}$

4 $W = \{(x, y): xy = 0\}$

5 $W = \{(x, y): |x| < 2, |y| > 1\}$

6 $W = \{(x, y): |x| > 1, |y| \le 2\}$

7 $W = \{(x, y): |x - 2| < 3, |y| \le 1\}$

8 $W = \{(x, y): |x - 1| < 5, |y + 1| < 2\}$

In each of Exercises 9–28 sketch the graph of the equation after plotting a sufficient number of points.

9 $y = 3x + 1$

10 $y = 4x - 3$

11 $y = -2x + 3$

12 $y = 2 - 3x$

13 $2y + 3x + 6 = 0$

14 $3y - 2x = 6$

15 $y = 2x^2$

16 $y = -x^2$

17 $y = 2x^2 - 1$

18 $y = -x^2 + 2$

19 $4y = x^2$

20 $3y + x^2 = 0$

21 $y = -\frac{1}{2}x^3$

22 $y = \frac{1}{2}x^3$

23 $y = x^3 - 2$

24 $y = 2 - x^3$

25 $y = \sqrt{x}$

26 $y = \sqrt{x} - 1$

27 $y = \sqrt{-x}$

28 $y = \sqrt{x - 1}$

Describe the graphs in Exercises 29–36.

29 $x^2 + y^2 = 16$

30 $x^2 + y^2 = 25$

31 $9x^2 + 9y^2 = 1$

32 $2x^2 + 2y^2 = 1$

33 $(x - 2)^2 + (y + 1)^2 = 4$

34 $(x + 3)^2 + (y - 4)^2 = 1$

35 $x^2 + (y - 3)^2 = 9$

36 $(x + 5)^2 + y^2 = 16$

In each of Exercises 37–46 find an equation of a circle satisfying the stated conditions.

37 Center $C(3, -2)$, radius 4

38 Center $C(-5, 2)$, radius 5

39 Center $C(\frac{1}{2}, -\frac{3}{2})$, radius 2

40 Center $C(\frac{1}{3}, 0)$, radius $\sqrt{3}$

41 Center at the origin, passing through $P(-3, 5)$

42 Center $C(-4, 6)$, passing through $P(1, 2)$

43 Center $C(-4, 2)$, tangent to the x-axis

44 Center $C(3, -5)$, tangent to the y-axis

45 Endpoints of a diameter $A(4, -3)$ and $B(-2, 7)$

46 Tangent to both axes, center in the first quadrant, radius 2

3   FUNCTIONS

The notion of **correspondence** is encountered frequently in everyday life. For example, to each book in a library there corresponds the number of pages in the book. As another example, to each human being there corresponds a birth date. To cite a third example, if the temperature of the air is recorded throughout a day, then at each instant of time there is a corresponding temperature.

The examples of correspondences we have given involve two sets X and Y. In our first example, X denotes the set of books in a library and Y the set of positive integers. For each book x in X there corresponds a positive integer y, namely the number of pages in the book. In the second example, if we let X denote the set of all human beings and Y the set of all possible dates, then to each person x in X there corresponds a birth date y.

We sometimes represent correspondences by diagrams of the type shown in Figure 3.17, where the sets X and Y are represented by points within regions in a plane. The curved arrow indicates that the element y of Y corresponds to the element x of X. We have pictured X and Y as different sets. However, X and Y may have elements in common. As a matter of fact, we often have $X = Y$.

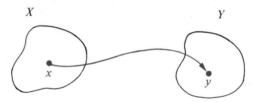

Figure 3.17

Our examples indicate that to each x in X there corresponds *one and only one* y in Y; that is, y *is unique* for a given x. However, the same element of Y may correspond to different elements of X. For example, two different books may have the same number of pages, two different people may have the same birthday, and so on.

In most of our work X and Y will be sets of numbers. To illustrate, let X and Y both denote the set $\mathbb{R}$ of real numbers, and to each real number x let us assign its square x^2. Thus to 3 we assign 9, to -5 we assign 25, to $\sqrt{2}$ the number 2, and so on. This gives us a correspondence from $\mathbb{R}$ to $\mathbb{R}$.

All the examples of correspondences we have given are *functions*, as defined below.

(3.7)   Definition

> A **function** f from a set X to a set Y is a correspondence that assigns to each element x of X a unique element y of Y. The element y is called the **image** of x under f and is denoted by $f(x)$. The set X is called the **domain** of the function. The **range** of the function consists of all images of elements of X.

In (3.7) we introduced the notation $f(x)$ for the element of Y which corresponds to x. This is usually read "f of x." We also call $f(x)$ the **value** of f at x. In terms of the pictorial representation given earlier, we may now sketch a diagram as in Figure 3.18. The curved arrows indicate that the elements $f(x), f(w),$

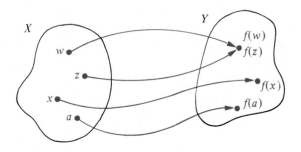

Figure 3.18

$f(z)$, and $f(a)$ of Y correspond to the elements $x, w, z,$ and a of X. Let us repeat the important fact that to each x in X there is assigned precisely one image $f(x)$ in Y; however, different elements of X such as w and z in Figure 3.18 may have the same image in Y.

Beginning students are sometimes confused by the symbols f and $f(x)$. Remember that f is used to represent the function. It is neither in X nor in Y. However, $f(x)$ is an element of Y, namely the element which f assigns to x.

Two functions f and g from X to Y are said to be **equal**, written

$$f = g \quad \text{if and only if} \quad f(x) = g(x)$$

for every x in X.

Example 1   Let f be the function with domain $\mathbb{R}$ such that $f(x) = x^2$ for every x in $\mathbb{R}$. Find $f(-6), f(\sqrt{3}),$ and $f(a),$ where a is any real number. What is the range of f?

Solution   Values of f (or images under f) may be found by substituting for x in the equation $f(x) = x^2$. Thus

$$f(-6) = (-6)^2 = 36, \quad f(\sqrt{3}) = (\sqrt{3})^2 = 3 \quad \text{and} \quad f(a) = a^2.$$

If T denotes the range of f, then by definition (3.7) T consists of all numbers of the form $f(a)$ where a is in $\mathbb{R}$. Hence T is the set of all squares a^2, where a is a real number. Since the square of any real number is nonnegative, T is contained in the set of all nonnegative real numbers. Moreover, every nonnegative real number c is an image under f, since $f(\sqrt{c}) = (\sqrt{c})^2 = c$. Hence the range of f is the set of all nonnegative real numbers.

If a function is defined as in the preceding example, the symbol used for the variable is immaterial; that is, expressions such as $f(x) = x^2$, $f(s) = s^2$, $f(t) = t^2$, and so on, all define the same function f. This is true because if a is any number in the domain of f, then the same image a^2 is obtained no matter which expression is employed.

Example 2   Let X denote the set of nonnegative real numbers and let f be the function from X to $\mathbb{R}$ defined by $f(x) = \sqrt{x} + 1$ for every x in X. Find $f(4)$ and $f(\pi)$. If b and c are in X find $f(b + c)$ and $f(b) + f(c)$.

Solution   As in Example 1, finding images under f is simply a matter of substituting the appropriate number for x in the expression for $f(x)$. Thus

$$f(4) = \sqrt{4} + 1 = 2 + 1 = 3$$
$$f(\pi) = \sqrt{\pi} + 1$$
$$f(b + c) = \sqrt{b + c} + 1$$
$$f(b) + f(c) = (\sqrt{b} + 1) + (\sqrt{c} + 1) = \sqrt{b} + \sqrt{c} + 2.$$

In certain branches of mathematics, such as calculus, it is important to carry out manipulations of the type given in the next example.

Example 3   Suppose $f(x) = x^2 + 3x - 2$ for every real number x. If a and t are real numbers, and $t \neq 0$, find

$$\frac{f(a + t) - f(a)}{t}.$$

Solution   We have

$$f(a + t) = (a + t)^2 + 3(a + t) - 2$$
$$= (a^2 + 2at + t^2) + (3a + 3t) - 2$$

and

$$f(a) = a^2 + 3a - 2.$$

Hence

$$\frac{f(a + t) - f(a)}{t} = \frac{(a^2 + 2at + t^2) + (3a + 3t) - 2 - (a^2 + 3a - 2)}{t}$$
$$= \frac{2at + t^2 + 3t}{t}$$
$$= 2a + t + 3.$$

Occasionally one of the notations

$$X \xrightarrow{f} Y, \quad f:X \to Y \quad \text{or} \quad f:x \to f(x)$$

is used to signify that f is a function from X to Y. It is not unusual in this event to say f **maps** X *into* Y or f *maps* x *into* $f(x)$. If f is the function in Example 1, then f maps x into x^2 and we may write $f:x \longrightarrow x^2$.

Many formulas which occur in mathematics and the sciences determine functions. As an illustration, the formula $A = \pi r^2$ for the area A of a circle of radius r assigns to each positive real number r a unique value of A. This determines a function f, where $f(r) = \pi r^2$, and we may write $A = f(r)$. The letter r, which represents an arbitrary number from the domain of f, is often called an **independent variable**. The letter A, which represents a number from the range of f, is called a **dependent variable**, since its value depends on the number assigned to r. When two variables r and A are related in this manner, it is customary to use the phrase A *is a function of* r. To cite another example, if an automobile travels at a uniform rate of 50 miles per hour, then the distance d (miles) traveled in time t (hours) is given by $d = 50t$ and hence the distance d is a function of time t.

We have seen that different elements in the domain of a function may have the same image. If images are always different, then as in the next definition the function is called one-to-one.

(3.8) **Definition**

> A function f from X to Y is a **one-to-one function** if, whenever $a \neq b$ in X, then $f(a) \neq f(b)$ in Y.

If f is one-to-one, then each $f(x)$ in the range is the image of *precisely one x in* X. The function illustrated in Figure 3.18 is not one-to-one since two different elements w and z of X have the same image in Y. A one-to-one function is often called a **one-to-one correspondence**. To illustrate, the association between real numbers and points on a coordinate line is an example of a one-to-one correspondence.

Example 4 (a) If $f(x) = 3x + 2$, where x is real, prove that f is one-to-one.

(b) If $g(x) = x^2 + 5$, where x is real, prove that g is not one-to-one.

Solution (a) If $a \neq b$, then $3a \neq 3b$ and hence $3a + 2 \neq 3b + 2$, or $f(a) \neq f(b)$. Hence f is one-to-one by (3.8).

(b) The function g is not one-to-one since different numbers in the domain may have the same image. For example, although $-1 \neq 1$, both $g(-1)$ and $g(1)$ are equal to 6.

The next two definitions introduce terms which are used to describe certain types of functions.

(3.9)   Definition

> The **identity function** f on a set X is defined by $f(x) = x$ for every x in X.

Note that if f is the identity function on X, then every x in X is mapped into itself.

(3.10)   Definition

> A function f from X to Y is a **constant function** if there is some (fixed) element c in Y such that $f(x) = c$ for every x in X.

The diagram in Figure 3.19 illustrates the fact that if f is constant function, then every arrow from X terminates at the same element in Y.

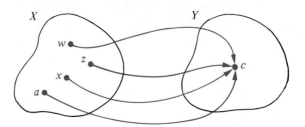

Figure 3.19

In the remainder of our work, unless specified otherwise, the phrase *f is a function* will mean that the domain and range are sets of real numbers. If a function is defined by means of some expression as in Examples 1–3, and the domain X is not stated explicitly, then X is considered to be the totality of real numbers for which the given expression is meaningful. To illustrate, if $f(x) = \sqrt{x}/(x-1)$, then the domain is assumed to be the set of nonnegative real numbers different from 1. If x is in the domain we sometimes say that f **is defined at** x, or that $f(x)$ **exists**. If a subset S is contained in the domain we often say that f **is defined on** S. The terminology f **is undefined at** x means that x is not in the domain of f.

The concept of ordered pair can be used to obtain an alternate approach to functions. We first observe that a function f from X to Y determines the following set W of ordered pairs:

$$W = \{(x, f(x)) : x \text{ is in } X\}.$$

Thus W is the totality of ordered pairs for which the first number is in X and the second number is the image of the first. In Example 2, where $f(x) = \sqrt{x} + 1$, W consists of all pairs of the form $(x, \sqrt{x} + 1)$, where x is a nonnegative real number.

It is important to note that for each x there is exactly one ordered pair (x, y) in W having x in the first position.

Conversely, if we begin with a set W of ordered pairs such that each x in X appears exactly once in the first position of an ordered pair, and numbers from Y appear in the second position, then W determines a function from X to Y. Specifically, for any x in X there is a unique pair (x, y) in W, and by letting y correspond to x, we obtain a function from X to Y.

It follows from the preceding discussion that the statement given below could also be used as a definition of function. We prefer, however, to think of it as an alternate approach to this concept.

(3.11)   **Alternate Definition of a Function**

> A function with domain X is a set W of ordered pairs such that for each x in X, there is exactly one ordered pair (x, y) in W having x in the first position.

In terms of the preceding definition the function f of Example 1, where $f(x) = x^2$, is the set of all ordered pairs of the form (x, x^2). Similarly, the ordered pairs $(x, x^2 + 3x - 2)$ determine the function of Example 3, where we had $f(x) = x^2 + 3x - 2$.

EXERCISES

1   If $f(x) = 2x^2 - 3x + 4$, find $f(1)$, $f(-1)$, $f(0)$, and $f(2)$.

2   If $f(x) = x^3 + 5x^2 - 1$, find $f(2)$, $f(-2)$, $f(0)$, and $f(-1)$.

3   If $f(x) = \sqrt{x - 1} + 2x$, find $f(1)$, $f(3)$, $f(5)$, and $f(10)$.

4   If $f(x) = \dfrac{x}{x - 2}$, find $f(1)$, $f(3)$, $f(-2)$, and $f(0)$.

In Exercises 5–8 find each of the following, where a, b, and h are real numbers:

(a) $f(a)$ (b) $f(-a)$

(c) $-f(a)$ (d) $f(a + h)$

(e) $f(a) + f(h)$ (f) $\dfrac{f(a + h) - f(a)}{h}$ provided $h \neq 0$

5    $f(x) = 5x - 2$ 6 $f(x) = 3 - 4x$

7    $f(x) = 2x^2 - x + 3$ 8 $f(x) = x^3 - 2x$

In Exercises 9–12 find the following:

(a) $g(1/a)$ (b) $1/g(a)$

(c) $g(a^2)$ (d) $(g(a))^2$

(e) $g(\sqrt{a})$ (f) $\sqrt{g(a)}$

9   $g(x) = 3x^2$ 10 $g(x) = 3x - 8$

11   $g(x) = \dfrac{2x}{x^2 + 1}$ 12 $g(x) = \dfrac{x^2}{x + 1}$

In each of Exercises 13–20 find the largest subset of $\mathbb{R}$ that can serve as the domain of the function f.

13   $f(x) = \sqrt{3x - 5}$ 14 $f(x) = \sqrt{7 - 2x}$

15   $f(x) = \sqrt{4 - x^2}$ 16 $f(x) = \sqrt{x^2 - 9}$

17   $f(x) = \dfrac{x + 1}{x^3 - 9x}$ 18 $f(x) = \dfrac{4x + 7}{6x^2 + 13x - 5}$

19   $f(x) = \dfrac{\sqrt{x}}{2x^2 - 11x + 12}$ 20 $f(x) = \dfrac{x^3 - 1}{x^2 - 1}$

In each of Exercises 21–26 find the number that maps into 4. If $a > 0$, what number maps into a? Find the range of f.

21   $f(x) = 7x - 5$ 22 $f(x) = 3x$

23   $f(x) = \sqrt{x - 3}$ 24 $f(x) = 1/x$

25   $f(x) = x^3$ 26 $f(x) = \sqrt[3]{x - 4}$

In each of Exercises 27–34 determine if the function f is one-to-one.

27   $f(x) = 2x + 9$ 28 $f(x) = 1/(7x + 9)$

29   $f(x) = 5 - 3x^2$ 30 $f(x) = 2x^2 - x - 3$

31   $f(x) = \sqrt{x}$ 32 $f(x) = x^3$

33   $f(x) = |x|$ 34 $f(x) = 4$

A function f with domain X is termed (i) **even** if $f(-a) = f(a)$ for every a in X, or (ii) **odd** if $f(-a) = -f(a)$ for every a in X. In each of Exercises 35–44 determine whether f is even, odd, or neither even nor odd.

35   $f(x) = 3x^3 - 4x$ 36 $f(x) = 7x^4 - x^2 + 7$

37   $f(x) = 9 - 5x^2$ 38 $f(x) = 2x^5 - 4x^3$

39   $f(x) = 2$ 40 $f(x) = 2x^3 + x^2$

41   $f(x) = 2x^2 - 3x + 4$ 42 $f(x) = \sqrt{x^2 + 1}$

43   $f(x) = \sqrt[3]{x^3 - 4}$ 44 $f(x) = |x| + 5$

45 If f is an even function, prove that the graph of f is symmetric with respect to the y-axis.

46 If f is an odd function, prove that the graph of f is symmetric with respect to the origin.

47 Find a formula which expresses the radius r of a circle as a function of its circumference C. If the circumference of *any* circle is increased by 12 inches, determine how much the radius increases.

48 Find a formula which expresses the volume of a cube as a function of its surface area. Find the volume if the surface area is 36 square inches.

49 An open box is to be made from a rectangular piece of cardboard having dimensions 20 inches by 30 inches by cutting out identical squares of area x^2 from each corner and turning up the sides. Express the volume V of the box as a function of x.

50 Find a formula which expresses the area A of an equilateral triangle as a function of the length s of a side.

51 Express the perimeter P of a square as a function of its area A.

52 Express the surface area S of a sphere as a function of its volume V.

53 A weather balloon is released at 1:00 P.M. and rises vertically at a rate of 2 meters per second. An observer is situated 100 meters from a point on the ground directly below the balloon. If t denotes the time (in seconds) after 1:00 P.M., express the distance d between the balloon and the observer as a function of t.

54 Two ships leave port at 9:00 A.M., one sailing south at a rate of 16 mph and the other west at a rate of 20 mph. If t denotes the time (in hours) after 9:00 A.M., express the distance d between the ships as a function of t.

55 A manufacturer sells a certain article to dealers at a rate of $20 each if less than 50 are ordered. If 50 or more are ordered (up to 600), the price per article is reduced at a rate of 2 cents times the number ordered. Let A denote the amount of money received when x articles are ordered. Express A as a function of x.

56 A travel agency offers a tour at a cost of $30 per person if not more than 60 take the tour. If more than 60 take the tour the cost is to be reduced 10 cents for each person in excess of 60. Let A denote the total cost if x people go on the tour. Express A as a function of x.

In each of Exercises 57–64 determine whether the set W of ordered pairs is a function in the sense of (3.11).

57 $W = \{(x, y): 2y = x^2 + 5\}$

58 $W = \{(x, y): x = 3y + 2\}$

59 $W = \{(x, y): x^2 + y^2 = 4\}$

60 $W = \{(x, y): y^2 - x^2 = 1\}$

61 $W = \{(x, y): y = 3\}$

62 $W = \{(x, y): x = y\}$

63 $W = \{(x, y): xy = 0\}$

64 $W = \{(x, y): x + y = 0\}$

4   GRAPHS OF FUNCTIONS

Graphs, or more precisely *sketches of graphs*, are often used to describe the variation of physical quantities. For example, a scientist may use Figure 3.20 to

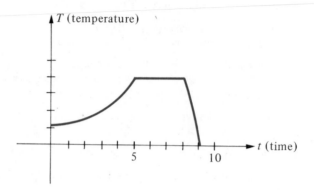

Figure 3.20

indicate the temperature T of a certain solution at various times t during an experiment. The sketch shows that the temperature increased gradually from time $t = 0$ to time $t = 5$, was steady between $t = 5$ and $t = 8$, and then decreased rapidly from $t = 8$ to $t = 9$. A visual aid of this type reveals the behavior of T more clearly than a long table of numerical values.

   In like manner, Figure 3.21 could represent the variation in atmospheric pressure y as the altitude x varies. One can see at a glance the manner in which the pressure decreases as the altitude increases.

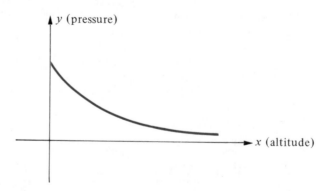

Figure 3.21

   If f is a function, we often use a sketch of a graph to exhibit the behavior of $f(x)$ as x varies through the domain of f. Such geometric representations of functions are often more enlightening than algebraic descriptions of the types considered in the examples of the previous section. By definition, the **graph of a function** f is the set of all points $(x, f(x))$ in a coordinate plane, where x is in the domain of f. The phrase "sketch the graph" has the same meaning as in Section 3. The graph of f can also be described as the set of all points $P(x, y)$ such that $y = f(x)$. Thus the graph of f is the same as the graph of the equation $y = f(x)$, and if $P(a, b)$ is on the graph of f, then the ordinate b is the functional value $f(a)$ as illustrated in Figure 3.22. It is important to note that since there is a unique $f(a)$ for each a in the domain, there is only *one* point on the graph with abscissa a.

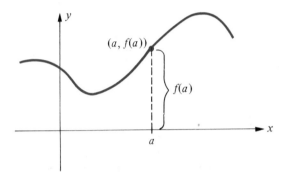

Figure 3.22

Consequently, for graphs of functions it is impossible to obtain a sketch such as that shown in Figure 3.12 where a vertical line pierces the graph in more than one point.

Example 1   Sketch the graph of f if $f(x) = 2x + 3$.

Solution   The graph consists of all points $(x, f(x))$, where $f(x) = 2x + 3$. The following table lists coordinates of several points on the graph, where $y = f(x)$.

x	-3	-2	-1	0	1	2
y	-3	-1	1	3	5	7

Plotting, it appears that the points lie on a straight line and we sketch the graph as in Figure 3.23. (In the next section we shall *prove* that the graph is a straight line.) The graph of f is the same as the graph of the equation $y = 2x + 3$. (Compare with Example 1 of Section 2.)

It is useful to determine the points at which the graph of a function f intersects the x-axis. The abscissas of these points are called the **x-intercepts** of the graph and are found by locating all points with zero ordinates, that is, all points $(x, f(x))$ such that $f(x) = 0$. In Example 1 the x-intercept is $-3/2$, since that number is the solution of $2x + 3 = 0$. A number a such that $f(a) = 0$ is also called a **zero of the function** f.

If the number 0 is in the domain of f, then $f(0)$ is called the **y-intercept** of the graph of f. It is the ordinate of the point at which the graph intersects the y-axis.

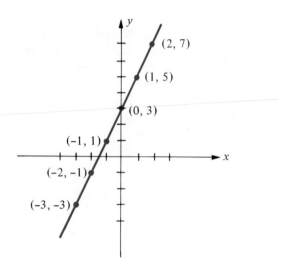

Figure 3.23. $f(x) = 2x + 3$

The graph of a function can have at most one y-intercept. The y-intercept in Example 1 is 3.

If f is the function in Example 1 and if $x_1 < x_2$, then $2x_1 + 3 < 2x_2 + 3$; that is, $f(x_1) < f(x_2)$. This means that as abscissas of points increase, ordinates also increase. The function f is then said to be *increasing*, according to (3.12) below. If f is increasing, then the graph rises as x increases. For certain functions, we have $f(x_1) > f(x_2)$ whenever $x_1 < x_2$. In this case the graph of f falls as x increases and the function is called a *decreasing* function. In general we shall speak of functions which increase or decrease on certain subsets of their domains, as in the following definition. Functions which neither increase nor decrease are called *constant*.

(3.12)   Definition

> If S is a subset of the domain of a function f, then
>
> (i) f is **increasing** on S if $f(x_1) < f(x_2)$ whenever $x_1 < x_2$ in S.
> (ii) f is **decreasing** on S if $f(x_1) > f(x_2)$ whenever $x_1 < x_2$ in S.
> (iii) f is **constant** on S if $f(x_1) = f(x_2)$ for every x_1, x_2 in S.

If we regard Figure 3.20 as the graph of a function f, then f is increasing on the closed interval $[0, 5]$, is constant on the closed interval $[5, 8]$, and is decreasing on $[8, 9]$. Figure 3.21 represents a function which is decreasing throughout its domain.

Example 2   Sketch the graph of f if $f(x) = x^2 - 3$.

Solution We list coordinates $(x, f(x))$ of some points on the graph of f in tabular form, as shown below.

x	-3	-2	-1	0	1	2	3
$f(x)$	6	1	-2	-3	-2	1	6

The x-intercepts are the solutions of the equation $f(x) = 0$, that is, of $x^2 - 3 = 0$. These are $\pm\sqrt{3}$. The y-intercept is $f(0) = -3$. Plotting the points given by the table and using the x-intercepts leads to the sketch in Figure 3.24. Compare the graph of f with that of the equation $y = x^2$ sketched in Figure 3.10. We could have shortened our work somewhat by observing that since $(-x)^2 = x^2$, the graph of $y = x^2 - 3$ is symmetric with respect to the y-axis.

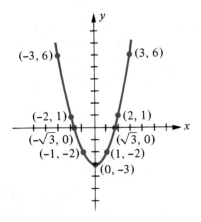

Figure 3.24. $f(x) = x^2 - 3$

In arriving at the sketch in Figure 3.24 we assumed that f is decreasing on the interval $(-\infty, 0]$ and increasing on the interval $[0, \infty)$. These facts can be proved using properties of inequalities. For example, consider any two positive real numbers x_1 and x_2 such that $x_1 < x_2$. Multiplying both sides of the inequality by x_1 and x_2, respectively, leads to

$$x_1^2 < x_1 x_2 \quad \text{and} \quad x_1 x_2 < x_2^2.$$

Applying (i) and (ii) of (2.4) together with the definition of f gives us

$$x_1^2 < x_2^2$$
$$x_1^2 - 3 < x_2^2 - 3$$
$$f(x_1) < f(x_2).$$

Hence by (i) of (3.12), f is increasing on the set of positive real numbers. If, in the previous discussion, $x_1 = 0$ and $x_1 < x_2$, then again $x_1^2 < x_2^2$ and $f(x_1) < f(x_2)$. Thus f is increasing on the interval $[0, \infty)$. It can be shown in similar fashion that f is decreasing on $(-\infty, 0]$. It follows that $f(x)$ takes on its least value at $x = 0$. This smallest value, -3, is called the **minimum value** of f. The corresponding point $(0, -3)$ is the lowest point on the graph. Clearly, $f(x)$ does not attain a **maximum value**, that is, a *largest* value. In the future we shall not justify our graphs in this way, but instead rely on sufficient point plotting to determine where functions are increasing or decreasing.

Example 3   Sketch the graph of f if $f(x) = |x|$.

Solution   If $x \geq 0$, then $f(x) = x$ and we obtain the set of all points (x, x) on the graph of f. Some special cases are $(0, 0)$, $(1, 1)$, $(2, 2)$, $(3, 3)$, and $(4, 4)$. Negative values of x give rise to the following table:

x	-1	-2	-3	-4
$f(x)$	1	2	3	4

More generally, we obtain all points of the form $(-a, a)$ where $a > 0$. Plotting points leads to the sketch shown in Figure 3.25. As in Example 2, this function decreases on $(-\infty, 0]$ and increases on $[0, \infty)$. Also note that the graph is symmetric with respect to the y-axis.

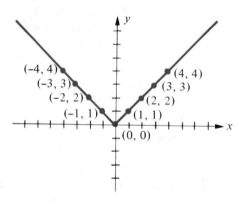

Figure 3.25. $f(x) = |x|$

No portion of the graph in Example 3 appears below the x-axis. A region of the coordinate plane in which there is no graph is called an **excluded region**.

Example 4   Sketch the graph of f if $f(x) = \sqrt{x - 1}$.

Solution   The domain of f does not include values of x such that $x - 1 < 0$, since $f(x)$ is not a real number in this case. Consequently the set of points (x, y) with $x < 1$ is an excluded region for the graph. The region below the x-axis is also excluded. (Why?) The following table lists some points $(x, f(x))$ on the graph.

x	1	2	3	4	5	6
$f(x)$	0	1	$\sqrt{2}$	$\sqrt{3}$	2	$\sqrt{5}$

Plotting points leads to the sketch shown in Figure 3.26. The function is increasing throughout its domain. The x-intercept is 1 and there is no y-intercept.

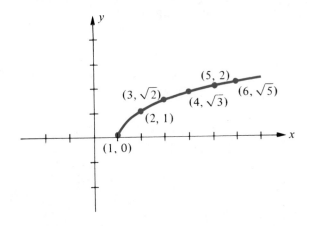

Figure 3.26. $f(x) = \sqrt{x - 1}$

Example 5   Sketch the graph of a constant function with domain $\mathbb{R}$.

Solution   From (3.10), f is a constant function if there is some (fixed) real number c such that $f(x) = c$ for every x. The graph of f consists of all points with coordinates (x, c), where x is any real number. In particular this includes $(-1, c), (0, c), (2, c)$, and so on. Since all the ordinates equal c, the graph is a line parallel to the x-axis with y-intercept c. A sketch for the case $c > 0$ is shown in Figure 3.27.

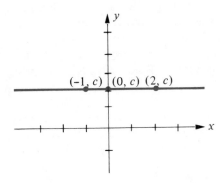

Figure 3.27. $f(x) = c$

Comparing Figures 3.24 and 3.10 we see that the graph of $f(x) = x^2 - 3$ can be obtained by lowering the graph of $y = x^2$ a distance of 3 units, that is, by decreasing the ordinate of each point by 3. Generally, if we know the graph of *any* function f and if c is a positive real number, then the graph of $y = f(x) + c$ can be obtained by raising the graph of $y = f(x)$ a distance c. For the graph of $y = f(x) - c$ we lower the graph of f a distance c.

Example 6   If $f(x) = x^2 + c$, sketch the graph of f if (a) $c = 4$; (b) $c = 1$; (c) $c = -2$.

Solution   We shall sketch all three graphs on the same coordinate axes. The graph of $y = x^2$ was sketched in Figure 3.10, and for reference we have represented it by dashes in Figure 3.28. To find the graph of $f(x) = x^2 + 4$ we simply increase the ordinate of

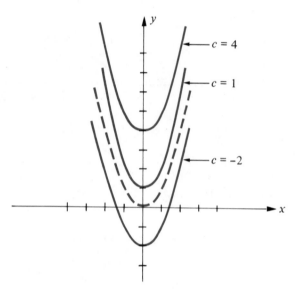

Figure 3.28. $f(x) = x^2 + c$

each point on the graph of $y = x^2$ by 4, as shown in the figure. Similarly, for $c = 1$ we increase ordinates by 1, whereas for $c = -2$ we decrease ordinates by 2. Note that each function is decreasing on the interval $(-\infty, 0]$ and increasing on $[0, \infty)$. Each graph is a parabola symmetric with respect to the y-axis. In order to get the correct position it is advisable to plot several points on each graph.

To obtain the graph of $y = cf(x)$ we may *multiply* the ordinates of points on the graph of $y = f(x)$ by c. Thus, if $y = 2f(x)$ we double ordinates, if $y = \frac{1}{2}f(x)$ we multiply each ordinate by 1/2, and so on.

Example 7   If $f(x) = x^2$, sketch the graphs of $y = cf(x)$ if $c = 4$, $c = 1/4$, and $c = -1$.

Solution   We wish to sketch the graphs of the equations

$$y = 4x^2, \quad y = \tfrac{1}{4}x^2, \quad \text{and} \quad y = -x^2.$$

To obtain the graph of $y = 4x^2$ we could refer to the graph of $y = x^2$ in Figure 3.10 and multiply the ordinate of each point by 4. This gives us a narrower parabola which is sharper at the vertex, as illustrated in (i) of Figure 3.29. In order

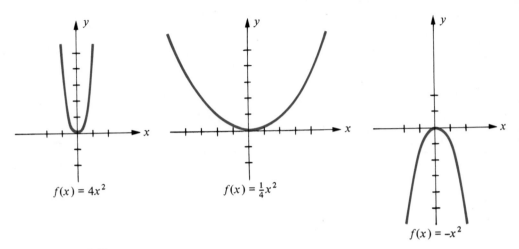

$f(x) = 4x^2$ $f(x) = \frac{1}{4}x^2$

$f(x) = -x^2$

Figure 3.29

to get the correct position, several points such as $(0, 0)$, $(1/2, 1)$, and $(1, 4)$ should be plotted.

Similarly, for the graph of $y = (1/4)x^2$ we multiply ordinates of points on the graph of $y = x^2$ by $1/4$. This gives us a wider parabola which is flatter at the vertex, as shown in (ii) of Figure 3.29.

Finally, for $y = -x^2$ we multiply ordinates in Figure 3.10 by -1. This amounts to reflecting that graph through the x-axis, as shown in (iii) of Figure 3.29.

Sometimes functions are described in terms of several expressions, as shown in the next example.

Example 8 Sketch the graph of the function f which is defined as follows:

$$f(x) = \begin{cases} 2x + 3 & \text{if } x < 0 \\ x^2 & \text{if } 0 \le x < 2 \\ 1 & \text{if } x \ge 2. \end{cases}$$

Solution If $x < 0$, then $f(x) = 2x + 3$. This means that when x is negative, the expression $2x + 3$ should be used to find functional values. Consequently, if $x < 0$, then the graph of f coincides with the graph in Figure 3.23 and we sketch that portion of the graph to the left of the y-axis as indicated in Figure 3.30.

If $0 \le x < 2$, we use x^2 to find functional values of f, and therefore this part of the graph of f coincides with the graph of the equation $y = x^2$ (see Figure 3.10). We then sketch the part of the graph of f between $x = 0$ and $x = 2$ as indicated in Figure 3.30.

Finally, if $x \ge 2$, the graph of f coincides with the constant function having values equal to 1. That part of the graph is the horizontal half-line illustrated in Figure 3.30.

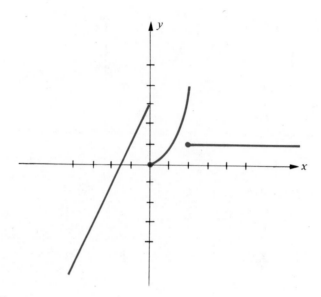

Figure 3.30

EXERCISES

In Exercises 1–20, sketch the graph of f and state where f is increasing, is decreasing, or is constant.

1   $f(x) = 5x$ 2 $f(x) = -3x$

3   $f(x) = -4x + 2$ 4 $f(x) = 4x - 2$

5   $f(x) = 3 - x^2$ 6 $f(x) = 4x^2 + 1$

7   $f(x) = 2x^2 - 4$ 8 $f(x) = \frac{1}{100}x^2$

9   $f(x) = \sqrt{x + 4}$ 10 $f(x) = \sqrt{4 - x}$

11   $f(x) = \sqrt{x} + 2$ 12 $f(x) = 2 - \sqrt{x}$

13   $f(x) = 4/x$ 14 $f(x) = 1/x^2$

15   $f(x) = |x - 2|$ 16 $f(x) = |x + 2|$

17   $f(x) = |x| - 2$ 18 $f(x) = |x| + 2$

19   $f(x) = \dfrac{x}{|x|}$ 20 $f(x) = x + |x|$

In Exercises 21–26, sketch the graph of the function f for the stated values of c.

21   $f(x) = 3x + c$; $c = 0, c = 2, c = -1$

22   $f(x) = -2x + c$; $c = 0, c = 1, c = -3$

23   $f(x) = x^3 + c$; $c = 0, c = 1, c = -2$

24   $f(x) = -x^3 + c$; $c = 0, c = 2, c = -1$

25 $f(x) = cx + 2;$ $c = 1, c = 2, c = -1/2$

26 $f(x) = cx - 1;$ $c = 3, c = 1/2, c = -2$

In each of Exercises 27–32 sketch the graph of f.

27 $f(x) = \begin{cases} 2 & \text{if } x < 0 \\ -1 & \text{if } x \geq 0 \end{cases}$

28 $f(x) = \begin{cases} -1 & \text{if } x \text{ is an integer} \\ 1 & \text{if } x \text{ is not an integer} \end{cases}$

29 $f(x) = \begin{cases} 3 & \text{if } x < -3 \\ -x & \text{if } -3 \leq x \leq 3 \\ -3 & \text{if } x > 3 \end{cases}$

30 $f(x) = \begin{cases} x & \text{if } x < 0 \\ -2 & \text{if } 0 \leq x < 1 \\ x^2 & \text{if } x \geq 1 \end{cases}$

31 $f(x) = \begin{cases} x^2 & \text{if } x \leq -1 \\ x^3 & \text{if } |x| < 1 \\ 2x & \text{if } x \geq 1 \end{cases}$

32 $f(x) = \begin{cases} x & \text{if } x \leq 1 \\ -x^2 & \text{if } 1 < x < 2 \\ x & \text{if } x \geq 2 \end{cases}$

33 If x is any real number, then there exist consecutive integers n and $n + 1$ such that $n \leq x < n + 1$. Let f be the function defined as follows: if x is a real number and $n \leq x < n + 1$, then $f(x) = n$. Sketch the graph of f. (The function f is called the **greatest integer function**.)

34 Explain why the graph of the equation $x^2 + y^2 = 1$ is not the graph of a function.

35 Sketch the graph of the identity function with domain $\mathbb{R}$.

36 Define a function f whose graph is the upper half of a circle with center at the origin and radius 1.

37 Define a function f whose graph is the lower half of a circle with center at the origin and radius 1.

38 If a function f is increasing throughout its domain, prove that f is one-to-one.

39 If a function f is decreasing throughout its domain, prove that f is one-to-one.

40 Prove that a function f is one-to-one if and only if every horizontal line pierces the graph of f in at most one point.

5   LINEAR FUNCTIONS

(3.13)   Definition

> A function f is a **linear function** if
> $$f(x) = ax + b$$
> where a and b are real numbers and $a \neq 0$.

The reason for the term "linear" is that the graph of f is a line, as we shall see later in this section.

Let us begin by introducing several fundamental concepts pertaining to lines. All lines referred to are considered to be in some fixed coordinate plane.

(3.14)   Definition

> If l is a line which is not parallel to the y-axis, and if $P_1(x_1, y_1)$ and $P_2(x_2, y_2)$ are distinct points on l, then the **slope m** of l is given by
>
> $$m = \frac{y_2 - y_1}{x_2 - x_1}.$$
>
> If l is parallel to the y-axis, then the slope is not defined.

The numerator $y_2 - y_1$ in the formula for m is sometimes called the **rise** from P_1 to P_2. It measures the vertical change in direction in proceeding from P_1 to P_2 and may be positive, negative, or zero. The denominator $x_2 - x_1$ is called the **run** from P_1 to P_2. It measures the amount of horizontal change in going from P_1 to P_2. The run may be positive or negative but is never zero because l is not parallel to the y-axis. Using this terminology, we could write (3.14) as

$$\text{slope of } l = \frac{\text{rise from } P_1 \text{ to } P_2}{\text{run from } P_1 \text{ to } P_2}.$$

In finding the slope of a line it is immaterial which point is labeled P_1 and which is labeled P_2, since

$$\frac{y_2 - y_1}{x_2 - x_1} = \frac{y_1 - y_2}{x_1 - x_2}.$$

Consequently, we may as well assume that the points are labeled so that $x_1 < x_2$, as in Figure 3.31. In this event $x_2 - x_1 > 0$, and hence the slope is positive, negative, or zero, depending on whether $y_2 > y_1, y_2 < y_1,$ or $y_2 = y_1$. The slope of

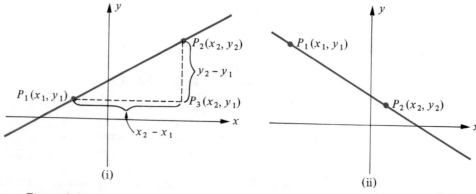

(i)   (ii)

Figure 3.31

the line shown in (i) of Figure 3.31 is positive, whereas the slope of the line shown in (ii) of the figure is negative. The slope is zero if and only if the line is horizontal.

Example 1   Sketch the lines through the following pairs of points and find their slopes.
(a) $A(-1,4)$ and $B(3,2)$
(b) $A(2,5)$ and $B(-2,-1)$
(c) $A(4,3)$ and $B(-2,3)$
(d) $A(4,-1)$ and $B(4,4)$.

Solution   The lines are sketched in Figure 3.32. Using (3.14) gives the slopes for parts (a)–(c).

(a) $$m = \frac{2-4}{3-(-1)} = \frac{-2}{4} = -\frac{1}{2}$$

(b) $$m = \frac{5-(-1)}{2-(-2)} = \frac{6}{4} = \frac{3}{2}$$

(c) $$m = \frac{3-3}{-2-4} = \frac{0}{-6} = 0$$

(d)   The slope is undefined since the line is vertical. This is also seen by noting that if (3.14) is used, then the denominator is zero.

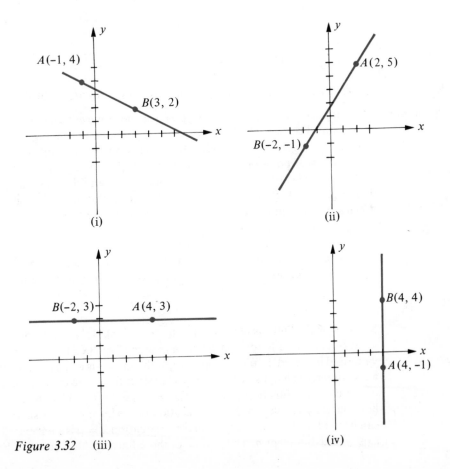

Figure 3.32   (iii)

It is important to note that the definition of slope is independent of the two points that are chosen on l, for if other points $P'_1(x'_1, y'_1)$ and $P'_2(x'_2, y'_2)$ are used, then as in Figure 3.33 the triangle with vertices P'_1, P'_2, and $P'_3(x'_2, y'_1)$ is similar to the triangle with vertices $P_1, P_2, P_3(x_2, y_1)$. Since the ratios of corresponding sides are equal, it follows that

$$\frac{y_2 - y_1}{x_2 - x_1} = \frac{y'_2 - y'_1}{x'_2 - x'_1}.$$

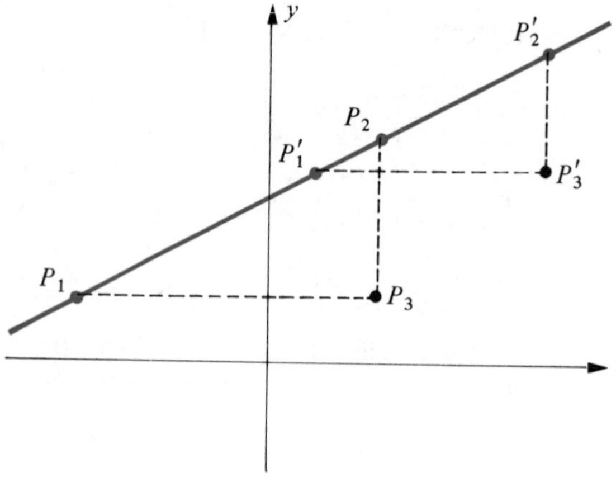

Figure 3.33

Example 2   Construct a line through $P(2, 1)$ that has slope   (a) $5/3$;   (b) $-5/3$.

Solution   If the slope of a line is a/b where b is positive, then for every change of b units in the horizontal direction, the line rises or falls $|a|$ units, depending on whether a is positive or negative, respectively. If $P(2, 1)$ is on the line and $m = 5/3$, we can obtain another point on the line by starting at P and moving 3 units to the right and 5 units upward. This gives us the point $Q(5, 6)$ and the line is determined (see (i) of Figure 3.34). Similarly, if $m = -5/3$ we move 3 units to the right and 5 units downward obtaining $Q(5, -4)$ as in (ii) of Figure 3.34.

The equation $y = b$, where b is a real number, may be considered as an equation in two variables x and y, since we can write it in the form

$$(0x) + y = b.$$

Some typical solutions of the equation are $(-2, b), (1, b)$, and $(3, b)$. Evidently, all solutions of the equation consist of pairs of the form (x, b) where x may have any value and b is fixed. It follows that the graph of $y = b$ is a line parallel to the x-axis with y-intercept b. This was to be expected since the graph is the same as the graph of the constant function f, where $f(x) = b$ (see Figure 3.27). Conversely, every horizontal line is the graph of an equation of the form $y = b$. A similar argument can be used to show that the graph of the equation $x = a$ is a line parallel to the y-axis with x-intercept a. The graphs of these lines are illustrated in Figure 3.35.

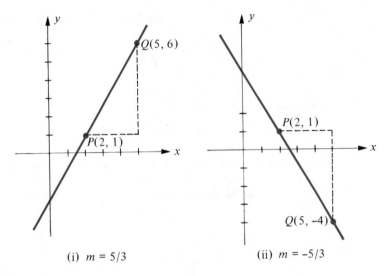

(i) $m = 5/3$ (ii) $m = -5/3$

Figure 3.34

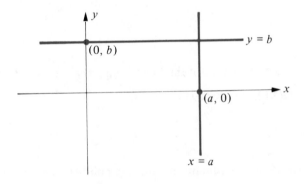

Figure 3.35

Let us now find an equation of a line l through a point $P_1(x_1, y_1)$ with slope m (only one such line exists). If $P(x, y)$ is any point with $x \neq x_1$ (see Figure 3.36),

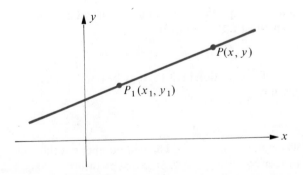

Figure 3.36

then P is on l if and only if the slope of the line through P_1 and P is m, that is, if and only if

$$\frac{y - y_1}{x - x_1} = m.$$

This equation may be written in the form

$$y - y_1 = m(x - x_1).$$

Note that (x_1, y_1) is also a solution of the latter equation and hence the points on l are precisely the points which correspond to the solutions. This equation for l is referred to as the **point-slope form**. Our discussion may be summarized as follows:

(3.15)   **Point-Slope Form for the Equation of a Line**

> An equation for the line through the point $P(x_1, y_1)$ with slope m is
>
> $$y - y_1 = m(x - x_1).$$

Example 3   Find an equation of the line through the points $A(1, 7)$ and $B(-3, 2)$.

Solution   By (3.14) the slope m of the line is

$$m = \frac{7 - 2}{1 - (-3)} = \frac{5}{4}.$$

Using the coordinates of A in the point-slope form (3.15) gives us

$$y - 7 = \frac{5}{4}(x - 1)$$

which is equivalent to

$$4y - 28 = 5x - 5 \quad \text{or} \quad 5x - 4y + 23 = 0.$$

The same equation would have been obtained if the coordinates of point B had been substituted in (3.15).

The equation in (3.15) may be rewritten as $y = mx - mx_1 + y_1$, which is of the form

$$y = mx + b$$

where $b = -mx_1 + y_1$. The real number b is the y-intercept of the graph, as may be seen by setting $x = 0$. Since the equation $y = mx + b$ displays the slope m and y-intercept b of l, it is called the **slope-intercept form** for the equation of a line.

Conversely, given an equation of that form we may change it to

$$y - b = m(x - 0).$$

Comparing with (3.15) we see that the graph is a line with slope m and passing through the point $(0, b)$. This gives us the next result.

(3.16)   Slope-Intercept Form for the Equation of a Line

> The graph of the equation $y = mx + b$ is a line having slope m and y-intercept b.

The work we have done shows that every line is the graph of an equation of the form

$$ax + by + c = 0$$

where a, b, and c are real numbers and a and b are not both zero. We call such an equation a **linear equation** in x and y. Let us show conversely that the graph of $ax + by + c = 0$ where a and b are not both zero is always a line. On the one hand, if $b \neq 0$ we may write the equation as

$$y = (-a/b)x + (-c/b)$$

which, by the slope-intercept form (3.16), is an equation of a line with slope $-a/b$ and y-intercept $-c/b$. On the other hand, if $b = 0$ but $a \neq 0$, then we may write the equation as $x = -c/a$, which is the equation of a line parallel to the y-axis with x-intercept $-c/a$. This establishes the following important theorem.

(3.17)   Theorem

> The graph of a linear equation $ax + by + c = 0$ is a line and, conversely, every line is the graph of a linear equation.

For simplicity we shall often use the terminology *the line* $ax + by + c = 0$ instead of the more accurate phrase *the line with equation* $ax + by + c = 0$.

Example 4   Sketch the graph of $2x - 5y = 8$.

Solution   We know from (3.17) that the graph is a line and hence it is sufficient to find two points on the graph. Let us find the x- and y-intercepts. Substituting $y = 0$ in the given equation we obtain the x-intercept 4. Substituting $x = 0$ we see that the y-intercept is $-8/5$. This leads to the graph in Figure 3.37.

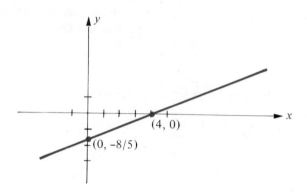

Figure 3.37. $2x - 5y = 8$

Another method of solution is to express the given equation in slope-intercept form. To do this we begin by isolating the term involving y on one side of the equals sign, obtaining

$$5y = 2x - 8.$$

Next, dividing both sides by 5 gives us

$$y = \frac{2}{5}x + \left(\frac{-8}{5}\right).$$

By comparison with (3.16) we obtain the slope $m = 2/5$ and the y-intercept $b = -8/5$. We may then sketch a line through the point $(0, -8/5)$ with slope $2/5$.

It can be shown geometrically that *two nonvertical lines are parallel if and only if they have the same slope.* We shall use this fact in the next example.

Example 5 Find an equation of a line through the point $(5, -7)$ which is parallel to the line $6x + 3y - 4 = 0$.

Solution Let us express the given equation in slope-intercept form. We begin by writing

$$3y = -6x + 4$$

and then divide both sides by 3, obtaining

$$y = -2x + \frac{4}{3}.$$

The last equation is in slope-intercept form (3.16) with $m = -2$, and hence the slope is -2. From the preceding remark the required line also has slope -2.

Applying the point-slope form (3.15) gives us

$$y + 7 = -2(x - 5)$$

or equivalently,

$$y + 7 = -2x + 10 \quad \text{or} \quad 2x + y - 3 = 0.$$

The results obtained in this section may be used to prove the following important theorem.

(3.18)   Theorem

> The graph of a linear function is a line.

Proof.   If f is linear, then as in (3.13) we may write

$$f(x) = mx + b$$

where $m \neq 0$. Since the graph of f is the same as the graph of the equation $y = mx + b$, the conclusion follows at once from (3.16).

EXERCISES

In each of Exercises 1–6 plot the points A and B and find the slope of the line through A and B.

1   $A(-3, 8)$, $B(-2, 16)$

2   $A(7, -1)$, $B(-4, 9)$

3   $A(\frac{1}{3}, \frac{2}{3})$, $B(-\frac{3}{2}, \frac{1}{2})$

4   $A(-5, -2)$, $B(6, 4)$

5   $A(5, 0)$, $B(-3, 2)$

6   $A(0, -4)$, $B(8, 3)$

7   Use slopes to show that $A(-4, -1)$, $B(2, 3)$, $C(1, -3)$, and $D(7, 1)$ are vertices of a parallelogram.

8   Use slopes to show that $A(-3, -9)$, $B(1, -1)$, and $C(4, 5)$ lie on a straight line.

In each of Exercises 9–22 find an equation for the line satisfying the given conditions.

9   Through $A(5, -2)$; slope $1/3$

10   Through $B(-3, 7)$; slope -2

11   Through $A(-8, -3)$ and $B(2, -5)$

12   Through $P(4, -1)$ and $Q(-6, 3)$

13   Slope -4, y-intercept 7

14   Slope $3/4$, x-intercept 5

15 x-intercept -3, y-intercept 7

16 Through $A(-7, -8)$; y-intercept 0

17 Through $A(7, -3)$, parallel to the   (a) x-axis   (b) y-axis

18 Through $A(-8, 2)$, perpendicular to the   (a) y-axis   (b) x-axis

19 Bisecting the second and fourth quadrants

20 Coinciding with the y-axis

21 Through $A(-4, 10)$, parallel to the line through $B(0, 5)$ and $C(-8, -8)$

22 Through $P(3/2, -1/4)$, parallel to the line with equation $2x + 4y = 5$

23 Find equations for the medians of the triangle with vertices $A(-4, -3)$, $B(2, 3)$, and $C(5, -1)$.

24 If $A(x_1, y_1)$, $B(x_2, y_2)$, $C(x_3, y_3)$, and $D(x_4, y_4)$ are vertices of an arbitrary quadrilateral, prove that the line segments joining the midpoints of adjacent sides form a parallelogram.

In each of Exercises 25–34 use the slope-intercept form to find the slope and y-intercept of the line with the given equation and sketch the graph.

25 $2x - 5y + 10 = 0$ **26** $2y - 3x = 4$

27 $3x + 5y = 0$ **28** $6x = 1 - 2y$

29 $y + 2 = 0$ **30** $3y = 8$

31 $5x = 20 - 6y$ **32** $9x - 4y = 0$

33 $y = 0$ **34** $x = (2/3)y + 4$

35 Find a real number k such that the point $P(-1, 2)$ is on the line $kx + 2y - 7 = 0$.

36 Find a real number k such that the line $5x + ky - 3 = 0$ has y-intercept -5.

37 Sketch the graph of f if $f(x) = -5x - 2$.

38 Sketch the graph of f if $f(x) = 4x - 3$.

39 Given $f(x) = ax + 3$, sketch the graph of f if:

    (a) $a = 0$ (b) $a = 2$ (c) $a = 5$ (d) $a = -3$

40 Given $f(x) = -5x + b$, sketch the graph of f if:

    (a) $b = 0$ (b) $b = 2$ (c) $b = 5$ (d) $b = -3$

41 If $f(x) = 3x + 7$ and $g(x) = -2x + 5$, find the abscissa of a point which is on the graphs of both f and g.

42 If the graph of a linear function f contains the points $(2, 7)$ and $(0, 1)$, find $f(x)$.

43 If f is a linear function such that $f(0) = -2$ and $f(-1) = 4$, find $f(x)$.

44 If the graph of a linear function f is parallel to the line $3x + 5y - 2 = 0$ and has x-intercept -4, find $f(x)$.

45 If $a > 0$, prove that the linear function f defined by $f(x) = ax + b$ is an increasing function throughout its domain. If $a < 0$, prove that f is decreasing throughout its domain.

46 Prove that the graph of the equation $ax + by = 0$, where a and b are not both zero, is a line passing through the origin.

47 If a line l has nonzero x- and y-intercepts a and b, respectively, prove that an equation for l is

$$\frac{x}{a} + \frac{y}{b} = 1.$$

(This is called the **intercept form** for the equation of a line.) Express the equation $4x - 2y = 6$ in intercept form.

48 Prove that an equation of the line through $P_1(x_1, y_1)$ and $P_2(x_2, y_2)$ is

$$(y - y_1)(x_2 - x_1) = (y_2 - y_1)(x - x_1).$$

(This is called the **two-point form** for the equation of a line.) Use the two-point form to find an equation of the line through $A(7, -1)$ and $B(4, 6)$.

49 Find all values of r such that the slope of the line through the points $(r, 4)$ and $(1, 3 - 2r)$ is less than 5.

50 Find all values of t such that the slope of the line through $(t, 3t + 1)$ and $(1 - 2t, t)$ is greater than 4.

6 COMPOSITE AND INVERSE FUNCTIONS

We shall now describe an important method of using two functions f and g to obtain a third function. Suppose X, Y, and Z are sets of real numbers and let f be a function from X to Y and g a function from Y to Z. In terms of the arrow notation introduced in Section 3 we have

$$X \xrightarrow{f} Y \xrightarrow{g} Z$$

that is, f maps X into Y and g maps Y into Z. A function from X to Z may be defined in a natural way. For every x in X, the number $f(x)$ is in Y. Since the domain of g is Y, we may then find the image of $f(x)$ under g. Of course, this element of Z is written as $g(f(x))$. By associating $g(f(x))$ with x, we obtain a function from X to Z called the *composite function* of g by f. This is illustrated pictorially in Figure 3.38, where the dashes indicate the correspondence we have

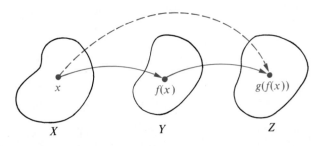

Figure 3.38

defined from X to Z. We sometimes use an operational symbol $\circ$ and denote the latter function by $g \circ f$. The following definition summarizes our discussion.

(3.19)   **Definition**

> If f is a function from X to Y and g is a function from Y to Z, then the **composite function** $g \circ f$ is the function from X to Z defined by
>
> $$(g \circ f)(x) = g(f(x))$$
>
> for every x in X.

Example 1   If $f(x) = x^3$ and $g(x) = 5x^2 + 2x + 1$, find $(g \circ f)(x)$.

Solution   Using (3.19) and the definition of f gives us

$$(g \circ f)(x) = g(f(x)) = g(x^3).$$

Since $g(x^3)$ means that x^3 should be substituted for x in the expression for $g(x)$ we have

$$g(x^3) = 5(x^3)^2 + 2(x^3) + 1.$$

Consequently

$$(g \circ f)(x) = 5x^6 + 2x^3 + 1.$$

In applying (3.19), it is not essential that the domain of g be all of Y, but merely that it *contain* the range of f. In certain cases we may wish to restrict x to some subset of X so that $f(x)$ is in the domain of g as illustrated in the next example.

Example 2   If the functions f and g are defined by $f(x) = x - 2$ and $g(x) = 5x + \sqrt{x}$, find $(g \circ f)(x)$.

Solution   Formal substitutions give us the following:

$$
\begin{aligned}
(g \circ f)(x) &= g(f(x)) & &\text{(definition of } g \circ f) \\
&= g(x - 2) & &\text{(definition of } f) \\
&= 5(x - 2) + \sqrt{x - 2} & &\text{(definition of } g) \\
&= 5x - 10 + \sqrt{x - 2} & &\text{(simplifying).}
\end{aligned}
$$

The domain X of f is the set of all real numbers; however, the last equality implies that $(g \circ f)(x)$ is a real number only if $x \geq 2$. Thus, when working with the composite function $g \circ f$ we must restrict x to the interval $[2, \infty)$.

If $X = Y = Z$ in (3.19), then it is possible to find $f(g(x))$. We first obtain the image of x under g and then apply f to $g(x)$. In this way we obtain a function from Z to X called the *composite function* of f by g and denoted by $f \circ g$. Thus, by definition,

$$(f \circ g)(x) = f(g(x))$$

for all x in Z.

Example 3 If $f(x) = x^2 - 1$ and $g(x) = 3x + 5$, find $(f \circ g)(x)$ and $(g \circ f)(x)$.

Solution We have the following identities:

$$
\begin{aligned}
(f \circ g)(x) &= f(g(x)) && \text{(definition of } f \circ g) \\
&= f(3x + 5) && \text{(definition of } g) \\
&= (3x + 5)^2 - 1 && \text{(definition of } f) \\
&= 9x^2 + 30x + 24 && \text{(simplifying).}
\end{aligned}
$$

Similarly,

$$
\begin{aligned}
(g \circ f) &= g(f(x)) && \text{(definition of } g \circ f) \\
&= g(x^2 - 1) && \text{(definition of } f) \\
&= 3(x^2 - 1) + 5 && \text{(definition of } g) \\
&= 3x^2 + 2 && \text{(simplifying).}
\end{aligned}
$$

We see from Example 3 that $f(g(x))$ and $g(f(x))$ are not always the same, that is, $f \circ g \neq g \circ f$. In certain cases it may happen that equality *does* occur. Of major importance is the case in which $f(g(x))$ and $g(f(x))$ are not only identical, but both are equal to x. Needless to say, f and g must be very special functions in order for this to happen. In the following discussion we indicate the manner in which they will be restricted.

Suppose f is a *one-to-one function* with domain X and range Y. As mentioned in Section 3 this implies that each element of Y is the image of precisely one element of X. Another way of phrasing this is to say that *each element of Y can be written in one and only one way in the form* $f(x)$, where x is in X. We may then define a function g from Y to X by demanding that

$$g(f(x)) = x \quad \text{for every } x \text{ in } X.$$

This amounts to *reversing* the correspondence given by f. If f is represented geometrically by drawing arrows as in (i) of Figure 3.39, then g can be represented by simply *reversing* these arrows as illustrated in (ii) of the figure. It follows that g is a one-to-one function with domain Y and range X. (See Exercise 38.)

As illustrated in Figure 3.39, if $f(x) = y$, then $x = g(y)$. This means that

$$f(g(y)) = y \quad \text{for every } y \text{ in } Y.$$

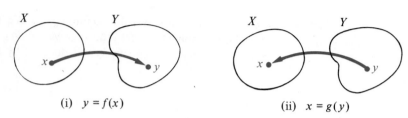

(i) $y = f(x)$ (ii) $x = g(y)$

Figure 3.39

Since the notation used for the variable is immaterial, we may write

$$f(g(x)) = x \quad \text{for every } x \text{ in } Y.$$

The functions f and g are called *inverse functions* of one another, according to the following definition.

(3.20)   Definition

> If f is a one-to-one function with domain X and range Y, then a function g with domain Y and range X is called the **inverse function of f** if
>
> $$f(g(x)) = x \quad \text{for every } x \text{ in } Y \text{ and}$$
> $$g(f(x)) = x \quad \text{for every } x \text{ in } X.$$

There can be only one inverse function of f (see Exercise 39). Moreover, if g is the inverse function of f, then by (3.20) $(g \circ f)(x) = x$, that is, $g \circ f$ is the identity function on X (see (3.9)). Similarly, since $(f \circ g)(x) = x$, for all x in Y, $f \circ g$ is the identity function on Y. For this reason the symbol f^{-1} is often used to denote the inverse function of f. Employing this notation,

(3.21)

> $$f^{-1}(f(x)) = x \quad \text{for every } x \text{ in } X$$
> $$f(f^{-1}(x)) = x \quad \text{for every } x \text{ in } Y.$$

The symbol -1 in (3.21) should not be mistaken for an exponent; that is, $f^{-1}(x)$ does not mean $1/f(x)$. The reciprocal $1/f(x)$ may be denoted by $[f(x)]^{-1}$.

Inverse functions are very important in the study of trigonometry. In Chapter Five we shall discuss two other important classes of inverse functions. It is important to remember that in order to define the inverse of a function f it is *absolutely essential that f be one-to-one.* The most common examples of one-to-one functions are those which are either increasing or decreasing on their domains, for in this case if $a \neq b$ in the domain, then $f(a) \neq f(b)$ in the range.

An algebraic method can sometimes be used to find the inverse of a one-to-one function f with domain X and range Y. Given any x in X, its image y in Y may be found by means of the equation $y = f(x)$. In order to determine the inverse function f^{-1} we wish to *reverse* this procedure, in the sense that given y, the element x may be found Since x and y are related by means of $y = f(x)$, it follows that if the latter equation can be solved for x in terms of y, we may arrive at the inverse function f^{-1}. This technique is illustrated in the following examples.

Example 4   If $f(x) = 3x - 5$ for every real number x, find the inverse function of f.

Solution   The graph of the given linear function f is a line of slope 3 (why?) and hence f is increasing for all x. It follows that f is a one-to-one function with domain and range $\mathbb{R}$, and hence the inverse function g exists. If we let

$$y = 3x - 5$$

and then solve for x in terms of y, we get

$$x = \frac{y + 5}{3}.$$

The last equation enables us to find x when given y. Letting

$$g(y) = \frac{y + 5}{3}$$

gives us a function g from Y to X that reverses the correspondence determined by f. Since the symbol used for the independent variable is immaterial, we may replace y by x in the expression for g, obtaining

$$g(x) = \frac{x + 5}{3}.$$

To verify that g is actually the inverse function of f, we m. ist verify that the two conditions stated in (3.20) are fulfilled. Thus

$$f(g(x)) = f\left(\frac{x + 5}{3}\right) \qquad \text{(definition of } g)$$

$$= 3\left(\frac{x + 5}{3}\right) - 5 \qquad \text{(definition of } f)$$

$$= x \qquad \text{(simplifying).}$$

Also,

$$g(f(x)) = g(3x - 5) \qquad \text{(definition of } f)$$

$$= \frac{(3x - 5) + 5}{3} \qquad \text{(definition of } g)$$

$$= x \qquad \text{(simplifying).}$$

This proves that g is the inverse function of f. Using the notation of (3.21),

$$f^{-1}(x) = \frac{x + 5}{3}.$$

Example 5   Find the inverse function of f if the domain X is the interval $[0, \infty)$ and $f(x) = x^2 - 3$ for all x in X.

Solution   The domain has been restricted so that f is increasing and hence is one-to-one. The range of f is the interval $[-3, \infty)$. As in Example 4 we begin by considering the equation

$$y = x^2 - 3.$$

Solving for x gives us

$$x = \pm\sqrt{y + 3}.$$

Since x is nonnegative we reject $x = -\sqrt{y + 3}$ and, as in the preceding example, we let

$$g(y) = \sqrt{y + 3}$$

or equivalently,

$$g(x) = \sqrt{x + 3}.$$

We now check the two conditions in (3.20), obtaining

$$f(g(x)) = f(\sqrt{x + 3}) = (\sqrt{x + 3})^2 - 3$$
$$= (x + 3) - 3 = x$$

and

$$g(f(x)) = g(x^2 - 3) = \sqrt{(x^2 - 3) + 3} = x.$$

This proves that

$$f^{-1}(x) = \sqrt{x + 3} \quad \text{where } x \geq -3.$$

There is an interesting relationship between the graphs of a function f and its inverse function f^{-1}. We first note that f maps a into b if and only if f^{-1} maps b into a; that is, $b = f(a)$ means the same thing as $a = f^{-1}(b)$. These equations imply that the point (a, b) is on the graph of f if and only if the point (b, a) is on the graph of f^{-1}. As an illustration, in Example 5 we found that the functions f and f^{-1} given by

$$f(x) = x^2 - 3 \quad \text{and} \quad f^{-1}(x) = \sqrt{x + 3}$$

are inverse functions of one another, provided x is suitably restricted. Some points on the graph of f are $(0, -3), (1, -2), (2, 1),$ and $(3, 6)$. Corresponding points on the graph of f^{-1} are $(-3, 0), (-2, 1), (1, 2),$ and $(6, 3)$. The graphs of f and f^{-1} are sketched on the same coordinate axes in Figure 3.40. If the page is folded along the line l which bisects quadrants I and III (as indicated by the dashes in the figure), then the graphs of f and f^{-1} coincide. The two graphs are said to be *reflections* of one another through the line *l*. This is typical of the graph of every function *f* that has an inverse function f^{-1}.

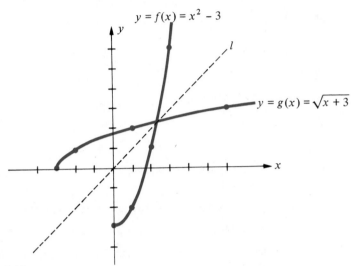

Figure 3.40

EXERCISES

In Exercises 1–18 find $(f \circ g)(x)$ and $(g \circ f)(x)$.

1   $f(x) = 3x + 2, \; g(x) = 2x - 1$

2   $f(x) = x + 4, \; g(x) = 5x - 3$

3   $f(x) = 4x^2 - 5, \; g(x) = 3x$

4   $f(x) = 7x + 1, \; g(x) = 2x^2$

5   $f(x) = 3x^2 + 2x, \; g(x) = 2x - 1$

6   $f(x) = 5x + 7, \; g(x) = 4 - x^2$

7   $f(x) = x - 1, \; g(x) = x^3$

8   $f(x) = x^3 + 5x, \; g(x) = 4x$

9   $f(x) = x^2 + 9x, \; g(x) = \sqrt{x + 9}$

10   $f(x) = \sqrt[3]{x^2 + 1}, \; g(x) = x^3 + 1$

11   $f(x) = \dfrac{1}{2x - 5}, \; g(x) = 1/x^2$

12   $f(x) = \dfrac{x}{x + 1}, \; g(x) = x - 1$

13   $f(x) = |x|, \; g(x) = -5$

14   $f(x) = 7, \; g(x) = 10$

15   $f(x) = x^2, \; g(x) = 1/x^2$

16   $f(x) = \dfrac{1}{x + 1}, \; g(x) = x + 1$

17   $f(x) = 2x - 3, \; g(x) = \dfrac{x + 3}{2}$

18   $f(x) = x^3 - 1, \; g(x) = \sqrt[3]{x + 1}$

In each of Exercises 19–22 prove that *f* and *g* are inverse functions of one another and sketch the graphs of *f* and *g* on the same coordinate plane.

19   $f(x) = 7x + 5; \; g(x) = (x - 5)/7$

20   $f(x) = x^2 - 1, \; x \geq 0; \; g(x) = \sqrt{x + 1}, \; x \geq -1$

21   $f(x) = \sqrt{2x - 4}, \; x \geq 2; \; g(x) = (x^2 + 4)/2, \; x \geq 0$

22   $f(x) = x^3 + 1; \; g(x) = \sqrt[3]{x - 1}$

In Exercises 23–36 find the inverse function of f.

23   $f(x) = 4x - 3$

24   $f(x) = 9 - 7x$

25   $f(x) = \dfrac{1}{2x + 5}, x > -5/2$

26   $f(x) = \dfrac{1}{3x - 1}, x > 1/3$

27   $f(x) = 9 - x^2, x \geq 0$

28   $f(x) = 4x^2 + 1, x \geq 0$

29   $f(x) = 5x^3 - 2$

30   $f(x) = 7 - 2x^3$

31   $f(x) = \sqrt{3x - 5}, x \geq 5/3$

32   $f(x) = \sqrt{4 - x^2}, 0 \leq x \leq 2$

33   $f(x) = \sqrt[3]{x} + 8$

34   $f(x) = (x^3 + 1)^5$

35   $f(x) = x$

36   $f(x) = -x$

37   (a)   Prove that the function f defined by $f(x) = ax + b$, where $a \neq 0$, has an inverse function, and find $f^{-1}(x)$.

   (b)   Does a constant function have an inverse? Explain.

   (c)   Does an identity function have an inverse? Explain.

38   If f is a one-to-one function with domain X and range Y, prove that f^{-1} is a one-to-one function with domain Y and range X.

39   Prove that a one-to-one function can have at most one inverse function.

40   If $f(x) = ax + b$ and $g(x) = cx + d$, find conditions on c and d in terms of a and b which will guarantee that $f \circ g = g \circ f$. Discuss the case where $a = b = 1$.

7   VARIATION

In this section we shall introduce terminology which is used in science to describe relationships among variable quantities. In the following definitions we shall use the letters u, v, w, and s for variables. The domains and ranges of the variables will not be specified. In any particular problem they should be evident.

(3.22)   **Definition**

> The phrase u **varies directly as** v, or u **is directly proportional to** v, means that $u = kv$ for some real number k.

The number k is sometimes called the **constant of variation** or the **constant of proportionality**. For example, if an automobile is moving at a rate of 50 miles per hour, then the distance d it travels in t hours is given by $d = 50t$. Hence the distance d is directly proportional to the time t and the constant of proportionality is 50.

(3.23)   **Definition**

> If n is a positive real number, then the phrase u **varies directly as the nth**
> **power of** v, or u **is directly proportional to the nth power of** v, means that
> $u = kv^n$ for some real number k.

To illustrate, the formula $A = \pi r^2$ for the area of a circle states that the area
A varies directly as the square of the radius r. The constant of proportionality is π.
As another illustration, the formula $V = \frac{4}{3}\pi r^3$ for the volume of a sphere of radius
r states that the volume V is directly proportional to the cube of the radius. The
constant of proportionality in this case is $\frac{4}{3}\pi$.

(3.24)   **Definition**

> The phrase u **varies inversely as** v means that $u = k/v$, where k is a real
> number. If $u = k/v^n$ for some positive real number n, then u **varies inversely**
> **as the nth power of** v.

Note that in direct variation the dependent variable increases numerically
as the independent variable increases, whereas in inverse variation the absolute
value of the dependent variable decreases as the independent variable increases.
In many applied problems the constant of proportionality can be determined by examining experimental facts. This is illustrated in the following
example.

Example 1   If the temperature remains constant, then the pressure of an enclosed gas is
inversely proportional to the volume. The pressure of a certain gas within a
spherical balloon of radius 9 inches is 20 pounds per square inch. If the radius of
the balloon increases to 12 inches, find the new pressure of the gas.

Solution   The original volume is $\frac{4}{3}\pi(9)^3 = 972\pi$ cubic inches. If we denote the pressure by P
and the volume by V, then from (3.24) we have

$$P = \frac{k}{V}$$

for some real number k. Since $P = 20$ when $V = 972\pi$,

$$20 = \frac{k}{972\pi}$$

and hence $k = 20(972\pi) = 19440\pi$. Consequently a formula for P is

$$P = \frac{19440\pi}{V}.$$

If the radius is 12 inches, then $V = \frac{4}{3}(12)^3\pi = 2304\pi$ cubic inches. Substituting this number for V in the previous equation gives us

$$P = \frac{19440\pi}{2304\pi} = \frac{135}{16} = 8.4375.$$

Thus the pressure is 8.4375 pounds per square inch when the radius is 12 inches.

Several independent variables often occur in the same problem. Some of these situations are considered below.

(3.25)   **Definition**

> The phrase w **varies jointly as** u **and** v means $w = kuv$ for some real number k. If $w = ku^nv^m$ for positive real numbers n and m, then w **varies jointly as the** n**th power of** u **and the** m**th power of** v.

It should be obvious how (3.25) may be extended to the case of more than two variables.

Finally, combinations of the above types of variation may occur. It is inconvenient to discuss all possible situations, but perhaps the following illustration will suffice. If a variable s varies jointly as u and the cube of v and inversely as the square of w, then

$$s = k\frac{uv^3}{w^2}$$

where k is some real number.

Example 2   The weight that can be safely supported by a beam with a rectangular cross section varies jointly as the width and square of the depth of the cross section and inversely as the length of the beam. If a 2-inch by 4-inch beam which is 8 feet long safely supports a load of 500 pounds, what weight can be safely supported by a 2-inch by 8-inch beam which is 10 feet long? (Assume that the width is the *shorter* dimension of the cross section.)

Solution   If the width, depth, length, and weight are denoted by w, d, l, and W respectively, then

$$W = k\frac{wd^2}{l}.$$

According to the given data,

$$500 = k\frac{2(4^2)}{8}.$$

Solving for k we obtain $k = 125$, and hence the formula for W is

$$W = 125\left(\frac{wd^2}{l}\right).$$

To answer the question we substitute $w = 2$, $d = 8$, and $l = 10$, obtaining

$$W = 125\left(\frac{2 \cdot 8^2}{10}\right) = 1600 \text{ pounds.}$$

EXERCISES

In Exercises 1–10 express each statement as a formula and determine the constant of proportionality from the given conditions.

1 a is directly proportional to v. If $v = 30$, then $a = 12$.

2 s varies directly as t. If $t = 10$, then $s = 18$.

3 r varies directly as s and inversely as t. If $s = -2$ and $t = 4$, then $r = 7$.

4 w varies directly as z and inversely as the square root of u. If $z = 2$ and $u = 9$, then $w = 6$.

5 y is directly proportional to the square of x and inversely proportional to the cube of z. If $x = 5$ and $z = 3$, then $y = 25$.

6 q is inversely proportional to the sum of x and y. If $x = 0.5$ and $y = 0.7$, then $q = 1.4$.

7 c varies jointly as the square of a and the cube of b. If $a = 7$ and $b = -2$, then $c = 16$.

8 r varies jointly as s and v and inversely as the cube of p. If $s = 2$, $v = 3$, and $p = 5$, then $r = 40$.

9 F varies jointly as m_1 and m_2 and inversely as the square of d. If $m_1 = 100$, $m_2 = 50$, and $d = 25$, then $F = 225$.

10 z is directly proportional to the sum of the squares of x and y. If $x = 4$ and $y = 1$, then $z = 50$.

11 The pressure acting at a point in a liquid is directly proportional to the distance from the surface of the liquid to the point. In a certain oil tank the pressure at a depth of 2 feet is 118 pounds per square foot. Find the pressure at a depth of 5 feet.

12 The volume V of a gas varies directly as the temperature T and inversely as the pressure P. The temperature, pressure, and volume of a certain gas are 240°, 20 pounds per square inch, and 30 cubic feet, respectively. If the temperature drops 40° and the pressure increases by 2 pounds per square inch, find the new volume.

13 Hooke's Law states that the force required to stretch a spring x units beyond its natural length is directly proportional to x. If a weight of 4 pounds stretches a spring from its natural length of 10 inches to a length of 10.3 inches, what weight will stretch it to a length of 11.5 inches?

14 If an object is dropped from a position near the surface of the earth, then the distance s that it falls in t seconds varies as the square of t. If $s = 36$ when $t = 1.5$, find a formula for s as a function of t. If an object is dropped from a height of 1,000 feet, how long does it take to reach the ground?

15   The volume V of a gas varies directly as the temperature T and inversely as the pressure P. Find a formula for V in terms of T and P, if $V = 50$ when $P = 40$ and $T = 300$. What is V when $T = 320$ and $P = 30$?

16   The kinetic energy K of an object in motion varies jointly as its mass m and the square of its velocity v. If an object weighing 25 pounds and moving with a velocity of 90 feet per second has a kinetic energy of 400 foot-pounds, find its kinetic energy when the velocity is 130 feet per second.

17   The electrical resistance of a wire varies directly as its length and inversely as the square of its diameter. If a wire 100 feet long of diameter 0.01 inches has a resistance of 25 ohms, find the resistance in a wire made of the same material which has a diameter of 0.015 inches and is 50 feet long.

18   The intensity of illumination I from a source of light varies inversely as the square of the distance d from the source. If a searchlight has an intensity of 1,000,000 candlepower at 50 feet, what is the intensity at a distance of 1 mile?

19   The period of a simple pendulum, that is, the time required for one complete oscillation, varies directly as the square root of its length. If a pendulum 2 feet long has a period of 1.5 seconds, find the period of a pendulum 6 feet long.

20   The distance a ball rolls down an inclined plane varies as the square of the time. If a ball rolls 5 feet in the first second, how far will it roll in 4 seconds?

8   REVIEW

Concepts

Define or discuss each of the following.

1   Ordered pair

2   Rectangular coordinate system in a plane

3   Coordinate axes

4   Quadrants

5   The abscissa and ordinate of a point

6   The distance formula

7   The midpoint formula

8   The graph of an equation in x and y

9   Tests for symmetry

10   The standard equation of a circle

11   Unit circle

12   Function

13   The domain and range of a function

14   The graph of a function

15   Increasing function

16   Decreasing function

17   One-to-one function

18   Identity function

19   Constant function

20   Linear function

21   Slope of a line

22   The point-slope form for the equation of a line

23   The slope-intercept form

24   Linear equation in x and y

25   The composite function of two functions

26   The inverse of a function

27   Direct variation

28   Inverse variation

Exercises

1   Plot the points $A(3, 1)$, $B(-5, -3)$, $C(4, -1)$ and prove that they are vertices of a right triangle. What is the area of the triangle?

2   Given the points $P(-5, 9)$ and $Q(-8, -7)$,   (a) find the midpoint of the segment PQ (b) find a point T such that Q is the midpoint of PT.

3   Describe the set of all points (x, y) in a coordinate plane such that $xy < 0$.

4   Find the slope of the line through $C(11, -5)$ and $D(-8, 6)$.

5   Prove that the points $A(-3, 1)$, $B(1, -1)$, $C(4, 1)$, and $D(3, 5)$ are vertices of a trapezoid.

6   Find an equation of the circle with center $C(7, -4)$ and passing through the point $Q(-3, 3)$.

7   Find an equation of the circle with center $C(-5, -1)$ which is tangent to the line $x = 4$.

8   Express the equation $8x + 3y - 24 = 0$ in slope-intercept form.

9   Find an equation of the line through $A(1/2, -1/3)$ that is parallel to the line $6x + 2y + 5 = 0$.

10   Find an equation of the line having x-intercept -3 and passing through the center of the circle $(x - 2)^2 + (y + 5)^2 = 3$.

Sketch the graphs of the equations in Exercises 11–19.

11   $2y + 5x - 8 = 0$

12   $x = 3y + 4$

13   $x + 5 = 0$

14   $2y - 7 = 0$

15   $y = \sqrt{1 - x}$

16   $y - x^2 = 1$

17   $y^2 - x = 1$

18   $y^2 = 16 - x^2$

19   $(x + 2)^2 + (y - 8)^2 = 4$

In each of Exercises 20–26 sketch the graph of f and determine the intervals in which f is increasing or decreasing.

20 $f(x) = |x + 3|$

21 $f(x) = \dfrac{1 - 3x}{2}$

22 $f(x) = \sqrt{2 - x}$

23 $f(x) = 1 - \sqrt{x + 1}$

24 $f(x) = 9 - x^2$

25 $f(x) = 1000$

26 $f(x) = \begin{cases} x^2 & \text{if } x < 0 \\ 3x & \text{if } 0 \le x < 2 \\ 6 & \text{if } x \ge 2 \end{cases}$

27   If $f(x) = x/\sqrt{x + 3}$ find

    (a) $f(1)$ (b) $f(-1)$ (c) $f(0)$

    (d) $f(-x)$ (e) $-f(x)$ (f) $f(x^2)$ (g) $(f(x))^2$

28   Find the domain and range of f if

    (a) $f(x) = \sqrt{3x - 4}$ (b) $f(x) = 1/(x + 3)^2$

In Exercises 29 and 30 find $(f \circ g)(x)$ and $(g \circ f)(x)$.

29 $f(x) = 2x^2 - 5x + 1$, $g(x) = 3x + 2$

30 $f(x) = \sqrt{3x + 2}$, $g(x) = 1/x^2$

In Exercises 31 and 32 find $f^{-1}(x)$ and sketch the graphs of f and f^{-1} on the same coordinate plane.

31 $f(x) = 10 - 15x$

32 $f(x) = 9 - 2x^2$, $x \le 0$

33   If the altitude and radius of a right circular cylinder are equal, express the volume V as a function of the circumference C of the base.

34   Find a formula which expresses the fact that w varies jointly as x and the square of y, and inversely as z, if $w = 30$ when $x = 3$, $y = 2$, and $z = 5$.

CHAPTER 4

Polynomial Functions, Rational Functions, and Conic Sections

Polynomial functions are the most basic functions in algebra, since they are defined only in terms of additions and multiplications of real numbers and variables. Techniques for sketching their graphs are discussed in the first two sections. Next we consider quotients of polynomial functions, that is, rational functions. The chapter concludes with a brief survey of an important class of graphs called conic sections.

1 QUADRATIC FUNCTIONS

Among the most important functions in mathematics are those defined as follows.

(4.1) Definition

> A function f is a **polynomial function** if
> $$f(x) = a_n x^n + a_{n-1} x^{n-1} + \cdots + a_1 x + a_0$$
> where the coefficients $a_0, a_1, \ldots, a_n$ are real numbers and the exponents are nonnegative integers.

According to (1.30), f is a polynomial function if $f(x)$ is a polynomial, and we say that f has **degree n** if $f(x)$ has degree n. Note that a polynomial function of degree 1 is a linear function. If the degree is 2, then as in the next definition, f is called a *quadratic function.*

(4.2) Definition

> A function f is a **quadratic function** if
> $$f(x) = ax^2 + bx + c$$
> where a, b, and c are real numbers and $a \neq 0$.

155

If $b = c = 0$ in (4.2), then $f(x) = ax^2$ and, as pointed out in Chapter Three, the graph is a parabola with vertex at the origin, opening upward if $a > 0$ or downward if $a < 0$ (see Figure 3.29). If $b = 0$ and $c \neq 0$, then

$$f(x) = ax^2 + c$$

and the graph is a parabola with vertex at the point $(0, c)$ on the y-axis. Some typical graphs are illustrated in Figure 4.1.

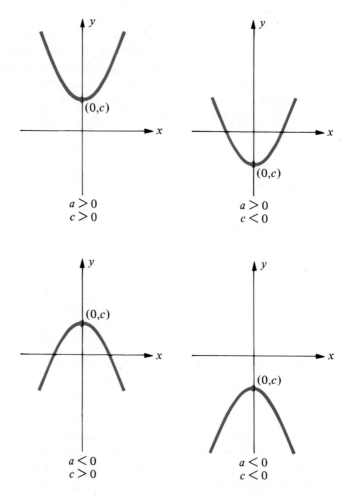

Figure 4.1. $f(x) = ax^2 + c$

Example 1   Sketch the graph of $f(x) = -\frac{1}{2}x^2 + 4$.

Solution   The graph of $f(x) = -\frac{1}{2}x^2$ is similar to the graph of $f(x) = -x^2$ sketched in (iii) of Figure 3.29 but is somewhat wider. The graph of $f(x) = -\frac{1}{2}x^2 + 4$ may be found by raising the graph of $f(x) = -\frac{1}{2}x^2$ four units. To help with the sketch we may plot several points from the following table and use symmetry with respect to the y-axis (see Figure 4.2).

x	0	1	2	$\sqrt{8}$	3
$f(x)$	4	7/2	2	0	$-1/2$

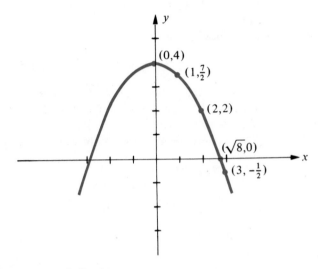

Figure 4.2. $f(x) = -\frac{1}{2}x^2 + 4$

If $b \neq 0$ in (4.2), then a technique called **completing the square** can be used to change the form of $f(x)$ to

(4.3) $f(x) = a(x - h)^2 + k$

where h and k are real numbers. The right-hand side of (4.3) may be written $a(x')^2 + k$, where $x' = x - h$, and as we shall see later, the graph will be similar to that of $f(x) = ax^2 + k$. The process of completing the square is illustrated in the next example.

Example 2   Express $f(x)$ in the form (4.3) if:
(a) $f(x) = x^2 - 6x - 5$,   (b) $f(x) = 3x^2 + 24x + 50$.

Solution   (a)   We begin by writing

$$f(x) = (x^2 - 6x \qquad) - 5.$$

We now complete the square by determining the number to add to the expression within the parentheses so that it takes on the form $(x - h)^2$ for some h. It is not difficult to show that *the desired number is always the square of one-half the coefficient of x.* In the present example the coefficient of x is -6, and consequently the square of one-half of -6 is $(-3)^2$, or 9. Adding 9 to $x^2 - 6x$ gives $x^2 - 6x + 9$

or $(x - 3)^2$. Of course, if we add 9 to $f(x)$, then we must also subtract 9, so that $f(x)$ does not change. Accordingly we obtain

$$f(x) = (x^2 - 6x \quad) - 5$$
$$= (x^2 - 6x + 9) - 5 - 9$$
$$= (x - 3)^2 - 14$$

which has form (4.3) with $a = 1$, $h = 3$, and $k = -14$.

(b)   Given $f(x) = 3x^2 + 24x + 50$, we begin by removing a factor 3 from the first two terms as follows:

$$f(x) = 3(x^2 + 8x \quad) + 50.$$

As in part (a), to complete the square for the expression in parentheses we add the square of one-half the coefficient of x, namely $(8/2)^2$, or 16. However, if we add 16 to the expression within parentheses, then because of the factor 3 we are actually adding 48 to $f(x)$. Hence we must compensate by subtracting 48 as follows:

$$f(x) = 3(x^2 + 8x \quad) + 50$$
$$= 3(x^2 + 8x + 16) + 50 - 48$$
$$= 3(x + 4)^2 + 2$$

which has form (4.3) with $a = 3$, $h = -4$, and $k = 2$.

The graph of f in (4.3) can be obtained by shifting the graph of the equation

(4.4) $$y = ax^2 + k$$

either to the right or left, depending on the sign of h. This is true because the ordinate of the point with abscissa x_1 on the graph of (4.3) is the same as the ordinate of the point with abscissa $x_1 - h$ on the graph of (4.4). It follows that the graph of (4.4) and, therefore, the graph of a quadratic function, is a parabola. If $a > 0$, the parabola opens upward and the vertex is obtained by finding the smallest value of $f(x)$ in (4.3). Since $(x - h)^2 \geq 0$, this smallest value occurs if $(x - h)^2 = 0$, that is, if $x = h$. The ordinate of the point with abscissa h is

$$f(h) = a(h - h)^2 + k = k$$

and hence the vertex is the point (h, k). If $a < 0$ the vertex is again (h, k), but the parabola opens downward.

If we let $y = f(x)$ in (4.3) we obtain

$$y = a(x - h)^2 + k$$

or equivalently,

(4.5) $$\boxed{y - k = a(x - h)^2}$$

Equation (4.5) is called the **standard equation of a parabola** with vertex (h, k) and axis parallel to the y-axis.

Example 3   Sketch the graph of $f(x) = 2x^2 - 6x + 4$.

Solution   The graph of f is the same as the graph of the equation $y = 2x^2 - 6x + 4$. From the preceding discussion we know that the graph is a parabola, and we may obtain form (4.5) by completing the square as follows:

$$
\begin{aligned}
y &= 2x^2 - 6x + 4 \\
&= 2(x^2 - 3x \qquad\;) + 4 \\
&= 2\left(x^2 - 3x + \frac{9}{4}\right) + \left(4 - \frac{9}{2}\right) \\
&= 2\left(x - \frac{3}{2}\right)^2 - \frac{1}{2}
\end{aligned}
$$

If we rewrite the last equation as

$$
y + \frac{1}{2} = 2\left(x - \frac{3}{2}\right)^2
$$

and compare with the standard equation (4.5), we see that $h = 3/2$ and $k = -1/2$. Consequently the vertex (h, k) of the parabola is $(3/2, -1/2)$. Since $a = 2 > 0$, the parabola opens upward. The y-intercept is $f(0) = 4$. To find the x-intercepts we solve $2x^2 - 6x + 4 = 0$ or the equivalent equation $(2x - 2)(x - 2) = 0$, obtaining $x = 2$ and $x = 1$. The vertex together with the x- and y-intercepts are sufficient for a reasonably accurate sketch (see Figure 4.3).

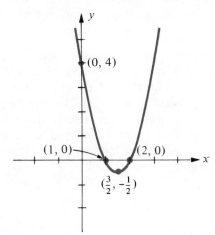

Figure 4.3. $f(x) = 2x^2 - 6x + 4$

Example 4   Sketch the graph of the equation $y = 8 - 2x - x^2$.

Solution   We know from the discussion in this section that the graph is a parabola. Let us find the vertex by completing the square as follows:

$$
\begin{aligned}
y &= -x^2 - 2x + 8 \\
&= -(x^2 + 2x) + 8 \\
&= -(x^2 + 2x + 1) + 8 + 1 \\
&= -(x + 1)^2 + 9.
\end{aligned}
$$

If we now write

$$y - 9 = -(x + 1)^2$$

and compare with the standard equation (4.5) we see that $h = -1$, $k = 9$ and hence the vertex is $(-1, 9)$. Since $a = -1 < 0$, the parabola opens downward. To find the x-intercepts we solve the equation $8 - 2x - x^2 = 0$. Either by factoring or by using the quadratic formula we obtain $x = -4$ and $x = 2$. The y-intercept is 8. Using this information gives us the sketch in Figure 4.4.

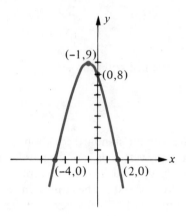

Figure 4.4. $f(x) = 8 - 2x - x^2$

Parabolas (and hence quadratic functions) are very useful in applications of mathematics to the physical world. It can be shown that if a projectile is fired and is acted upon only by the force of gravity—that is, if air resistance and other outside factors are inconsequential—then the path of the projectile is a parabola. Properties of parabolas are used in the design of mirrors for telescopes and searchlights. They are also employed in the design of field microphones used in television broadcasts of football games. These are only a few of many physical applications.

EXERCISES

1   Given $f(x) = ax^2 + 2$, sketch the graph of f if:
    (a) $a = 2$ (b) $a = 5$ (c) $a = \frac{1}{2}$ (d) $a = -3$

2   Given $f(x) = 4x^2 + c$, sketch the graph of f if:
    (a) $c = 2$ (b) $c = 5$ (c) $c = \frac{1}{2}$ (d) $c = -3$

In each of Exercises 3–6 sketch the graph of f and find the vertex.

3   $f(x) = 4x^2 - 9$ 4 $f(x) = 2x^2 + 3$

5   $f(x) = 9 - 4x^2$ 6 $f(x) = 16 - 9x^2$

In Exercises 7–14 use the quadratic formula (2.2) to find all real numbers a such that $f(a) = 0$.

7 $f(x) = 4x^2 - 11x - 3$ **8** $f(x) = 25x^2 + 10x + 1$

9 $f(x) = 9x^2 - 12x + 4$ **10** $f(x) = 10x^2 + x - 21$

11 $f(x) = 2x^2 + x - 5$ **12** $f(x) = 2x^2 + 4x - 3$

13 $f(x) = \dfrac{2x + 1}{x - 3} - \dfrac{x - 1}{x + 2}$ **14** $f(x) = 9x^2 + 13x$

In Exercises 15–20 use the technique of completing the square to express $f(x)$ in the form (4.3).

15 $f(x) = 2x^2 - 16x + 23$ **16** $f(x) = 3x^2 - 12x + 7$

17 $f(x) = -5x^2 - 10x + 3$ **18** $f(x) = -2x^2 - 12x - 12$

19 $f(x) = 4x^2 - 4x + 2$ **20** $f(x) = 9x^2 + 12x - 8$

In each of Exercises 21–30 sketch the graph of f. (Find the vertex by completing the square.)

21 $f(x) = x^2 + 5x + 4$ **22** $f(x) = x^2 - 6x$

23 $f(x) = 8x - 12 - x^2$ **24** $f(x) = 10 + 3x - x^2$

25 $f(x) = x^2 + x + 3$ **26** $f(x) = x^2 + 2x + 5$

27 $f(x) = 3x^2 - 12x + 16$ **28** $f(x) = 2x^2 + 12x + 13$

29 $f(x) = -5x^2 - 10x - 4$ **30** $f(x) = -3x^2 + 24x - 42$

31   If $f(x) = x^2 + kx - 5 + 3k$, find two different values for k such that f has only one zero.

32   If $f(x) = 3x^2 + kx - 4k$, find all values of k such that f has no real zeros.

33   If one zero of $f(x) = kx^2 + 3x + k$ is -2, find the other zero.

34   Find a real number k such that the graph of $y = x^2 + kx + 3$ contains the point $P(-1, -2)$.

2   GRAPHS OF POLYNOMIAL FUNCTIONS OF DEGREE GREATER THAN 2

As in (4.1), f is a polynomial function of degree n if

$$f(x) = a_n x^n + a_{n-1} x^{n-1} + \cdots + a_1 x + a_0$$

where $a_n \neq 0$. If all the coefficients except a_n are zero, then we may write

(4.6) $$f(x) = ax^n, \quad \text{where } a = a_n \neq 0.$$

We know from our previous work that if $n = 1$ the graph of (4.6) is a line passing through the origin, whereas if $n = 2$ the graph is a parabola with vertex at the origin. The special case in which $a = 1$ and $n = 3$, that is, $f(x) = x^3$, is sketched in Figure 3.14. Several other illustrations of (4.6) with $n = 3$ are given in the next example.

Example 1   Sketch the graph of f if   (a) $f(x) = \frac{1}{2}x^3$,   (b) $f(x) = -\frac{1}{2}x^3$.

Solution   (a)   The table below gives several points on the graph.

x	0	$\frac{1}{2}$	1	$\frac{3}{2}$	2	$\frac{5}{2}$
$f(x)$	0	$\frac{1}{16}$	$\frac{1}{2}$	$\frac{27}{16}$	4	$\frac{125}{16} \approx 7.8$

It follows from (3.4) that the graph of f is symmetric with respect to the origin and hence the points $(-\frac{1}{2}, -\frac{1}{16})$, $(-1, -\frac{1}{2})$, etc., are also on the graph. The graph is sketched in (i) of Figure 4.5.
(b)   If $f(x) = -\frac{1}{2}x^3$, the graph can be obtained from that in part (a) by multiplying all ordinates by -1. This amounts to reflecting the graph through the x-axis as shown in (ii) of Figure 4.5.

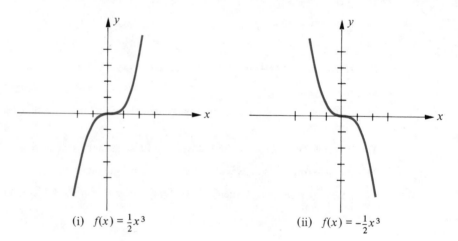

(i) $f(x) = \frac{1}{2}x^3$ (ii) $f(x) = -\frac{1}{2}x^3$

Figure 4.5

In general, if $f(x) = ax^3$, then increasing the numerical value of the coefficient a causes the graph to rise or fall more sharply. For example, if $f(x) = 10x^3$, then $f(1) = 10$, $f(2) = 80$, $f(-2) = -80$, and so on. The effect of using a larger exponent in (4.6) is illustrated in Figure 4.6 (with $a = 1$).

If n is an *even* integer, then the graph of $f(x) = ax^n$ is symmetric with respect to the y-axis. (Why?) Several illustrations (with $a = 1$) are given in Figure 4.7. Note that as the exponent increases, the graph becomes much flatter at the origin. It also rises (or falls) more rapidly if we let x increase through positive (or negative) values. Using a negative coefficient such as -1 inverts the graphs.

As usual, the graph of $f(x) = ax^n + c$, where $c \neq 0$, can be found by raising or lowering the graph of $f(x) = ax^n$. For example, if $f(x) = 1 - x^6$ we obtain the sketch in Figure 4.8.

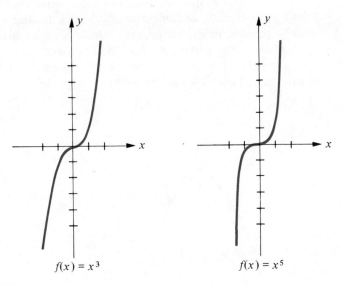

$f(x) = x^3$ $f(x) = x^5$

Figure 4.6

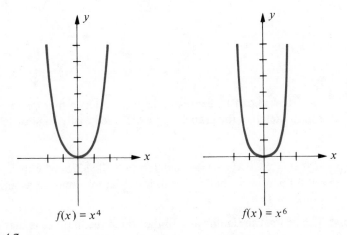

$f(x) = x^4$ $f(x) = x^6$

Figure 4.7

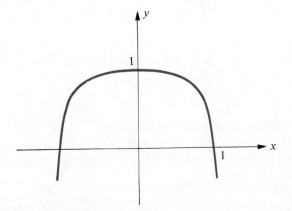

Figure 4.8. $f(x) = 1 - x^6$

A complete analysis of graphs of polynomial functions of degree greater than 2 requires methods studied in calculus. As the degree increases, the graphs usually become more complicated. However, they always have a smooth appearance with a number of "hills" and "valleys," as illustrated in Figure 4.9. This number may be large when the degree is large, but it may also be small, as illustrated in Figures 4.6 through 4.8.

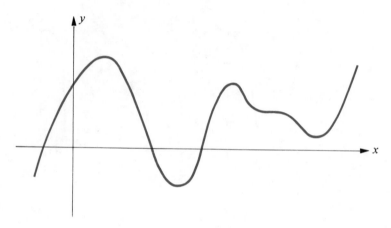

Figure 4.9

If $f(x)$ can be expressed as a product of first-degree polynomials, then a rough sketch of the graph can be obtained easily, as shown in Examples 2 and 3.

Example 2   If $f(x) = x^3 + x^2 - 4x - 4$, determine all values of x such that $f(x) > 0$, and all values of x such that $f(x) < 0$. Use that information as an aid in sketching the graph of f.

Solution   The geometric significance of the stated inequalities is that the graph lies above the x-axis for values of x such that $f(x) > 0$, whereas the graph lies below the x-axis if $f(x) < 0$. We may factor $f(x)$ by grouping terms as follows:

$$f(x) = (x^3 + x^2) - (4x + 4)$$
$$= x^2(x + 1) - 4(x + 1)$$
$$= (x^2 - 4)(x + 1)$$
$$= (x + 2)(x - 2)(x + 1).$$

It follows that $f(x)$ is zero if x equals -2, -1, or 2. These are the x-intercepts of the graph. The corresponding points on the graph divide the x-axis into four parts, as illustrated in Figure 4.10. We next consider values of x such that

(a) $x < -2$ (b) $-2 < x < -1$ (c) $-1 < x < 2$ (d) $2 < x$.

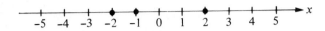

Figure 4.10

The signs of $f(x)$ for the values of x indicated in (a)–(d) may be determined by investigating the signs of the three factors $x + 2, x + 1$, and $x - 2$. The diagram in Figure 4.11 displays the signs of these factors for values of x corresponding to

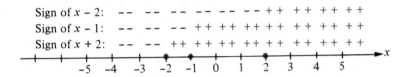

Figure 4.11

each of the four inequalities listed in (a)–(d) above. We see from the diagram that all three factors are negative if $x < -2$. Since $f(x)$ is the product of the three factors, it follows that $f(x)$ is negative if $x < -2$. Similarly, if $-2 < x < -1$, then two factors are negative and one is positive and hence $f(x)$ is positive. If $-1 < x < 2$, then only one factor is negative and, therefore, $f(x)$ is negative. Finally, if $x > 2$, then all factors are positive and hence so is $f(x)$. The following table summarizes these remarks.

Domain of x	Sign of $f(x)$	Position of graph
$x < -2$	$-$	Below x-axis
$-2 < x < -1$	$+$	Above x-axis
$-1 < x < 2$	$-$	Below x-axis
$x > 2$	$+$	Above x-axis

Coordinates of several points on the graph are listed below.

x	-3	$-\frac{3}{2}$	0	1	$\frac{5}{2}$
$f(x)$	-10	$\frac{7}{8}$	-4	-6	$\frac{63}{8} \approx 7.9$

Using the information provided by the last two tables gives us the graph in Figure 4.12. In order to find the high and low points on the graph it is necessary to use methods developed in calculus.

The graph of every polynomial function of degree 3 has an S-shaped appearance similar to that shown in Figure 4.12 or an inverted version of that graph if the coefficient of x^3 is negative. However, sometimes there may be only one x-intercept, or the S may be elongated, as in Figures 4.5 and 4.6.

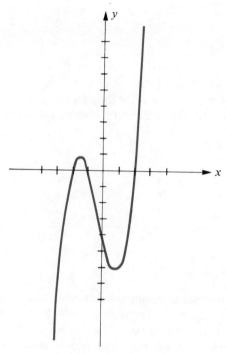

Figure 4.12. $f(x) = x^3 + x^2 - 4x - 4$

Example 3   If $f(x) = x^4 - 4x^3 + 3x^2$, determine all values of x such that $f(x) > 0$, and all x such that $f(x) < 0$. Sketch the graph of f.

Solution   We begin by factoring $f(x)$. Thus

$$f(x) = x^2(x^2 - 4x + 3)$$
$$= x^2(x - 1)(x - 3).$$

The x-intercepts of the graph are 0, 1, and 3. (Why?) We next plot the corresponding points on a coordinate line and display the signs of the factors as in the solution of the preceding example (see Figure 4.13). Using the fact that $f(x)$ is

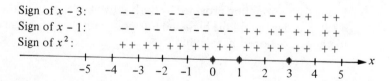

Figure 4.13

positive if two factors are negative and one is positive, or if all three factors are positive, we arrive at the following table.

Domain of x	Sign of $f(x)$	Position of graph
$x < 0$	$+$	Above x-axis
$0 < x < 1$	$+$	Above x-axis
$1 < x < 3$	$-$	Below x-axis
$x > 3$	$+$	Above x-axis

Making use of the above information and plotting several points gives us the sketch in Figure 4.14.

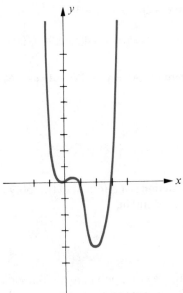

Figure 4.14. $f(x) = x^4 - 4x^3 + 3x^2$

As a final remark, if $f(x)$ cannot be factored, we may always employ the technique of plotting many points and fitting a smooth curve to the resulting configuration. This is, of course, an extremely tedious procedure and should be used only as a last resort.

EXERCISES

1 If $f(x) = ax^3 + 2$, sketch the graph of f if:
 (a) $a = 2$ (b) $a = 4$ (c) $a = \frac{1}{4}$ (d) $a = -2$

2 If $f(x) = 2x^3 + c$, sketch the graph of f if:
 (a) $c = 2$ (b) $c = 4$ (c) $c = \frac{1}{4}$ (d) $c = -2$

In each of Exercises 3–18 determine all x such that $f(x) > 0$, and all x such that $f(x) < 0$. Sketch the graph of f.

3 $f(x) = \frac{1}{2}x^3 - 4$

4 $f(x) = -\frac{1}{4}x^3 - 16$

5 $f(x) = \frac{1}{8}x^4 + 2$

6 $f(x) = 1 - x^5$

7   $f(x) = x^3 - 9x$

8   $f(x) = 16x - x^3$

9   $f(x) = -x^3 - x^2 + 2x$

10   $f(x) = x^3 + x^2 - 12x$

11   $f(x) = (x + 4)(x - 1)(x - 5)$

12   $f(x) = (x + 2)(x - 3)(x - 4)$

13   $f(x) = x^4 - 16$

14   $f(x) = 16 - x^4$

15   $f(x) = -x^4 - 3x^2 + 4$

16   $f(x) = x^4 - 7x^2 - 18$

17   $f(x) = x(x - 2)(x + 1)(x + 3)$

18   $f(x) = x(x + 1)^2(x - 3)(x - 5)$

19   Find a number k such that the graph of $f(x) = 3x^3 - kx^2 + x - 5k$ contains the point $(-1, 4)$.

20   If one zero of $f(x) = x^3 - 2x^2 - 16x + 16k$ is 2, find two other zeros.

3   RATIONAL FUNCTIONS

A **rational function** is a quotient of two polynomial functions. Thus f is rational if for all x in its domain,

(4.7)
$$f(x) = \frac{g(x)}{h(x)}$$

where $g(x)$ and $h(x)$ are polynomials. Since division by zero is not permissible, the domain of a rational function consists of all real numbers except those for which the denominator $h(x)$ is zero. In order to obtain the graph of f in (4.7), it is extremely important to examine the behavior of $f(x)$ when x is near a number c such that $h(c) = 0$ and $g(c) \neq 0$.

Example 1   Sketch the graph of f if $f(x) = \dfrac{1}{x}$.

Solution   The function f is rational since $f(x)$ has the form (4.7), with $g(x) = 1$ and $h(x) = x$. The domain of f is the set of all nonzero real numbers. (Why?) Before constructing a table, let us make some general observations. If x is positive, so is $1/x$, and hence quadrant IV is an excluded region. Quadrant II is also excluded since if $x < 0$, then $1/x < 0$. If x is close to zero, then $1/x$ is very large numerically. As x increases through positive values, $1/x$ decreases and is close to zero when x is large. The variation of $f(x)$ is brought out in the following table.

x	$\frac{1}{100}$	$\frac{1}{10}$	$\frac{1}{4}$	$\frac{1}{2}$	1	2	4	10	100
$f(x)$	100	10	4	2	1	$\frac{1}{2}$	$\frac{1}{4}$	$\frac{1}{10}$	$\frac{1}{100}$

By plotting some points and using the previous discussion, we obtain the first quadrant part of the graph in Figure 4.15. Since the graph of f is the same as the graph of $y = 1/x$, we may use (3.4) to show that the graph is symmetric with respect to the origin. This fact gives us the third quadrant part of the graph in Figure 4.15.

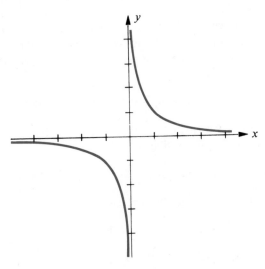

Figure 4.15. $f(x) = \dfrac{1}{x}$

If, in the previous example, we assign values to x which are very close to 0, then $f(x)$ becomes very large numerically. Thus,

$$f(0.0001) = 10{,}000, \quad f(0.000001) = 1{,}000{,}000$$

and so on. Indeed, $f(x)$ can be made as large as desired by taking a positive number x sufficiently close to 0. This is sometimes referred to by the statement:

$f(x)$ increases without bound as x approaches 0 through positive values

and is denoted symbolically by

$$f(x) \longrightarrow \infty \quad \text{as} \quad x \longrightarrow 0^{+}.$$

The phrase "$f(x)$ becomes positively infinite" is sometimes used to describe the variation described above. It is important to remember that the symbol ∞ (read "infinity") does not represent a real number, but is used merely as an abbreviation for certain types of functional behavior.

In like manner, if x is *negative* and close to 0, then $f(x)$ is a numerically large negative number. For example,

$$f(-0.001) = -1{,}000, \quad f(-0.00001) = -100{,}000$$

and so on. In this case we say that

$f(x)$ decreases without bound as x approaches 0 through negative values.

and write

$$f(x) \longrightarrow -\infty \quad \text{as} \quad x \longrightarrow 0^-.$$

The phrase "$f(x)$ becomes negatively infinite" is sometimes used in place of the phrase "decreases without bound." The line $x = 0$ in Figure 4.15, that is, the y-axis, is called a **vertical asymptote** for the graph of f. Vertical asymptotes are common characteristics of graphs of rational functions.

    If, in Example 1, we assign very *large* values to x, then $f(x)$ is close to 0. For example,

$$f(1{,}000) = 0.001 \quad \text{and} \quad f(1{,}000{,}000) = 0.000001.$$

Moreover, we can make $f(x)$ as close to 0 as desired by choosing x sufficiently large. In this case we say that

$$f(x) \text{ approaches 0 as } x \text{ becomes positively infinite}$$

and denote this fact by writing

$$f(x) \longrightarrow 0 \quad \text{as} \quad x \longrightarrow \infty.$$

Similarly, in Example 1,

$$f(x) \text{ approaches 0 as } x \text{ becomes negatively infinite,}$$

written

$$f(x) \longrightarrow 0 \quad \text{as} \quad x \longrightarrow -\infty.$$

Since the graph gets closer and closer to the line $y = 0$ as $|x|$ increases, we call the x-axis a **horizontal asymptote** for the graph.

Example 2   Sketch the graph of f if $f(x) = 1/x^2$.

Solution   The graph here differs from that in Example 1 since $f(x)$ is never negative. Indeed, we see that

$$f(x) \longrightarrow \infty \quad \text{as} \quad x \longrightarrow 0^+$$

and

$$f(x) \longrightarrow \infty \quad \text{as} \quad x \longrightarrow 0^-.$$

In addition, the present graph is steeper near the origin and approaches the x-axis more rapidly as x increases. These facts can be seen by comparing the following table with the corresponding table in Example 1.

x	$\frac{1}{100}$	$\frac{1}{10}$	$\frac{1}{4}$	$\frac{1}{2}$	1	2	10	100
$f(x)$	10000	100	16	4	1	$\frac{1}{4}$	$\frac{1}{100}$	$\frac{1}{10000}$

Plotting leads to the first quadrant part of the graph in Figure 4.16. Since the graph is symmetric with respect to the y-axis (why?), it is unnecessary to tabulate points with negative abscissas. We merely reflect the first quadrant part through the y-axis. Note that the y-axis is a vertical asymptote and the x-axis is a horizontal asymptote.

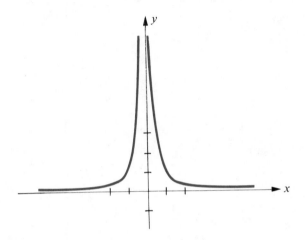

Figure 4.16. $f(x) = \dfrac{1}{x^2}$

Examples 1 and 2 can be generalized to $f(x) = 1/x^n$ where n is *any* positive integer. If n is odd, as for $1/x^3$ and $1/x^5$, the graph has the general appearance shown in Figure 4.15, except that there is a sharper corner near the origin. If n is even, as for $1/x^4$ and $1/x^6$, the graph is similar to that in Figure 4.16, except that the graph again has a sharper corner near the origin.

Let us extend the previous discussion to the case in which x approaches *any* real number a. In particular, the symbol $x \longrightarrow a^+$ will signify that x approaches a through values *greater* than a, whereas $x \longrightarrow a^-$ will mean that x approaches a through values *less* than a. Some illustrations of the manner in which a function may increase or decrease without bound, together with the notation used, are shown in Figure 4.17.

Of major importance in sketching the graph of a rational function is the special case

$$f(x) = \frac{1}{(x-a)^n}$$

where n is a positive integer and a is any real number. If n is even, then $(x - a)^n$ is always positive when $x \neq a$, and therefore $f(x) \longrightarrow \infty$ if either $x \longrightarrow a^-$ or $x \longrightarrow a^+$, as illustrated in (i) of Figure 4.18. However, if n is *odd* and if $x < a$, then $x - a < 0$ and hence $(x - a)^n < 0$. In this event $f(x)$ is negative and we see that $f(x) \longrightarrow -\infty$ as $x \longrightarrow a^-$ (see (ii) of Figure 4.18).

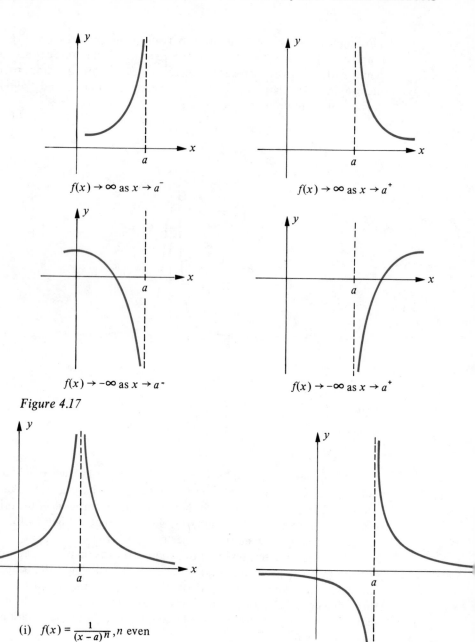

$f(x) \to \infty$ as $x \to a^-$

$f(x) \to \infty$ as $x \to a^+$

$f(x) \to -\infty$ as $x \to a^-$

$f(x) \to -\infty$ as $x \to a^+$

Figure 4.17

(i) $f(x) = \dfrac{1}{(x-a)^n}$, n even

Figure 4.18

(ii) $f(x) = \dfrac{1}{(x-a)^n}$, n odd

The remarks made in the preceding paragraph about the variation of $f(x)$ as $x \to a^-$ or $x \to a^+$ are also true if

$$f(x) = \frac{k}{(x-a)^n}$$

where k is any *positive* real number.

To complete the discussion, let us consider what happens if k is *negative*. In this case if n is even, then again $(x - a)^n$ is always positive when $x \neq a$, and hence $f(x)$ is always negative. Consequently, $f(x) \rightarrow -\infty$ as $x \rightarrow a^+$ or $x \rightarrow a^-$. The general shape of the graph can be obtained by *inverting* the graph in (i) of Figure 4.18. Similarly, if n is odd, the shape of the graph may be found by inverting the graph in (ii) of Figure 4.18.

Sketches of graphs of $f(x) = k/(x - a)^n$ for several values of k, a, and n are shown in Figure 4.19. The reader should check each graph by plotting several points and carefully noting the behavior of $f(x)$ if x is near a.

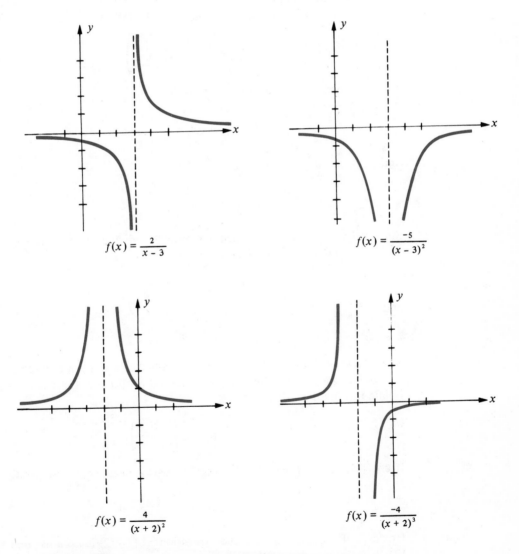

$$f(x) = \frac{2}{x - 3}$$

$$f(x) = \frac{-5}{(x - 3)^2}$$

$$f(x) = \frac{4}{(x + 2)^2}$$

$$f(x) = \frac{-4}{(x + 2)^3}$$

Figure 4.19

In the following examples we consider several other types of rational functions.

Example 3   Sketch the graph of f if $f(x) = \dfrac{x^2}{x^2 - x - 2}$.

Solution   Let us begin by factoring the denominator, obtaining

$$f(x) = \frac{x^2}{(x + 1)(x - 2)}.$$

Since the denominator is zero at 2 and -1, it is necessary to investigate the behavior of $f(x)$ as x approaches either of these two numbers. In order to consider $x \to 2^+$ and $x \to 2^-$, we isolate the factor $x - 2$ follows:

$$f(x) = \left(\frac{x^2}{x + 1}\right)\left(\frac{1}{x - 2}\right).$$

From the previous discussion we know that

$$\frac{1}{x - 2} \to \infty \quad \text{as} \quad x \to 2^+$$

and

$$\frac{1}{x - 2} \to -\infty \quad \text{as} \quad x \to 2^-.$$

Since $x^2/(x + 1)$ is positive when x is near 2 and since $f(x)$ is the *product* of $x^2/(x + 1)$ and $1/(x - 2)$, it follows that

$$f(x) \to \infty \quad \text{as} \quad x \to 2^+$$

and

$$f(x) \to -\infty \quad \text{as} \quad x \to 2^-.$$

These facts are illustrated in the partial graph shown in (i) of Figure 4.20.

In like manner, to determine the variation of $f(x)$ as x approaches -1, we isolate the factor $x + 1$ as follows:

$$f(x) = \left(\frac{x^2}{x - 2}\right)\left(\frac{1}{x + 1}\right).$$

If $x < -1$, then $x + 1 < 0$ and hence $1/(x + 1)$ is negative. Consequently,

$$\frac{1}{x + 1} \to -\infty \quad \text{as} \quad x \to -1^-.$$

However, if x is close to -1, the expression $x^2/(x - 2)$ is *negative*, and since $1/(x + 1) \to -\infty$, we see that the *product* of $x^2/(x - 2)$ and $1/(x + 1)$ becomes

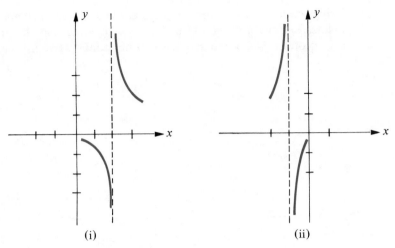

(i)   (ii)

Figure 4.20

positively infinite; that is,

$$f(x) \longrightarrow \infty \quad \text{as} \quad x \longrightarrow -1^-.$$

Similarly, if $x > -1$, then $x + 1 > 0$. Hence

$$\frac{1}{x + 1} \longrightarrow \infty \quad \text{as} \quad x \longrightarrow -1^+.$$

Again, since $x^2/(x - 2)$ is *negative* when x is near -1, it follows that the product of $x^2/(x - 2)$ and $1/(x + 1)$ becomes *negatively* infinite; that is,

$$f(x) \longrightarrow -\infty \quad \text{as} \quad x \longrightarrow -1^+.$$

This behavior of $f(x)$ near $x = -1$ is illustrated in (ii) of Figure 4.20.

In order to determine what happens as $x \longrightarrow \infty$ or $x \longrightarrow -\infty$, it is helpful to divide numerator and denominator of the given expression for $f(x)$ by x^2, obtaining

$$f(x) = \frac{x^2/x^2}{(x^2 - x - 2)/x^2} = \frac{1}{1 - (1/x) - (2/x^2)}.$$

If we assign very large numerical values to x, the expressions $1/x$ and $2/x^2$ are very close to 0, and hence

$$f(x) \approx \frac{1}{1 - 0 - 0} = 1.$$

This indicates that

$$f(x) \longrightarrow 1 \quad \text{as} \quad x \longrightarrow \infty$$

and

$$f(x) \longrightarrow 1 \quad \text{as} \quad x \longrightarrow -\infty.$$

Using the previous information together with Figure 4.20 and plotting several points, we obtain the graph in Figure 4.21. Note that the graph has vertical asymptotes $x = -1$ and $x = 2$, and a horizontal asymptote $y = 1$.

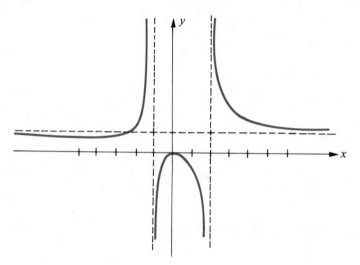

Figure 4.21. $f(x) = \dfrac{x^2}{x^2 - x - 2}$

Example 4 Sketch the graph of f if $f(x) = \dfrac{-x}{x^2 - x - 6}$.

Solution The factored form of $f(x)$ is

$$f(x) = \frac{-x}{(x + 2)(x - 3)}.$$

The zeros of the denominator are -2 and 3. As in Example 3, we begin by isolating the factor $x - 3$ as follows:

$$f(x) = \left(\frac{-x}{x + 2}\right)\left(\frac{1}{x - 3}\right).$$

Since

$$\frac{1}{x - 3} \longrightarrow \infty \quad \text{as} \quad x \longrightarrow 3^+$$

and since $-x/(x + 2)$ is *negative* when x is close to 3, we see that

$$f(x) \longrightarrow -\infty \quad \text{as} \quad x \longrightarrow 3^+.$$

Similarly, since

$$\frac{1}{x - 3} \longrightarrow -\infty \quad \text{as} \quad x \longrightarrow 3^-$$

and since $-x/(x + 2)$ is negative if x is close to 3, we have

$$f(x) \longrightarrow \infty \quad \text{as} \quad x \longrightarrow 3^{-}.$$

This behavior is illustrated in (i) of Figure 4.22.

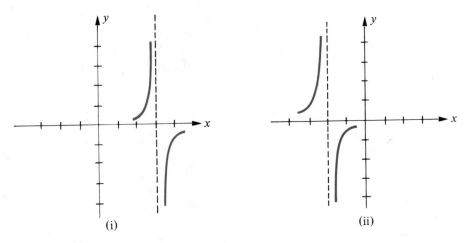

(i)   (ii)

Figure 4.22

In order to investigate the situation near $x = -2$, we write

$$f(x) = \left(\frac{-x}{x - 3}\right)\left(\frac{1}{x + 2}\right).$$

Since

$$\frac{1}{x + 2} \longrightarrow -\infty \quad \text{as} \quad x \longrightarrow -2^{-}$$

and since $-x/(x - 3)$ is *negative* if x is close to -2, it follows that

$$f(x) \longrightarrow \infty \quad \text{as} \quad x \longrightarrow -2^{-}.$$

It is left to the reader to show that

$$f(x) \longrightarrow -\infty \quad \text{as} \quad x \longrightarrow -2^{+}.$$

The sketch in (ii) of Figure 4.22 illustrates the variation of $f(x)$ near $x = -2$.
    To determine what is true if $x \longrightarrow \infty$ or $x \longrightarrow -\infty$, we divide numerator and denominator of $f(x)$ by x^2, obtaining

$$f(x) = \frac{(-1/x)}{1 - (1/x) - (6/x^2)}.$$

If x is numerically very large, then the expressions in parentheses are close to 0, and hence

$$f(x) \approx \frac{0}{1 - 0 - 0} = 0.$$

This indicates that

$$f(x) \to 0 \quad \text{as} \quad x \to \infty$$

and

$$f(x) \to 0 \quad \text{as} \quad x \to -\infty.$$

Using this information together with Figure 4.22 and plotting several points gives us the sketch in Figure 4.23. The graph has vertical asymptotes $x = -2$ and $x = 3$ and a horizontal asymptote $y = 0$ (the x-axis).

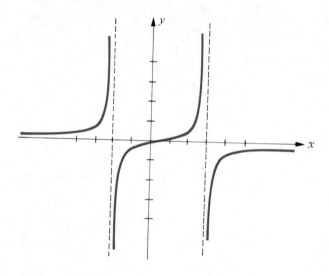

Figure 4.23. $f(x) = \dfrac{-x}{x^2 - x - 6}$

Example 5   Sketch the graph of f if $f(x) = \dfrac{2x^4}{x^4 + 1}$.

Solution   Since $f(-x) = f(x)$, it follows from Test (3.3) that the graph is symmetric with respect to the y-axis. Since the denominator of $f(x)$ is never 0, there are no vertical asymptotes. In order to examine what happens to $f(x)$ as $x \to \infty$ or $x \to -\infty$ we divide numerator and denominator by x^4, obtaining

$$f(x) = \frac{2}{1 + (1/x^4)}.$$

Since

$$\frac{1}{x^4} \to 0 \quad \text{as} \quad x \to \infty$$

it follows that

$$f(x) \longrightarrow \frac{2}{1+0} = 2 \quad \text{as} \quad x \longrightarrow \infty.$$

Using this fact, plotting several points, and making use of the symmetry with respect to the y-axis leads to the graph in Figure 4.24. The line $y = 2$ is a horizontal asymptote.

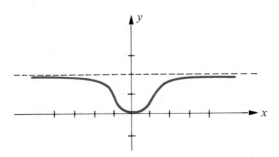

Figure 4.24. $f(x) = \dfrac{2x^4}{x^4 + 1}$

It is worth noting that when we investigated the situation $x \longrightarrow \infty$ in Examples 3 and 4, we divided numerator and denominator of $f(x)$ by x^2, whereas in Example 5 we divided by x^4. In general, if we wish to investigate $f(x) = g(x)/h(x)$ as $x \longrightarrow \infty$ or as $x \longrightarrow -\infty$, and if the degree of $g(x)$ is not higher than the degree of $h(x)$, then we divide numerator and denominator by x^k, where k is the degree of $h(x)$. If the degree of $g(x)$ is higher than the degree of $h(x)$, then it can be shown that $f(x)$ either increases or decreases without bound as $x \longrightarrow \infty$ or as $x \longrightarrow -\infty$.

Graphs of rational functions may become very complicated as the degrees of the polynomials in the numerator and denominator increase. For a thorough treatment it is necessary to employ techniques developed in calculus.

EXERCISES

Sketch the graph of f in each of Exercises 1–20.

1 $f(x) = \dfrac{1}{x - 2}$

2 $f(x) = \dfrac{1}{x + 3}$

3 $f(x) = \dfrac{-2}{x + 4}$

4 $f(x) = \dfrac{-3}{x - 1}$

5 $f(x) = \dfrac{x}{x - 5}$

6 $f(x) = \dfrac{x}{3x + 2}$

7 $f(x) = \dfrac{4}{(x - 1)^2}$

8 $f(x) = \dfrac{-1}{(x + 2)^2}$

9 $f(x) = \dfrac{1}{x^2 - 4}$

10 $f(x) = \dfrac{2}{x^2 + x - 2}$

11 $f(x) = \dfrac{5x}{4 - x^2}$

12 $f(x) = \dfrac{x^2}{x^2 - 4}$

13 $f(x) = \dfrac{x^2}{x^2 - 7x + 10}$

14 $f(x) = \dfrac{x}{x^2 - x - 6}$

15 $f(x) = \dfrac{3x + 2}{x}$

16 $f(x) = \dfrac{x^2 - 4}{x^2}$

17 $f(x) = \dfrac{4}{x^2 + 4}$

18 $f(x) = \dfrac{3x}{x^2 + 1}$

19 $f(x) = \dfrac{1}{x^3 + x^2 - 6x}$

20 $f(x) = \dfrac{x^2 - x}{16 - x^2}$

4   CONIC SECTIONS

Each of the geometric figures to be discussed in this section can be obtained by the intersection of a double-napped right circular cone and a plane, as illustrated in Figure 4.25. For this reason they are called **conic sections**, or simply **conics**. If, as in (i) of the figure, the plane cuts through one nappe and is perpendicular to the axis

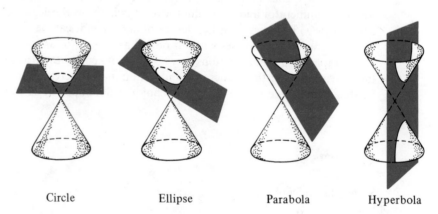

Circle   Ellipse   Parabola   Hyperbola

Figure 4.25

of the cone, then the curve of intersection is a circle. If the plane cuts entirely across one nappe and is not perpendicular to the axis, as in (ii) of the figure, then an *ellipse* is obtained. If, as in (iii), the plane does not cut completely across one nappe and does not intersect both nappes, then the curve of intersection is a *parabola*. Finally, if the plane cuts through both nappes, as in (iv), then the resulting figure is called a *hyperbola*.

The conic sections were studied extensively by the early Greek mathematicians who used the methods of Euclidean geometry. A remarkable fact about conic sections is that although they were studied thousands of years ago, they are far from obsolete. Indeed, they are important tools for current investigations in outer space and for the study of the behavior of atomic particles. Several applications of parabolas were mentioned in Section 1. There are also numerous applications involving ellipses and hyperbolas. For example, orbits of planets are ellipses. If the ellipse is very flat, the curve resembles the path of a comet. Elliptic gears or cams are sometimes used in machines. The hyperbola is useful for describing the path of an alpha particle in the electric field of the nucleus of an atom. The interested person can find many other applications of conic sections.

In Chapter Three (see (3.5)) we derived the following standard equation for a circle with center $C(h, k)$ and radius r:

$$(x - h)^2 + (y - k)^2 = r^2.$$

If we square the expressions on the left and simplify, we obtain an equation of the form

$$x^2 + y^2 + ax + by + c = 0$$

where a, b, and c are real numbers. Conversely, if we begin with an equation of that form it may be possible, by completing the squares in x and y, to obtain the standard equation, as illustrated in the next example.

Example 1   Describe the graph of the equation

$$x^2 + y^2 - 4x + 6y - 3 = 0.$$

Solution   We begin by grouping terms as follows:

$$(x^2 - 4x) + (y^2 + 6y) = 3.$$

Next we complete the squares by adding appropriate numbers within the parentheses. Of course, to obtain an equivalent equation we must add the numbers to both sides of the equation. This leads to

$$(x^2 - 4x + 4) + (y^2 + 6y + 9) = 3 + 4 + 9$$

or

$$(x - 2)^2 + (y + 3)^2 = 16.$$

Comparing with the standard equation we see that the graph is a circle with center $(2, -3)$ and radius 4.

In certain cases, completing squares as in Example 1 may lead to an equation of the form $(x - h)^2 + (y - k)^2 = 0$, and the graph consists of only one point (h, k). In other cases we could end up with a negative number on the right-hand side of the equation, and no graph exists. (Why?) Generally, we may state the following fact.

> If the equation
>
> $$x^2 + y^2 + ax + by + c = 0$$
>
> has a graph, then it is either a circle or a point.

We know from our work in Section 1 that if $a \neq 0$, then the graph of the equation $y = ax^2 + bx + c$ is a parabola which opens upward if $a > 0$ or downward if $a < 0$. By interchanging the variables x and y we obtain the equation

(4.8)
$$x = ay^2 + by + c.$$

If $b = c = 0$, then $x = ay^2$ and the graph is a parabola with vertex at the origin, symmetric to the x-axis, and opening to the right if $a > 0$ or to the left if $a < 0$, as illustrated in Figure 4.26.

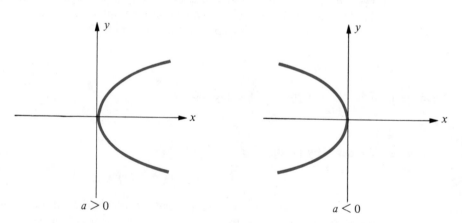

$$a > 0 \qquad\qquad a < 0$$

Figure 4.26. $x = ay^2$

If $b = 0$ and $c \neq 0$ in (4.8), then $x = ay^2 + c$ and the graph may be obtained by shifting one of the graphs in Figure 4.26 to the right or left, depending on the sign of c.

In the most general case of (4.8) we may complete the square in y, obtaining an equation of the form

(4.9)
$$x - h = a(y - k)^2.$$

In analogy with our discussion of (4.5), the graph of (4.9) is a parabola with vertex (h, k), opening to the right if $a > 0$ or to the left if $a < 0$.

Example 2 Sketch the graph of the equation $x = 3y^2 + 8y - 3$.

Solution The equation has the form (4.8) with $a = 3$, $b = 8$, and $c = -3$. Since $a > 0$ the graph is a parabola which opens to the right. We complete the square in y as follows:

$$x = 3y^2 + 8y - 3$$

$$= 3\left(y^2 + \frac{8}{3}y\right) - 3$$

$$= 3\left(y^2 + \frac{8}{3}y + \frac{16}{9}\right) - 3 - \frac{16}{3}$$

$$= 3\left(y + \frac{4}{3}\right)^2 - \frac{25}{3}.$$

If we write the last equation as

$$x + \frac{25}{3} = 3\left(y + \frac{4}{3}\right)^2$$

and compare with (4.9), we obtain $h = -25/3$ and $k = -4/3$, and hence the vertex is $(-25/3, -4/3)$. To find the y-intercepts of the graph we let $x = 0$ in the given equation and obtain

$$0 = 3y^2 + 8y - 3 = (3y - 1)(y + 3).$$

Consequently, the y-intercepts are $1/3$ and -3. The x-intercept -3 is found by setting $y = 0$ in the equation of the parabola. Plotting the vertex and the points corresponding to the intercepts leads to the sketch in Figure 4.27.

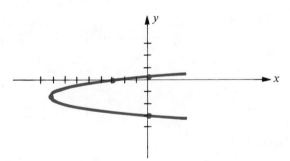

Figure 4.27. $x = 3y^2 + 8y - 3$

The graph of the equation $x^2 + y^2 = a^2$, where $a > 0$, is a circle of radius a with center at the origin (see (3.6)). A generalization of this equation is

(4.10)
$$\frac{x^2}{a^2} + \frac{y^2}{b^2} = 1$$

where a and b are positive real numbers. Note that (4.10) reduces to the circle equation if $a = b$. To find the x-intercepts of the graph we let $y = 0$ in (4.10), obtaining $\pm a$. Similarly, letting $x = 0$ gives us the y-intercepts $\pm b$. We may solve (4.10) for y in terms of x as follows:

$$\frac{y^2}{b^2} = 1 - \frac{x^2}{a^2} = \frac{a^2 - x^2}{a^2}$$

$$y^2 = \frac{b^2}{a^2}(a^2 - x^2)$$

(4.11)
$$y = \pm\frac{b}{a}\sqrt{a^2 - x^2}.$$

In order to obtain points on the graph, the radicand $a^2 - x^2$ must be non-negative. This will be true if $-a \le x \le a$. Consequently the entire graph of (4.10) lies between the vertical lines $x = -a$ and $x = a$.

For each permissible value of x in (4.11) there correspond two values for y. Let us consider the nonnegative values given by

$$y = \frac{b}{a}\sqrt{a^2 - x^2}.$$

If we let x vary from $-a$ to 0, we see that y increases from 0 to b. As x varies from 0 to a, y decreases from b to 0. This gives us the upper half of the graph in Figure 4.28. The lower half is the graph of $y = (-b/a)\sqrt{a^2 - x^2}$. The graph of (4.10) is called an **ellipse** with center at the origin.

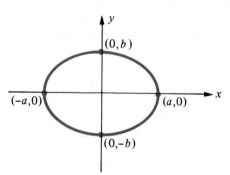

Figure 4.28. $\dfrac{x^2}{a^2} + \dfrac{y^2}{b^2} = 1$

Multiplying both sides of (4.10) by a^2b^2 we obtain

$$b^2x^2 + a^2y^2 = a^2b^2$$

which may be written in the form

(4.12)
$$Ax^2 + By^2 = C$$

where A, B, and C are positive real numbers. Knowing that (4.12) is the equation of an ellipse, the graph may be readily sketched by using the x- and y-intercepts and plotting several points.

Example 3   Sketch the graph of $9x^2 + 16y^2 = 144$.

Solution   The equation has form (4.12) with $A = 9$, $B = 16$, and $C = 144$. Hence the graph is an ellipse with center at the origin. The x-intercepts (found by letting $y = 0$) are ± 4, and the y-intercepts (found by letting $x = 0$) are ± 3. Let us next locate the points on the graph with abscissa 2. Substituting $x = 2$ in the given equation we obtain

$$9(2)^2 + 16y^2 = 144 \quad \text{or} \quad y^2 = \tfrac{108}{16} = \tfrac{27}{4}$$

and hence

$$y = \pm \frac{\sqrt{27}}{2} \approx 2.6.$$

Consequently the points $(2, \pm\sqrt{27}/2)$ are on the graph. Similarly $(-2, \pm\sqrt{27}/2)$ are solutions of the given equation. The graph is sketched in Figure 4.29.

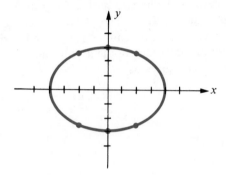

Figure 4.29. $9x^2 + 16y^2 = 144$

The horizontal and vertical line segments which join the x- and y-intercepts in Figure 4.29 are called the **axes** of the ellipse. The longer of the two axes may be on the y-axis, as in the graph of $16x^2 + 9y^2 = 144$ sketched in Figure 4.30.

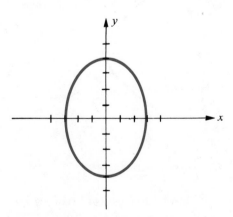

Figure 4.30. $16x^2 + 9y^2 = 144$

An equation similar to (4.10) is

(4.13)
$$\frac{x^2}{a^2} - \frac{y^2}{b^2} = 1.$$

The x-intercepts of the graph are $\pm a$; however, there are no y-intercepts since the equation $-y^2/b^2 = 1$ has no solutions. (Why?) Solving (4.13) for y in terms of x we obtain

(4.14)
$$y = \pm\frac{b}{a}\sqrt{x^2 - a^2}.$$

To obtain points on the graph we must have $x^2 - a^2 \geq 0$. (Why?) This will be true if $x \geq a$ or $x \leq -a$. We shall not discuss the details of determining the shape of the graph of (4.13). For any specific equation we could plot points; however, this is a cumbersome and time-consuming task. It can be shown that the graph has the general appearance illustrated in Figure 4.31. The graph is called a **hyperbola** with center at the origin. The two dashed lines in the figure are referred to as the **asymptotes of the hyperbola**. They serve as excellent guidelines for sketching the graph since as $|x|$ increases, the corresponding points on the graph approach the asymptotes. Equations of the asymptotes may be found by replacing the 1 in (4.13) by 0. This gives us

$$\frac{x^2}{a^2} - \frac{y^2}{b^2} = 0 \quad \text{or} \quad \frac{y^2}{b^2} = \frac{x^2}{a^2}$$

and hence

$$y = \pm\frac{b}{a}x.$$

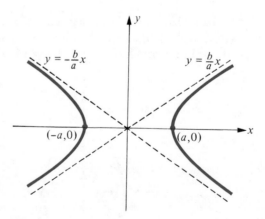

Figure 4.31. $\dfrac{x^2}{a^2} - \dfrac{y^2}{b^2} = 1$

If we multiply both sides of (4.13) by a^2b^2, we obtain an equation of the form

(4.15)
$$Ax^2 - By^2 = C$$

where A, B, and C are positive real numbers. The x-intercepts and the equations of the asymptotes (obtained from $Ax^2 - By^2 = 0$) can be used to obtain a rough sketch of the graph.

Example 4   Sketch the graph of $4x^2 - 9y^2 = 36$.

Solution   Since the equation has form (4.15), the graph is a hyperbola with center at the origin. The x-intercepts (obtained by letting $y = 0$) are ± 3 and there are no y-intercepts. The equations of the asymptotes may be found by replacing the number 36 by 0 in the given equation. This gives us

$$4x^2 - 9y^2 = 0 \quad \text{or} \quad y^2 = \tfrac{4}{9}x^2$$

and hence

$$y = \pm\tfrac{2}{3}x.$$

Plotting the x-intercepts and using the asymptotes as guidelines leads to the sketch in Figure 4.32. The student may find it instructive to plot several other points on the graph.

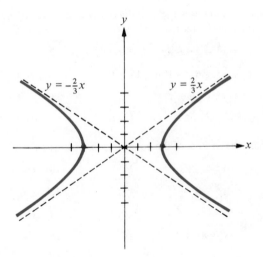

Figure 4.32. $4x^2 - 9y^2 = 36$

The graph of the equation

(4.16)
$$Ay^2 - Bx^2 = C$$

where A, B, and C are positive real numbers, is also a hyperbola. To sketch the graph we begin by finding the y-intercepts. There are no x-intercepts. (Why?) Equations for the asymptotes are obtained by replacing C by 0 in (4.16). It is left to the reader to show that the graph of $4y^2 - 9x^2 = 36$ has the shape illustrated in Figure 4.33.

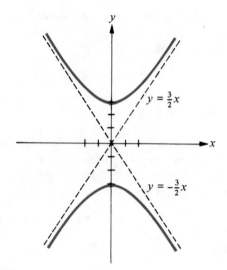

Figure 4.33. $4y^2 - 9x^2 = 36$

EXERCISES

In each of Exercises 1–8 find the center and radius of the circle having the given equation (provided a circle exists).

1   $x^2 + y^2 - 2x - 4y + 1 = 0$ 2 $x^2 + y^2 - 6x - 8y + 16 = 0$

3   $x^2 + y^2 + 6x - 10y + 18 = 0$ 4 $x^2 + y^2 + 4x + 2y + 5 = 0$

5   $x^2 + y^2 + 6y = 1$ 6 $x^2 + y^2 - 4x = 12$

7   $2x^2 + 2y^2 - 2x - 6y + 3 = 0$ 8 $9x^2 + 9y^2 - 6x + 6y + 1 = 0$

Sketch the graphs of the equations in Exercises 9–40.

9   $x = 4y^2$ 10 $x = -2y^2$

11   $x = 4y^2 - 3$ 12 $x = 8 - 2y^2$

13   $x = 9 - y^2$ 14 $x = y^2 + 4$

15   $x = y^2 + 4y + 1$ 16 $x = y^2 + 6y + 5$

17   $x = -2y^2 + 12y - 13$ 18 $x = 4y - y^2$

19   $y = x^2 - 4x - 6$ 20 $y = 3x^2 - 30x + 75$

21   $x^2 + 5y^2 = 25$ 22 $4x^2 + 25y^2 = 100$

23   $9x^2 + 4y^2 = 36$ 24 $9x^2 + y^2 = 81$

25   $5x^2 + 3y^2 = 15$ 26 $4x^2 + 3y^2 = 24$

27   $25x^2 + 9y^2 = 16$ 28 $x^2 + 8y^2 = 4$

29   $\dfrac{x^2}{4} + \dfrac{y^2}{9} = 2$ 30 $\dfrac{x^2}{4} + \dfrac{y^2}{4} = 1$

31   $9x^2 - 25y^2 = 225$ 32 $16x^2 - 9y^2 = 144$

33 $25y^2 - 9x^2 = 225$ **34** $16y^2 - 9x^2 = 144$

35 $y^2 - 36x^2 = 36$ **36** $x^2 - 5y^2 = 10$

37 $x^2 - y^2 = 1$ **38** $y^2 - x^2 = 4$

39 $4x^2 - 9y^2 = -1$ **40** $y^2 - 25x^2 = 1$

In Exercises 41 and 42 find an equation of an ellipse with center at the origin which has the given intercepts.

41 x-intercepts ± 10, y-intercepts ± 5

42 x-intercepts ± 8, y-intercepts $\pm 1/2$

43   An arch of a bridge is semielliptical with its longer axis horizontal. The base of the arch is 30 feet across and the highest part of the arch is 10 feet above the horizontal roadway. Find the height of the arch 6 feet from the center of the base.

44   Determine A so that the point $(2, -3)$ is on the conic $Ax^2 + 2y^2 = 4$. Is the conic an ellipse or a hyperbola?

45   If a square with sides parallel to the coordinate axes is inscribed in the ellipse with equation $x^2/a^2 + y^2/b^2 = 1$, express the area A of the square in terms of a and b.

46   The **eccentricity** e of the ellipse having equation (4.10) is defined as the ratio $\sqrt{a^2 - b^2}/a$. Prove that $0 < e < 1$. If a is fixed and b varies, describe the general shape of the ellipse when the eccentricity is close to 1 and when it is close to 0.

47   The graphs of the equations

$$\frac{x^2}{a^2} - \frac{y^2}{b^2} = 1 \quad \text{and} \quad \frac{x^2}{a^2} - \frac{y^2}{b^2} = -1$$

are called **conjugate hyperbolas**. Sketch the graphs of both equations on the same coordinate plane if $a = 5$ and $b = 3$. Describe the relationship between the two graphs.

48   Find an equation of a hyperbola having x-intercepts ± 2 and asymptotes $y = \pm 3x$.

5   REVIEW

Concepts

Define or discuss each of the following.

1   Polynomial function

2   Quadratic function

3   Graph of a quadratic function

4   Graphs of polynomial functions of degree greater than 2

5   Rational function

6   Vertical asymptote

7   Horizontal asymptote

8   Equations of parabolas, ellipses, and hyperbolas

Exercises

Sketch the graphs of the equations in Exercises 1–10.

1 $y^2 = 1 - x$

2 $4x^2 + y = 10$

3 $y = x^2 + 6x + 16$

4 $y^2 - 16y - x = 15$

5 $x^2 + y^2 - 8x = 0$

6 $x^2 + y^2 + 14x - 6y + 56 = 0$

7 $x^2 + 4y^2 = 64$

8 $4x^2 + y^2 = 64$

9 $x^2 - 4y^2 = 64$

10 $4y^2 - x^2 = 64$

In each of Exercises 11–20 sketch the graph of f.

11 $f(x) = (x - 4)^2$

12 $f(x) = 12x^2 + 5x - 3$

13 $f(x) = \dfrac{-2}{(x + 1)^2}$

14 $f(x) = \dfrac{1}{(x - 1)^3}$

15 $f(x) = \dfrac{3x^2}{16 - x^2}$

16 $f(x) = \dfrac{x}{(x + 5)(x^2 - 5x + 4)}$

17 $f(x) = (x + 2)^3$

18 $f(x) = 2x^2 + x^3 - x^4$

19 $f(x) = x^3 + 2x^2 - 8x$

20 $f(x) = x^6 - 32$

CHAPTER 5

Exponential
&
Logarithmic Functions

Exponential and logarithmic functions are very important in mathematics and its applications. In this chapter we shall define these functions and study some of their properties.

1 EXPONENTIAL FUNCTIONS

Throughout this section the letter a will denote a positive real number. In Chapter One we defined a^r, where r is any rational number. It is also possible to define a unique real number a^x for every real number x (rational or irrational) in such a way that the laws of exponents (1.18) remain valid. To illustrate, given a number such as a^π, we may use the nonterminating decimal representation 3.14159 ... for π and consider the numbers a^3, $a^{3.1}$, $a^{3.14}$, $a^{3.141}$, $a^{3.1415}$, ... We might expect that each successive power gets closer to a^π. This is precisely what happens if a^x is properly defined. However, the definition requires deeper concepts than are available to us and consequently it is better to leave it for a more advanced mathematics course such as calculus.

Although definitions and proofs are omitted we shall assume, henceforth, that formulas such as $a^x a^y = a^{x+y}$, $(a^x)^y = a^{xy}$, and so on, are valid for all real numbers x and y.

Before continuing the discussion of real exponents, we shall establish several results about rational exponents. Let us first show that if $a > 1$, then $a^r > 1$ for every positive rational number r. If $a > 1$, then multiplying both sides by a we obtain $a^2 > a$, and hence $a^2 > a > 1$. Multiplying by a again we obtain $a^3 > a^2 > a > 1$. Continuing, we see that $a^n > 1$ for every positive integer n. (A complete proof requires the method of mathematical induction discussed in Chapter Eleven.) Similarly, if $0 < a < 1$, it follows that $a^n < 1$ for every positive integer n. Next suppose that $a > 1$ and $r = p/q$, where p and q are positive integers. If it were true that $a^{p/q} \leq 1$, then from the previous discussion $(a^{p/q})^q \leq 1$, or $a^p \leq 1$, which contradicts the fact that $a^n > 1$ for every positive integer n. Consequently, $a^r > 1$ for every positive rational number r. We may use this fact to prove the following theorem.

(5.1) Theorem

If $a > 1$ and if r, s are rational numbers such that $r < s$, then $a^r < a^s$.

Proof. If $r < s$, then $s - r$ is a positive rational number and from the previous discussion, $a^{s-r} > 1$. Multiplying both sides by a^r we obtain

$$(a^{s-r})a^r > 1 \cdot a^r \quad \text{or} \quad a^s > a^r$$

which is what we wished to prove.

Since to each real number x there corresponds a unique real number a^x, we can define a function as follows.

(5.2)   Definition

> If $a > 0$, then the **exponential function** f **with base** a is defined by
>
> $$f(x) = a^x$$
>
> where x is any real number.

It is possible to extend Theorem (5.1) to the case of *real* exponents r and s. Specifically, if $a > 1$ and x_1 and x_2 are real numbers such that $x_1 < x_2$, then it can be shown that $a^{x_1} < a^{x_2}$, that is, $f(x_1) < f(x_2)$. This means that if $a > 1$, then the exponential function f with base a is increasing for all real numbers. It can also be shown that if $0 < a < 1$, then f is decreasing for all real numbers.

Example 1   Sketch the graph of f if $f(x) = 2^x$.

Solution   Coordinates of some points on the graph are listed in the following table.

x	-3	-2	-1	0	1	2	3	4
2^x	$\frac{1}{8}$	$\frac{1}{4}$	$\frac{1}{2}$	1	2	4	8	16

Plotting points and using the fact that f is increasing, gives us the sketch in Figure 5.1.

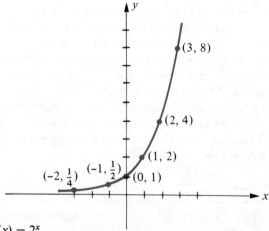

Figure 5.1. $f(x) = 2^x$

The graph in Figure 5.1 is typical of the exponential function (5.2) if $a > 1$. Since $a^0 = 1$, the y-intercept is always 1. Observe that as x decreases through negative values, the graph approaches the x-axis but never intersects it, since $a^x > 0$ for all x. This means that the x-axis is a **horizontal asymptote** for the graph. As x increases through positive values the graph rises very rapidly. Indeed, if we begin with $x = 0$ and consider successive unit changes in x, then the corresponding changes in y are 1, 2, 4, 8, 16, 32, 64, and so on. This type of variation is very common in nature and is characteristic of the **exponential law of growth**. At the end of this section we shall give several practical illustrations of this behavior.

Example 2   If $f(x) = (3/2)^x$ and $g(x) = 3^x$, sketch the graphs of f and g on the same coordinate plane.

Solution   The following table displays coordinates of several points on the graphs.

x	-2	-1	0	1	2	3	4
$(\tfrac{3}{2})^x$	$\tfrac{4}{9}$	$\tfrac{2}{3}$	1	$\tfrac{3}{2}$	$\tfrac{9}{4}$	$\tfrac{27}{8}$	$\tfrac{81}{16}$
3^x	$\tfrac{1}{9}$	$\tfrac{1}{3}$	1	3	9	27	81

With the aid of these points we obtain Figure 5.2, where dashes have been used for the graph of f in order to distinguish it from the graph of g.

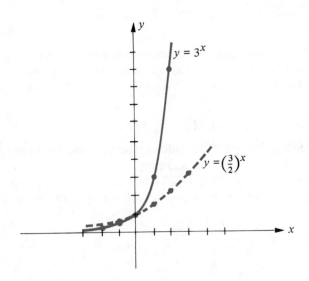

$y = 3^x$

$y = \left(\tfrac{3}{2}\right)^x$

Figure 5.2

Example 2 brings out the fact that if $1 < a < b$, then $a^x < b^x$ for positive values of x and $b^x < a^x$ for negative values of x. In particular, since $\tfrac{3}{2} < 2 < 3$, this tells us that the graph in Example 1 lies between the graphs of the functions in Example 2.

Example 3   Sketch the graph of the equation $y = (\frac{1}{2})^x$.

Solution   Some points on the graph may be obtained from the following table.

x	-3	-2	-1	0	1	2	3
$(\frac{1}{2})^x$	8	4	2	1	$\frac{1}{2}$	$\frac{1}{4}$	$\frac{1}{8}$

The graph is sketched in Figure 5.3. Since $(\frac{1}{2})^x = 2^{-x}$, the graph is the same as the graph of the equation $y = 2^{-x}$.

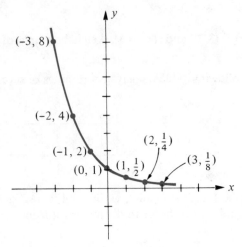

Figure 5.3. $f(x) = (\frac{1}{2})^x = 2^{-x}$

In advanced mathematics and applications it is often necessary to consider a function f such that $f(x) = a^p$, where p is some expression in x. We do not intend to study such functions in detail; however, let us consider one example.

Example 4   Sketch the graph of f if $f(x) = 2^{-x^2}$.

Solution   Since $f(x) = 1/2^{x^2}$, it follows that if x increases numerically, the corresponding point $(x, f(x))$ on the graph approaches the x-axis. Thus the x-axis is a horizontal asymptote for the graph. The maximum value of $f(x)$ occurs at $x = 0$. Tabulating some coordinates of points on the graph, we obtain the following table.

x	-2	-1	0	1	2
$f(x)$	$\frac{1}{16}$	$\frac{1}{2}$	1	$\frac{1}{2}$	$\frac{1}{16}$

The graph is sketched in Figure 5.4. Note the symmetry with respect to the y-axis. Functions of this type arise in the study of the branch of mathematics called *probability*.

The variation of many physical entities can be described by means of exponential functions. One of the most common examples occurs in the growth of

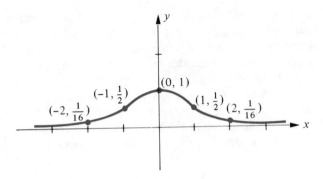

Figure 5.4. $f(x) = 2^{-x^2}$

certain populations. As an illustration, it might be observed experimentally that the number of bacteria in a culture doubles every hour. If there are 1,000 bacteria present at the start of the experiment, then the experimenter would obtain the readings listed below, where t is the time in hours and $f(t)$ is the bacteria count at time t.

t	0	1	2	3	4
$f(t)$	1,000	2,000	4,000	8,000	16,000

It appears that $f(t) = (1,000)2^t$. This formula makes it possible to predict the number of bacteria present at any time t. For example, at $t = 1.5$ we have

$$f(t) = (1,000)2^{3/2} = 1,000\sqrt{2^3} = 1,000\sqrt{8} \approx 2,828.$$

Exponential growth patterns of this type may also be observed in some human and animal populations.

Compound interest provides another illustration of exponential growth. If a sum of money P (called the **principal**) is invested at a simple interest rate of i per cent, then the interest at the end of one interest period is Pi. For example, if $P = \$100$ and i is 8% per year, then the interest at the end of one year is $\$100(0.08)$, or $\$8$. If the interest is reinvested at the end of this period then the new principal is

$$P + Pi, \quad \text{or} \quad P(1 + i).$$

Note that to find the new principal we multiply the original principal by $(1 + i)$. In the preceding illustration the new principal is $\$100(1.08)$, or $\$108$.

If another time period elapses, then the new principal may be found by multiplying $P(1 + i)$ by $(1 + i)$. Thus the principal after two time periods is $P(1 + i)^2$. If we reinvest and continue this process, the principal after three periods is $P(1 + i)^3$; after four it is $P(1 + i)^4$; and in general, after n time periods the principal P_n is given by

$$P_n = P(1 + i)^n.$$

Interest accumulated in this way is called **compound interest**. We see that the principal is given in terms of an exponential function whose base is $1 + i$ and exponent is n. The time period may vary, being measured in years, months, weeks, days, or any other suitable unit of time. When this formula for P_n is employed, it must be remembered that i is the interest rate per time period. For example, if the rate is stated as 9% *per year compounded monthly*, then the rate per month is $\frac{9}{12}\%$, or equivalently, $\frac{3}{4}\%$. In this case, $i = 0.75\% = 0.0075$ and n is the number of months. The sketch in Figure 5.5 illustrates the growth of $100 invested at this rate over a period of 15 years. We have connected the principal amounts by a

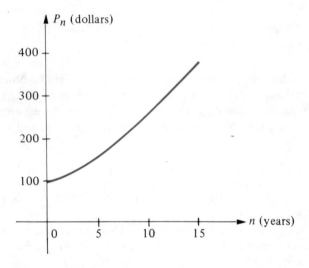

Figure 5.5. *Compound interest:* $P_n = 100(1.0075)^n$

smooth curve in order to indicate the growth during this time. Due to the complexity of the base, the use of the formula for P_n is extremely tedious unless tables or calculators are available. For those with calculators, several problems on compound interest have been included in the exercises.

Certain physical quantities may *decrease* exponentially. In such cases the base a of the exponential function is between 0 and 1. One of the most common examples is the decay of a radioactive substance. As an illustration, the polonium isotope ^{210}Po has a half-life of approximately 140 days; that is, given any amount, one-half of it will disintegrate in 140 days. If there is initially 20 mg present, then the following table indicates the amount remaining after various intervals of time.

t (days)	0	140	280	420	560
Amount remaining	20	10	5	2.5	1.25

The sketch in Figure 5.6 illustrates the exponential nature of the disintegration.

The behavior of an electrical condenser can be used to illustrate exponential decay. If the condenser is allowed to discharge, the initial rate of discharge is relatively high, but it then tapers off as in the preceding example of radioactive decay.

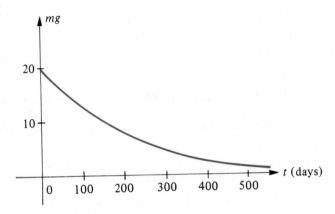

Figure 5.6. Decay of polonium

In calculus and its applications a certain irrational number, denoted by e, is often used as the base of an exponential function. To five decimal places,

$$e \approx 2.71828$$

The base e arises naturally in the discussion of many physical phenomena. For this reason the function f defined by $f(x) = e^x$ is called the **natural exponential function**. Since $2 < e < 3$, the graph of f lies "between" the graphs of the equations $y = 2^x$ and $y = 3^x$. Many hand-held calculators have an e^x key which can be used to approximate values of the natural exponential function. There may also be a key marked y^x that can be used for any positive base y. For those who have access to calculators we have included some exercises involving e^x and y^x.

EXERCISES

In each of Exercises 1–24 sketch the graph of the function f.

1 $f(x) = 4^x$

2 $f(x) = 5^x$

3 $f(x) = 10^x$

4 $f(x) = 8^x$

5 $f(x) = 3^{-x}$

6 $f(x) = 4^{-x}$

7 $f(x) = -2^x$

8 $f(x) = -3^x$

9 $f(x) = 4 - 2^{-x}$

10 $f(x) = 2 + 3^{-x}$

11 $f(x) = (2/3)^x$

12 $f(x) = (3/4)^{-x}$

13 $f(x) = (5/2)^{-x}$

14 $f(x) = (4/3)^x$

15 $f(x) = 2^{|x|}$

16 $f(x) = 2^{-|x|}$

17 $f(x) = 2^{x+3}$

18 $f(x) = 3^{x+2}$

19 $f(x) = 2^{3-x}$

20 $f(x) = 3^{-2-x}$

21 $f(x) = 3^{1-x^2}$

22 $f(x) = 2^{-(x+1)^2}$

23 $f(x) = 3^x + 3^{-x}$

24 $f(x) = 3^x - 3^{-x}$

25 If $f(x) = 2^x$ and $g(x) = x^2$, illustrate the difference in the rate of growth of f and g for $x \geq 0$ by sketching the graphs of both functions on the same coordinate plane.

26 Repeat Exercise 25 if $f(x) = 2^{-x}$ and $g(x) = x^{-2}$.

27 The half-life of radium is 1,600 years; that is, given any quantity, one-half of it will disintegrate in 1,600 years. If the initial amount is q_0 milligrams, then it can be shown that the quantity $q(t)$ remaining after t years is given by $q(t) = q_0 2^{kt}$. Find k.

28 The number of bacteria in a certain culture at time t is given by $Q(t) = 2(3^t)$, where t is measured in hours and $Q(t)$ in thousands. What is the initial number of bacteria? What is the number after 10 minutes? After 30 minutes? After 1 hour?

29 If $1,000 is invested at a rate of 12% per year compounded monthly, what is the principal after 1 month? 2 months? 6 months? 1 year?

30 If 10 grams of salt is added to a certain quantity of water, then the amount $q(t)$ which is undissolved after t minutes is given by $q(t) = 10(\frac{4}{5})^t$. Sketch a graph which shows the value $q(t)$ at any time from $t = 0$ to $t = 10$.

31 Why was $a < 0$ ruled out in the discussion of a^x?

32 Use (5.1) to prove that if $0 < a < 1$ and r and s are rational numbers such that $r < s$, then $a^r > a^s$.

33 How does the graph of $y = a^x$ compare with the graph of $y = -a^x$?

34 If $a > 1$, how does the graph of $y = a^x$ compare with the graph of $y = a^{-x}$?

CALCULATOR EXERCISES (*Optional*)

Solve Exercises 1–4 by using the compound interest formula $P_n = P(1 + i)^n$.

1 If $1,000 is invested at an interest rate of 6% per year compounded quarterly, find the principal at the end of
(a) one year. (b) two years. (c) five years. (d) ten years.
(*Hint:* $i = \frac{1}{4}(6\%) = 1.5\%$ and, at the end of one year, $n = 4$.)

2 Rework Exercise 1 if the interest rate is 6% per year compounded monthly.

3 If $10,000 is invested at a rate of 9% per year compounded semi-annually, how long will it take for the principal to exceed (a) $15,000? (b) $20,000? (c) $30,000?

4 A certain department store requires its credit card customers to pay interest at the rate of 18% per year compounded monthly on any unpaid bills. If a man buys a $500 television set on credit and then makes no payments for one year, how much does he owe?

5 Sketch the graph of f if $f(x)$ is given by:

(a) e^x (b) e^{-x} (c) $\dfrac{e^x + e^{-x}}{2}$ (d) $\dfrac{2}{e^x + e^{-x}}$

(e) $\dfrac{e^x - e^{-x}}{2}$ (f) $\dfrac{2}{e^x - e^{-x}}$

6 Sketch the graph of the equation $y = e^{-x^2/2}$.

7   Sketch the graph of $y = x(2^x)$.

8   If $f(x) = x^2(5^{\sqrt{x}})$, find $f(1.01) - f(1)$.

9   If $f(x) = (1 + x)^{1/x}$, approximate the following to four decimal places:

$$f(1), \quad f(0.1), \quad f(0.01), \quad f(0.001), \quad f(0.0001), \quad f(0.00001).$$

Compare your answers with the value of $e \approx 2.71828$ and arrive at a conjecture.

10   The population of a certain city is increasing at the rate of 5% per year and the present population is 500,000. Using methods developed in calculus it is estimated that the population after t more years will be approximately $500{,}000e^{0.05t}$. Estimate the population after   (a) 10 years.   (b) 20 years.   (c) 40 years.

11   Under certain conditions the atmospheric pressure p at altitude h ft is given by

$$p = 29e^{-0.000034h}.$$

What is the pressure at an altitude of 40,000 ft?

2   LOGARITHMS

Throughout this section and the next, it is assumed that a is a positive real number different from 1. Let us begin by examining the graph of f, where $f(x) = a^x$ (see Figure 5.7 for the case $a > 1$). It appears that every positive real number u is the ordinate of some point on the graph; that is, there is a number v such that $u = a^v$.

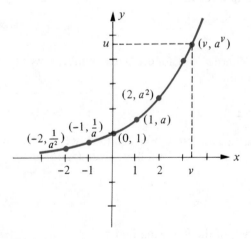

Figure 5.7. $f(x) = a^x$, $a > 1$

Indeed, v is the abscissa of the point where the line $y = u$ intersects the graph. Moreover, since f is increasing throughout its domain, the number u can occur as an ordinate only once. A similar situation exists if $0 < a < 1$. This makes the following theorem plausible.

(5.3)   Theorem

> For each positive real number u there is a unique real number v such that
> $$a^v = u.$$

A rigorous proof of (5.3) requires concepts studied in calculus.

(5.4)   Definition

> If u is any positive real number, then the (unique) exponent v such that $a^v = u$ is called the **logarithm of u with base a** and denoted by $\log_a u$.

A convenient way to memorize (5.4) is by means of the following statement:

(5.5)
$$v = \log_a u \quad \text{if and only if} \quad a^v = u$$

As illustrations of (5.5) we may write

$$3 = \log_2 8 \quad \text{since } 2^3 = 8$$
$$-2 = \log_5 \tfrac{1}{25} \quad \text{since } 5^{-2} = \tfrac{1}{25}$$
$$4 = \log_{10} 10{,}000 \quad \text{since } 10^4 = 10{,}000.$$

The fact that $\log_a u$ is the exponent v such that $a^v = u$ gives us the following important identity:

(5.6)
$$a^{\log_a u} = u$$

Since $a^r = a^r$ for every real number r, we may use (5.5) with $v = r$ and $u = a^r$ to obtain

(5.7)
$$r = \log_a a^r$$

In particular, letting $r = 1$ in (5.7), we see that $\log_a a = 1$.
Since $a^0 = 1$, it follows from (5.5) that

(5.8)
$$\log_a 1 = 0$$

Example 1   In each of the following find s.

(a) $s = \log_4 2$ (b) $\log_5 s = 2$ (c) $\log_s 8 = 3$

Solutions   We may use (5.5) to solve each part as follows.

(a)   If $s = \log_4 2$, then $4^s = 2$ and hence $s = \tfrac{1}{2}$.

(b)   If $\log_5 s = 2$, then $5^2 = s$ and hence $s = 25$.

(c)   If $\log_s 8 = 3$, then $s^3 = 8$ and hence $s = \sqrt[3]{8} = 2$.

Example 2   Solve the equation $\log_4 (5 + x) = 3$.

Solution   If $\log_4 (5 + x) = 3$, then by (5.5)

$$5 + x = 4^3 \quad \text{or} \quad 5 + x = 64.$$

Hence the solution is $x = 59$.

The following laws are fundamental for all work with logarithms, where it is assumed that u and w are positive real numbers.

(5.9)   **Laws of Logarithms**

> (i) $\log_a (uw) = \log_a u + \log_a w$
>
> (ii) $\log_a (u/w) = \log_a u - \log_a w$
>
> (iii) $\log_a (u^c) = c \log_a u$, for every real number c

Proof.   Let

(5.10) $$r = \log_a u \quad \text{and} \quad s = \log_a w.$$

By (5.5) this implies that $a^r = u$ and $a^s = w$. Consequently,

$$a^r a^s = uw$$

and then by a law of exponents

$$a^{r+s} = uw.$$

Again using (5.5), we see that the last equation is equivalent to

$$r + s = \log_a(uw).$$

Substituting for r and s from (5.10) gives us

$$\log_a u + \log_a w = \log_a uw.$$

This proves (i).

To prove (ii) we begin as in the proof of (i), but this time we divide a^r by a^s, obtaining

$$\frac{a^r}{a^s} = \frac{u}{w}, \quad \text{or} \quad a^{r-s} = \frac{u}{w}.$$

According to (5.5), the latter equation is equivalent to

$$r - s = \log_a (u/w).$$

Substituting for r and s from (5.10) gives us (ii).

Finally, if c is any real number, then using the same notation as above,

$$(a^r)^c = u^c, \quad \text{or} \quad a^{cr} = u^c.$$

According to the definition of logarithm, the last equality implies that

$$cr = \log_a u^c.$$

Substituting for r from (5.10), we obtain

$$c \log_a u = \log_a u^c.$$

This proves Law (iii).

Example 3   If $\log_a 3 = 0.4771$ and $\log_a 2 = 0.3010$, find each of the following.

(a) $\log_a 6$

(b) $\log_a (\tfrac{3}{2})$

(c) $\log_a \sqrt{2}$

(d) $\dfrac{\log_a 3}{\log_a 2}$

Solutions   (a)   Since $6 = 2 \cdot 3$, we may use (i) of (5.9) to obtain

$$\log_a 6 = \log_a (2 \cdot 3) = \log_a 2 + \log_a 3$$
$$= 0.4771 + 0.3010 = 0.7781.$$

(b)   By (ii) of (5.9),

$$\log_a (\tfrac{3}{2}) = \log_a 3 - \log_a 2$$
$$= 0.4771 - 0.3010 = 0.1761.$$

(c)   Using (iii) of (5.9), we get

$$\log_a \sqrt{2} = \log_a 2^{1/2} = (\tfrac{1}{2}) \log_a 2$$
$$= (\tfrac{1}{2})(0.3010) = 0.1505.$$

(d)   There is no law of logarithms which allows us to simplify $(\log_a 3)/(\log_a 2)$. Consequently, we *divide* 0.4771 by 0.3010, obtaining the approximation 1.585. It is important to notice the difference between this problem and the one stated in part (b).

Example 4   Solve each of the following equations.

(a) $\log_5 (2x + 3) = \log_5 11 + \log_5 3$

(b) $\log_4 (x + 6) - \log_4 10 = \log_4 (x - 1) - \log_4 2$

Solutions   (a)   Using (i) of (5.9), the given equation may be written as

$$\log_5 (2x + 3) = \log_5 (11 \cdot 3) = \log_5 33.$$

Consequently $2x + 3 = 33$, or $2x = 30$, and therefore the solution is $x = 15$.

(b)   The given equation is equivalent to

$$\log_4 (x + 6) - \log_4 (x - 1) = \log_4 10 - \log_4 2.$$

Applying (ii) of (5.9) gives us

$$\log_4 \left(\frac{x + 6}{x - 1}\right) = \log_4 \frac{10}{2} = \log_4 5$$

and hence

$$\frac{x + 6}{x - 1} = 5.$$

The last equation implies that

$$x + 6 = 5x - 5 \quad \text{or} \quad 4x = 11.$$

Thus the solution is $x = 11/4$.

Extraneous solutions sometimes occur in the process of solving logarithmic equations, as illustrated in the next example.

Example 5   Solve the equation $2 \log_7 x = \log_7 36$.

Solution   Applying (iii) of (5.9) we obtain $2 \log_7 x = \log_7 x^2$, and substitution in the given equation leads to

$$\log_7 x^2 = \log_7 36.$$

Consequently $x^2 = 36$ and hence either $x = 6$ or $x = -6$. However, $x = -6$ is not a solution of the original equation since x must be positive in order for $\log_7 x$ to exist. Thus there is only one solution, $x = 6$.

Sometimes it is necessary to *change the base* of a logarithm by expressing $\log_b u$ in terms of $\log_a u$, for some positive real number b different from 1. This can be accomplished as follows. We begin with the equivalent equations

$$v = \log_b u \quad \text{and} \quad b^v = u.$$

Taking the logarithm, base a, of both sides of the second equation gives us

$$\log_a b^v = \log_a u.$$

Applying (iii) of (5.9), we get

$$v \log_a b = \log_a u.$$

Solving for v (that is, $\log_b u$) we obtain

(5.11)
$$\log_b u = \frac{\log_a u}{\log_a b}$$

The most important special case of (5.11) is obtained by letting $u = a$. Since $\log_a a = 1$, this gives us

(5.12)
$$\log_b a = \frac{1}{\log_a b}$$

The laws of logarithms are often used as in the next example.

Example 6   Express $\log_a \dfrac{x^3 \sqrt{y}}{z^2}$ in terms of the logarithms of x, y, and z.

Solution   Using the three laws stated in (5.9), we have

$$\log_a \frac{x^3 \sqrt{y}}{z^2} = \log_a (x^3 \sqrt{y}) - \log_a z^2$$
$$= \log_a x^3 + \log_a \sqrt{y} - \log_a z^2$$
$$= 3 \log_a x + (\tfrac{1}{2}) \log_a y - 2 \log_a z.$$

The procedure in Example 6 can also be reversed; that is, beginning with the final equation we can retrace our steps to obtain the original expression (see Exercises 67–70).

As a final remark, note that there is no general law for expressing $\log_a (u + w)$ in terms of simpler logarithms. It is evident that it does not always equal $\log_a u + \log_a w$, since the latter equals $\log_a (uw)$.

EXERCISES

Use (5.5) to change the equations in Exercises 1–8 to logarithmic form.

1   $4^3 = 64$ 2 $3^5 = 243$

3   $2^7 = 128$ 4 $5^3 = 125$

5   $10^{-3} = 0.001$ 6 $10^{-2} = 0.01$

7   $t^r = s$ 8 $v^w = u$

Use (5.5) to change the equations in Exercises 9–16 to exponential form.

9   $\log_{10} 1000 = 3$ 10 $\log_3 81 = 4$

11   $\log_3 \dfrac{1}{243} = -5$ 12 $\log_4 \dfrac{1}{64} = -3$

13 $\log_7 1 = 0$

14 $\log_9 1 = 0$

15 $\log_t r = p$

16 $\log_v w = q$

Find the numbers in Exercises 17–30.

17 $\log_4 (1/16)$

18 $\log_2 32$

19 $\log_{10} 100$

20 $\log_8 64$

21 $10^{\log_{10} 5}$

22 $\log_{10} 0.0001$

23 $\log_7 \sqrt[3]{7}$

24 $10^{2 \log_{10} 3}$

25 $\log_{10} (1/10)$

26 $\log_{10} 100{,}000$

27 $\log_{1/2} 8$

28 $\log_3 (1/81)$

29 $3^{4 \log_3 2}$

30 $\log_6 \sqrt[5]{6}$

Given $\log_a 7 = 0.8$ and $\log_a 3 = 0.5$, change the expressions in Exercises 31–40 to decimal form.

31 $\log_a (7/3)$

32 $\log_a (3/7)$

33 $\log_a 7^3$

34 $\log_a (3^7)$

35 $\log_a (1/\sqrt{3})$

36 $\log_a \sqrt[4]{3}$

37 $\log_a 63$

38 $(\log_a 7)/(\log_a 3)$

39 $\log_7 a$

40 $(\log_3 a)(\log_a 3)$

Find the solutions of the equations in Exercises 41–58.

41 $\log_3 (x - 4) = 2$

42 $\log_2 (x - 5) = 4$

43 $\log_9 x = 3/2$

44 $\log_4 x = -3/2$

45 $\log_5 x^2 = -2$

46 $\log_{10} x^2 = -4$

47 $\log_2 (x^2 - 5x + 14) = 3$

48 $\log_2 (x^2 - 10x + 18) = 1$

49 $\log_x 7 = 3$

50 $5^{\log_5 x} = 10$

51 $\log_6 (2x - 3) = \log_6 12 - \log_6 3$

52 $2 \log_3 x = 3 \log_2 5$

53 $\log_2 x - \log_2 (x + 1) = 3 \log_2 4$

54 $\log_5 x + \log_5 (x + 6) = \frac{1}{2} \log_5 9$

55 $\log_{10} x^2 = \log_{10} x$

56 $\log_x 10 = 10$

57 $\frac{1}{2} \log_5 (x - 2) = 3 \log_5 2 - \frac{3}{2} \log_5 (x - 2)$

58 $\log_{10} 5^x = \log_3 1$

In each of Exercises 59–66, express the logarithm in terms of logarithms of x, y, and z.

59 $\log_a \dfrac{x^2 y}{z^3}$

60 $\log_a \dfrac{x^3 y^2}{z^5}$

61 $\log_a \dfrac{\sqrt{xz^2}}{y^4}$ **62** $\log_a x \sqrt[3]{\dfrac{y^2}{z^4}}$

63 $\log_a \sqrt[3]{\dfrac{x^2}{yz^5}}$ **64** $\log_a \dfrac{\sqrt{xy^6}}{\sqrt[3]{z^2}}$

65 $\log_a \sqrt{x\sqrt{yz^3}}$ **66** $\log_a \sqrt[3]{x^2 y\sqrt{z}}$

In each of Exercises 67–70 write the expression as one logarithm.

67 $2\log_a x + \frac{1}{3}\log_a (x - 2) - 5\log_a (2x + 3)$

68 $5\log_a x - \frac{1}{2}\log_a (3x - 4) + 3\log_a(5x + 1)$

69 $\log_a (y^2 x^3) - 2\log_a x \sqrt[3]{y} + 3\log_a \left(\dfrac{x}{y}\right)$

70 $2\log_a \dfrac{y^3}{x} - 3\log y + \frac{1}{2}\log_a x^4 y^2$

3 LOGARITHMIC FUNCTIONS

We shall now use the concept of logarithm to introduce a new function whose domain is the set of positive real numbers.

(5.13) Definition

> The function f defined by
>
> $$f(x) = \log_a x$$
>
> for all positive real numbers x is called the **logarithmic function with base a**.

The graph of f is the same as the graph of the equation $y = \log_a x$ which, by (5.5), is equivalent to

$$x = a^y.$$

In order to find some pairs which are solutions of the preceding equation, we may substitute for y and find the corresponding values of x, as illustrated in the following table.

y	-3	-2	-1	0	1	2	3
x	$\dfrac{1}{a^3}$	$\dfrac{1}{a^2}$	$\dfrac{1}{a}$	1	a	a^2	a^3

If $a > 1$, we obtain the sketch in Figure 5.8. In this case f is an increasing function throughout its domain. If $0 < a < 1$, then the graph has the general shape shown in Figure 5.9, and hence f is a decreasing function. Note that for every a under consideration the region to the left of the y-axis is excluded. There is no y-intercept and the x-intercept is 1.

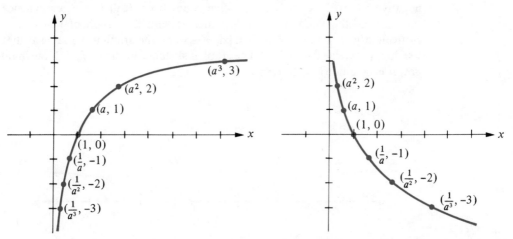

Figure 5.8. $f(x) = \log_a x, a > 1$ Figure 5.9. $f(x) = \log_a x, 0 < a < 1$

Functions defined by expressions of the form $\log_a p$, where p is some expression in x, often occur in mathematics and its applications. Functions of this type are classified as members of the logarithmic family; however, the graphs may differ from those sketched in Figures 5.8 and 5.9, as is illustrated in the following examples.

Example 1   Sketch the graph of f if $f(x) = \log_3(-x)$, $x < 0$.

Solution   If $x < 0$, then $-x > 0$ and hence $\log_3(-x)$ is defined. We wish to sketch the graph of the equation $y = \log_3(-x)$, or equivalently, $3^y = -x$. The following table displays coordinates of some points on the graph, which is sketched in Figure 5.10.

y	-2	-1	0	1	2
x	$-\frac{1}{9}$	$-\frac{1}{3}$	-1	-3	-9

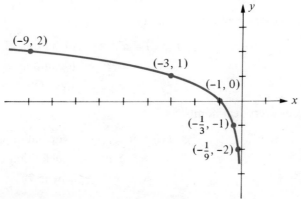

Figure 5.10. $f(x) = \log_3(-x)$

Example 2   Sketch the graph of the equation $y = \log_3 |x|, x \neq 0$.

Solution   Since $|x| > 0$ for all $x \neq 0$, there are points on the graph corresponding to negative values of x as well as to positive values. If $x > 0$, then $|x| = x$ and hence to the right of the y-axis the graph coincides with the graph of $y = \log_3 x$, or equivalently, $x = 3^y$. If $x < 0$, then $|x| = -x$ and the graph is the same as that of $y = \log_3(-x)$ (see Example 1). The graph is sketched in Figure 5.11. Note that the graph is symmetric with respect to the y-axis.

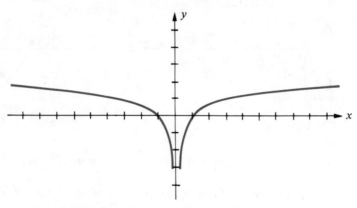

Figure 5.11. $f(x) = \log_3 |x|$

There is a close relationship between exponential and logarithmic functions. This is to be expected since the logarithmic function was defined in terms of the exponential function. A precise description of the relationship is stated in the following theorem.

(5.14)   **Theorem**

> The exponential and logarithmic functions with base a are inverse functions of one another.

Proof.   If $f(x) = a^x$ and $g(x) = \log_a x$, then according to (3.20) we must show that

$$f(g(x)) = x \quad \text{and} \quad g(f(x)) = x.$$

For every positive real number x,

$$
\begin{aligned}
f(g(x)) &= a^{g(x)} && \text{(definition of } f) \\
&= a^{\log_a x} && \text{(definition of } g) \\
&= x && \text{by (5.6).}
\end{aligned}
$$

For every real number x,

$$
\begin{aligned}
g(f(x)) &= \log_a f(x) && \text{(definition of } g) \\
&= \log_a a^x && \text{(definition of } f) \\
&= x && \text{by (5.7).}
\end{aligned}
$$

This proves the theorem.

Logarithmic functions occur frequently in applications. Indeed, if two variables u and v are related so that u is an exponential function of v, then v is a logarithmic function of u. As a specific example, if \$1.00 is invested at a rate of 9% per year compounded monthly, then from the discussion in Section 2 the principal P after n interest periods is

$$P = (1.0075)^n.$$

By (5.5), the logarithmic form of this equation is

$$n = \log_{1.0075} P.$$

Thus, the number of periods n required for \$1.00 to grow to an amount P is related logarithmically to P.

Another example we considered in Section 2 was that of population growth. In particular, the equation for the number N of bacteria in a certain culture after t hours was

$$N = (1000)2^t \quad \text{or} \quad 2^t = N/1000.$$

Changing to logarithmic form we obtain

$$t = \log_2 (N/1000).$$

Hence the time t is a logarithmic function of N.

In Section 1 we defined the natural exponential function f by means of the equation $f(x) = e^x$. The inverse of f, that is, the logarithmic function with base e, is called the **natural logarithmic function**. The notation **ln** x is used as an abbreviation for $\log_e x$. Since $e \approx 3$, the graph of $y = \ln x$ is similar in appearance to the graph of $y = \log_3 x$. For natural logarithms a law such as (i) of (5.9) is written

$$\boxed{\ln (uv) = \ln u + \ln v}$$

In like manner, letting $a = e$ in (5.6) and (5.7) we obtain

$$\boxed{e^{\ln u} = u \quad \text{and} \quad r = \ln e^r}$$

Example 3 According to Newton's Law of Cooling, the rate at which an object cools is directly proportional to the difference in temperature between the object and the surrounding medium. Newton's Law can be used to show that under certain conditions the temperature T of an object is given by

$$T = 75e^{-2t}$$

where t is time. Express t as a function of T.

Solution The given equation may be rewritten

$$e^{-2t} = T/75.$$

Using logarithms with base e yields

$$-2t = \log_e (T/75) = \ln (T/75).$$

Consequently,

$$t = -\tfrac{1}{2} \ln (T/75) \quad \text{or} \quad t = -\tfrac{1}{2}[\ln T - \ln 75].$$

We shall not have many occasions to work with natural logarithms in this text. For computational purposes the base 10 is used. Base 10 gives us the system of common logarithms discussed in subsequent sections.

Many hand-held calculators have a key labeled ln x which can be used to approximate values of the natural logarithmic function. Several exercises are included at the end of this section for those who have a calculator of that type.

EXERCISES

Sketch the graph of f in each of Exercises 1–10.

1 (a) $f(x) = \log_2 x$ (b) $f(x) = \log_4 x$

2 (a) $f(x) = \log_5 x$ (b) $f(x) = \log_{10} x$

3 (a) $f(x) = \log_3 x$ (b) $f(x) = \log_3 (x - 2)$

4 (a) $f(x) = \log_2 (-x)$ (b) $f(x) = \log_2 (3 - x)$

5 (a) $f(x) = \log_3 (3x)$ (b) $f(x) = 3 \log_3 x$

6 (a) $f(x) = \log_2 (x^2)$ (b) $f(x) = \log_2 (x^3)$

7 (a) $f(x) = \log_2 \sqrt{x}$ (b) $f(x) = \log_2 \sqrt[3]{x}$

8 (a) $f(x) = \log_2 |x|$ (b) $f(x) = |\log_2 x|$

9 (a) $f(x) = \log_3 (1/x)$ (b) $f(x) = 1/(\log_3 x)$

10 (a) $f(x) = \log_3 (2 + x)$ (b) $f(x) = 2 + \log_3 x$

11 What is the geometric relationship between the graphs of $y = \log_a x$ and $y = a^x$? Why does this relationship exist?

12 Describe the relationship of the graph of $y = \ln x$ to the graphs of $y = \log_2 x$ and $y = \log_3 x$.

13 If one begins with q_0 milligrams of pure radium, then the amount q remaining after t years is given by the formula

$$q = q_0(2)^{-t/1600}$$

Use logarithms with base 2 to solve for t in terms of q and q_0.

14 The number of bacteria in a certain culture at time t is given by

$$N = 10^4(3)^t.$$

Use logarithms with base 3 to solve for t in terms of N.

15 An electrical condenser with initial charge Q_0 is allowed to discharge. After t seconds the charge Q is given by

$$Q = Q_0 e^{kt}$$

where k is a constant of proportionality. Use natural logarithms to solve for t in terms of Q_0, Q, and k.

16 Under certain conditions the atmospheric pressure p at altitude h is given by

$$p = 29e^{-0.000034h}.$$

Use natural logarithms to solve for h as a function of p.

17 The loudness of sound, as experienced by the human ear, is based upon intensity levels. A formula used for finding the intensity level α which corresponds to a sound intensity I is

$$\alpha = 10\log_{10}(I/I_0) \text{ decibels}$$

where I_0 is a special value of I agreed to be the weakest sound that can be detected by the ear under certain conditions. Find α if:

(a) I is 10 times as great as I_0.

(b) I is 1,000 times as great as I_0.

(c) I is 10,000 times as great as I_0. (This is the intensity level of the average voice.)

CALCULATOR EXERCISES (*Optional*)

1 Sketch the graph of

(a) $y = \ln x$.   (b) $y = \ln|x|$.   (c) $y = |\ln x|$.

2 Sketch the graph of f if $f(x) = x^2 \ln x$.

3 Verify that $\ln e^x = x$ and $e^{\ln x} = x$ for the following values of x:

(a) 1   (b) 3.4   (c) 0.056   (d) 8.143

4 Refer to Calculator Exercise 10 of Section 1. For what value of t will the population of the city be 1,000,000?

5 From (5.11), $\log_b u = (\log_a u)/(\log_a b)$. Letting $a = e$ we obtain

$$\log_b u = \frac{\ln u}{\ln b}.$$

Use this formula to approximate

(a) $\log_3 4$ (b) $\log_5(1.64)$ (c) $\log_7\sqrt{2}$

(d) $\log_{10} 382$ (e) $\log_{10}(47/62)$

4   COMMON LOGARITHMS

Logarithms with base 10 are useful for certain numerical computations. It is customary to refer to such logarithms as **common logarithms** and to use the symbol **log x** as an abbreviation for $\log_{10} x$.

Base 10 is used in computational work because every positive real number can be written in the scientific form $c \cdot 10^k$, where $1 \le c < 10$ and k is an integer (see Section 6 of Chapter One). For example,

$$513 = (5.13)10^2 \qquad 2375 = (2.375)10^3 \qquad 720000 = (7.2)10^5$$
$$0.641 = (6.41)10^{-1} \qquad 0.00000438 = (4.38)10^{-6} \qquad 4.601 = (4.601)10^0.$$

If x is any positive real number and we write

$$x = c \cdot 10^k$$

where $1 \le c < 10$ and k is an integer, then applying (i) of (5.9) gives us

$$\log x = \log c + \log 10^k.$$

Next, using (5.7) with $a = 10$ and $r = k$ we obtain

(5.15) $$\log x = \log c + k.$$

Equation (5.15) tells us that to find $\log x$ for any positive real number x it is sufficient to know the logarithms of numbers between 1 and 10. The number $\log c$, where $1 \le c < 10$, is called the **mantissa**, and the integer k is called the **characteristic** of $\log x$.

If $1 \le c < 10$, then since $\log x$ increases as x increases,

$$\log 1 \le \log c < \log 10,$$

or, by (5.8) and (5.7),

$$0 \le \log c < 1.$$

Hence the mantissa of a logarithm is a number between 0 and 1. When working numerical problems it is usually necessary to approximate logarithms. For example, it can be shown that

$$\log 2 = 0.3010299957\ldots$$

where the decimal is nonrepeating and nonterminating. We shall round off such logarithms to four decimal places and write

$$\log 2 \approx 0.3010.$$

If a number between 0 and 1 is written as a finite decimal, it is sometimes referred to as a **decimal fraction**. Thus (5.15) tells us that if x is any positive real number, then *log x may be approximated by the sum of a positive decimal fraction (the mantissa) and an integer k (the characteristic).* We shall refer to this representation as the **standard form** for $\log x$.

Logarithms of many of the numbers between 1 and 10 have been calculated. Table 2 contains four-decimal-place approximations for logarithms of numbers between 1.00 and 9.99 at intervals of 0.01. This table can be used to find the logarithm of any three-digit number to four-decimal-place accuracy. There exist far more extensive tables which provide logarithms of many additional numbers to much greater accuracy than four decimal places. The use of Table 2 is illustrated in the following examples.

Example 1   Find each of the following.

(a) $\log 43.6$ (b) $\log 43,600$ (c) $\log 0.0436$

Solutions   (a)   Since $43.6 = (4.36)10^1$, the characteristic of $\log 43.6$ is 1. Referring to Table 2, we find that the mantissa of $\log 4.36$ may be approximated by 0.6395. Hence, as in (5.15),

$$\log 43.6 \approx 0.6395 + 1 = 1.6395.$$

(b)   Since $43,600 = (4.36)10^4$, the mantissa is the same as in part (a); however the characteristic is 4. Consequently,

$$\log 43,600 \approx 0.6395 + 4 = 4.6395.$$

(c)   If we write $0.0436 = (4.36)10^{-2}$, then

$$\log 0.0436 = \log 4.36 + (-2).$$

Hence

(5.16) $$\log 0.0436 \approx 0.6395 + (-2).$$

We could subtract 2 from 0.6395 and obtain

$$\log 0.0436 \approx -1.3605$$

but this is not standard form, since $-1.3605 = -0.3605 + (-1)$, a number in which the decimal fraction is *negative*. A common error is to write (5.16) as -2.6395. This is incorrect since $-2.6395 = -0.6395 + (-2)$, which is not the same as $0.6395 + (-2)$.

If a logarithm has a negative characteristic, it is customary either to leave it in standard form as in (5.16) or to rewrite the logarithm, keeping the decimal part positive. To illustrate the latter technique, let us add and subtract 8 on the right side of (5.16). This gives us

$$\log 0.0436 \approx 0.6395 + (8 - 8) + (-2),$$

or

$$\log 0.0436 \approx 8.6395 - 10.$$

We could also write

$$\log 0.0436 \approx 18.6395 - 20 = 43.6395 - 45,$$

and so on, as long as the *integral part* of the logarithm is -2.

Example 2   Find each of the following.

(a) $\log (0.00652)^2$

(b) $\log (0.00652)^{-2}$

(c) $\log (0.00652)^{1/2}$

Solutions   (a)   By (iii) of (5.9),

$$\log (0.00652)^2 = 2 \log 0.00652.$$

Since $0.00652 = (6.52)10^{-3}$,

$$\log 0.00652 = \log 6.52 + (-3).$$

Referring to Table 2, we see that $\log 6.52$ is approximately 0.8142 and therefore

$$\log 0.00652 \approx 0.8142 + (-3).$$

Hence

$$\begin{aligned}
\log (0.00652)^2 &= 2 \log 0.00652 \\
&\approx 2[0.8142 + (-3)] \\
&= 1.6284 + (-6).
\end{aligned}$$

The standard form is $0.6284 + (-5)$.

(b)   Using (iii) of (5.9) and the value for $\log 0.00652$ found in part (a),

$$\begin{aligned}
\log (0.00652)^{-2} &= -2 \log 0.00652 \\
&\approx -2[0.8142 + (-3)] \\
&= -1.6284 + 6.
\end{aligned}$$

It is important to note that -1.6284 means $-0.6284 + (-1)$ and consequently the decimal part is negative. To obtain the standard form, we may write

$$\begin{aligned}
-1.6284 + 6 &= 6.0000 - 1.6284 \\
&= 4.3716.
\end{aligned}$$

This shows that the mantissa is 0.3716 and the characteristic is 4.

(c)   By (iii) of (5.9),

$$\begin{aligned}
\log (0.00652)^{1/2} &= (\tfrac{1}{2}) \log 0.00652 \\
&\approx \tfrac{1}{2}[0.8142 + (-3)].
\end{aligned}$$

If we multiply by $1/2$, the standard form is not obtained since neither number in the resulting sum is the characteristic. In order to avoid this, we may adjust the expression within brackets by adding and subtracting a suitable number. If we use 1 in this way, we obtain

$$\log(0.00652)^{1/2} \approx \tfrac{1}{2}[1.8142 + (-4)]$$
$$= 0.9071 + (-2)$$

which is in standard form. We could also have added and subtracted a number other than 1. For example,

$$\tfrac{1}{2}[0.8142 + (-3)] = \tfrac{1}{2}[17.8142 + (-20)]$$
$$= 8.9071 + (-10).$$

Table 2 can be used to find an approximation to x if $\log x$ is given, as illustrated in the following example.

Example 3   Find a decimal approximation to x if $\log x$ is as follows.

(a) $\log x = 1.7959$ (b) $\log x = -3.5918$

Solutions   (a) The mantissa 0.7959 determines the sequence of digits in x and the characteristic determines the position of the decimal point. Referring to the *body* of Table 2, we see that the mantissa 0.7959 is the logarithm of 6.25. Since the characteristic is 1, x lies between 10 and 100. Consequently, $x \approx 62.5$.

(b) In order to find x from Table 2, $\log x$ must be written in standard form. In order to change $\log x = -3.5918$ to standard form, we may add and subtract 4, obtaining

$$\log x = (4 - 3.5918) - 4$$
$$= 0.4082 - 4.$$

Referring to Table 2, we see that the mantissa 0.4082 is the logarithm of 2.56. Since the characteristic of $\log x$ is -4, it follows that $x \approx 0.000256$.

In each of the previous examples the mantissas appeared in Table 2. If this is not the case, then x can be approximated by the method of interpolation discussed in Section 6.

If a hand-held calculator with a log key is used to determine common logarithms, then the standard form for $\log x$ is obtained only if $x \geq 1$. For example, to find log 463 on a typical calculator we enter 463 and press the log key, obtaining the standard form

However, if we find log 0.0463 in similar fashion, then the following number appears on the display panel:

This is not the standard form for the logarithm, but is similar to that which occurred in the solutions to Example 1(c) and Example 3(b). To find the standard form we could add 2 to the logarithm (using a calculator) obtaining 0.665580991 and then append -2. Thus

$$\log 0.0463 \approx 0.665580991 - 2.$$

EXERCISES

In Exercises 1–16 use Table 2 and laws of logarithms to approximate the common logarithms of the given numbers.

1   347;   0.00347;   3.47

2   86.2;   8,620;   0.862

3   0.54;   540;   540,000

4   208;   2.08;   20,800

5   60.2;   0.0000602;   602

6   5;   0.5;   0.0005

7   $(44.9)^2$; $(44.9)^{1/2}$; $(44.9)^{-2}$

8   $(1810)^4$; $(1810)^{40}$; $(1810)^{1/4}$

9   $(0.943)^3$; $(0.943)^{-3}$; $(0.943)^{1/3}$

10   $(0.017)^{10}$; $10^{0.017}$; $10^{1.43}$

11   $(638)(17.3)$

12   $\dfrac{(2.73)(78.5)}{621}$

13   $\dfrac{(47.4)^3}{(29.5)^2}$

14   $\dfrac{(897)^4}{\sqrt{17.8}}$

15   $\sqrt[3]{20.6(371)^3}$

16   $\dfrac{(0.0048)^{10}}{\sqrt{0.29}}$

In Exercises 17–30 use Table 2 to find a decimal approximation to x.

17   $\log x = 3.6274$

18   $\log x = 1.8965$

19   $\log x = 0.9469$

20   $\log x = 0.5729$

21   $\log x = 5.2095$

22   $\log x = 6.7300 - 10$

23   $\log x = 9.7348 - 10$

24   $\log x = 7.6739 - 10$

25   $\log x = 8.8306 - 10$

26   $\log x = 4.9680$

27   $\log x = 2.2765$

28   $\log x = 3.0043$

29   $\log x = -1.6253$

30   $\log x = -2.2118$

31   Chemists use a number denoted by pH to describe quantitatively the acidity or basicity of solutions. By definition,

$$pH = -\log[H^+]$$

where $[H^+]$ is the hydrogen ion concentration in moles per liter. Given the following, with the indicated $[H^+]$, approximate the pH of each substance.

   (a)   vinegar: $[H^+] \approx 6.3 \times 10^{-3}$

   (b)   carrots: $[H^+] \approx 1.0 \times 10^{-5}$

   (c)   sea water: $[H^+] \approx 5.0 \times 10^{-9}$

32   Approximate the hydrogen ion concentration $[H^+]$ in each of the following substances. (Refer to Exercise 31 and use the closest entry in Table 2 to approximate $[H^+]$.)

   (a)   apples: pH ≈ 3.0

   (b)   beer: pH ≈ 4.2

   (c)   milk: pH ≈ 6.6

33   A solution is considered acidic if $[H^+] > 10^{-7}$ or basic if $[H^+] < 10^{-7}$. What are the corresponding inequalities involving pH?

34   Many solutions have a pH between 1 and 14. What is the corresponding range of the hydrogen ion content $[H^+]$?

35   The sound intensity level formula considered in Exercise 17 of Section 3 may be written

$$\alpha = 10\log(I/I_0).$$

Find α if I is 285,000 times as great as I_0.

36   A sound intensity level of 140 decibels produces pain in the average human ear. Approximately how many times greater than I_0 must I be in order for α to reach this level? (Refer to Exercise 35 and use the nearest entry in Table 2 to find the approximation.)

5   EXPONENTIAL AND LOGARITHMIC EQUATIONS

The variables in certain equations appear as exponents or logarithms, as illustrated in the following examples.

Example 1   Solve the equation $3^x = 21$.

Solution   Taking the common logarithm of both sides and using (iii) of (5.9), we obtain

$$\log(3^x) = \log 21$$
$$x\log 3 = \log 21$$
$$x = \frac{\log 21}{\log 3}.$$

If an approximation is desired, we may use Table 2 to obtain

$$x \approx \frac{1.3222}{0.4771} \approx 2.77$$

where the last number is obtained by dividing 1.3222 by 0.4771. A partial check on the solution is to note that since $3^2 = 9$ and $3^3 = 27$, the number x such that $3^x = 21$ should lie between 2 and 3, somewhat closer to 3 than to 2.

Example 2   Solve $5^{2x+1} = 6^{x-2}$.

Solution   Taking the common logarithm of both sides and using (iii) of (5.9), we obtain

$$(2x + 1)\log 5 = (x - 2)\log 6.$$

We may now solve for x as follows:

$$2x\log 5 + \log 5 = x\log 6 - 2\log 6$$
$$2x\log 5 - x\log 6 = -\log 5 - 2\log 6$$
$$x(2\log 5 - \log 6) = -(\log 5 + \log 6^2)$$
$$x = \frac{-(\log 5 + \log 36)}{2\log 5 - \log 6}$$
$$x = \frac{-\log(5 \cdot 36)}{\log 5^2 - \log 6}$$
$$x = \frac{-\log 180}{\log(25/6)}.$$

If an approximation to the solution is desired, we may proceed as in Example 1.

Example 3   Solve the equation $\log(5x - 1) - \log(x - 3) = 2$.

Solution   The given equation may be written as

$$\log\frac{5x - 1}{x - 3} = 2.$$

Using the definition of logarithm (5.5) with $a = 10$ gives us

$$\frac{5x - 1}{x - 3} = 10^2.$$

Consequently,

$$5x - 1 = 10^2(x - 3) = 100x - 300, \quad \text{or} \quad 299 = 95x.$$

Hence

$$x = \frac{299}{95}.$$

Example 4   Solve the equation $\dfrac{5^x - 5^{-x}}{2} = 3$.

Solution   Multiplying both sides of the given equation first by 2 and then by 5^x gives us

$$5^x - 5^{-x} = 6$$
$$5^{2x} - 1 = 6(5^x)$$

which may be written

$$(5^x)^2 - 6(5^x) - 1 = 0.$$

Letting $u = 5^x$ gives us a quadratic equation in the variable u. Applying the quadratic formula (2.2)

$$5^x = \frac{6 \pm \sqrt{36 + 4}}{2} = 3 \pm \sqrt{10}.$$

Since 5^x is never negative, the number $3 - \sqrt{10}$ must be discarded; therefore

$$5^x = 3 + \sqrt{10}.$$

Taking the common logarithm of both sides and using (iii) of (5.9),

$$x \log 5 = \log(3 + \sqrt{10}) \quad \text{or} \quad x = \frac{\log(3 + \sqrt{10})}{\log 5}.$$

To obtain an approximate solution we may write $3 + \sqrt{10} \approx 6.16$ and use Table 2. This gives us

$$x \approx \frac{\log 6.16}{\log 5} \approx \frac{0.7896}{0.6990} \approx 1.13.$$

EXERCISES

Find the solutions of the equations in Exercises 1–20.

1   $10^x = 7$

2   $5^x = 8$

3   $4^x = 3$

4   $10^x = 6$

5   $3^{4-x} = 5$

6   $(1/3)^x = 100$

7   $3^{x+4} = 2^{1-3x}$

8   $4^{2x+3} = 5^{x-2}$

9   $2^{-x} = 8$

10   $2^{-x^2} = 5$

11 $\log x = 1 - \log(x - 3)$

12 $\log(5x + 1) = 2 + \log(2x - 3)$

13 $\log(x^2 + 4) - \log(x + 2) = 3 + \log(x - 2)$

14 $\log(x - 4) - \log(3x - 10) = \log(1/x)$

15 $\log(x^2) = (\log x)^2$

16 $\log\sqrt{x} = \sqrt{\log x}$

17 $\log(\log x) = 2$

18 $\log\sqrt{x^3 - 9} = 2$

19 $x^{\sqrt{\log x}} = 10^8$

20 $\log(x^3) = (\log x)^3$

In Exercises 21–24 solve for x in terms of y.

21 $y = \dfrac{10^x + 10^{-x}}{2}$

22 $y = \dfrac{10^x - 10^{-x}}{2}$

23 $y = \dfrac{10^x - 10^{-x}}{10^x + 10^{-x}}$

24 $y = \dfrac{1}{10^x - 10^{-x}}$

In Exercises 25–28 use natural logarithms to solve for x in terms of y.

25 $y = \dfrac{e^x - e^{-x}}{2}$

26 $y = \dfrac{e^x + e^{-x}}{2}$

27 $y = \dfrac{e^x - e^{-x}}{e^x + e^{-x}}$

28 $y = \dfrac{e^x + e^{-x}}{e^x - e^{-x}}$

29 The current i in a certain electrical circuit is given by

$$i = \frac{E}{R}(1 - e^{-Rt/L}).$$

Use natural logarithms to solve for t in terms of the remaining symbols.

30 If a sum of money P_0 is invested at an interest rate of $100r$ per cent per year, compounded m times per year, then the principal P at the end of t years is given by

$$P = P_0\left(1 + \frac{r}{m}\right)^{mt}.$$

Solve for t in terms of the other symbols.

31 In Exercise 17 of Section 3 we considered the sound intensity level formula

$$\alpha = 10\log(I/I_0).$$

(a) Solve for I in terms of α and I_0.

(b) Show that a one-decibel rise in the intensity level corresponds to a 26% increase in the intensity.

32 The chemical formula $\text{pH} = -\log[\text{H}^+]$ was introduced in Exercise 31 of Section 4. Solve for $[\text{H}^+]$ in terms of pH.

CALCULATOR EXERCISES (*Optional*)

Use natural logarithms to approximate the solutions of the equations in Exercises 1–3.

1 $5^x = 7$ (*Hint*: $\ln 5^x = \ln 7$, or $x \ln 5 = \ln 7$)

2 $4^{-x} = 10$

3 $e^{-x^2} = 0.163$

4 If $10,000 is invested at an interest rate of 8% per year compounded quarterly, when will the principal exceed $20,000? (*Hint*: Use Exercise 30.)

5 How long will it take money to double if it is invested at a rate of 6% per year compounded monthly?

6 LINEAR INTERPOLATION

The only logarithms that can be found *directly* from Table 2 are logarithms of numbers that contain at most three nonzero digits. If *four* nonzero digits are involved, then it is possible to obtain an approximation by using the method of linear interpolation described in this section. The terminology *linear interpolation* is used because, as we shall see, the method is based upon approximating portions of the graph of $y = \log x$ by line segments. It should be pointed out that there are quicker, more mechanical, ways to find logarithms accurately. Indeed, if a hand-held calculator is available, it is usually only necessary to press a few keys to obtain accuracy to many decimal places. Nonetheless, linear interpolation is a very useful skill, since it can be applied to any tabular data in which entries vary in a manner similar to that of a linear function, at least over small portions of the table. Consequently, as you proceed through this section keep in mind that you are learning a process that can be used with many tables other than those for logarithms.

In order to illustrate the process of linear interpolation, and at the same time give some justification for it, let us consider the specific example $\log 12.64$. Since the logarithmic function with base 10 is increasing, this number lies between $\log 12.60 \approx 1.1004$ and $\log 12.70 \approx 1.1038$. Examining the graph of $y = \log x$, we have the situation shown in Figure 5.12, where we have distorted the units on the x- and y-axes and also the portion of the graph shown. A more accurate drawing would indicate that the graph of $y = \log x$ is much closer to the line segment joining $P(12.60, 1.1004)$ to $Q(12.70, 1.1038)$ than is shown in the figure. Since $\log 12.64$ is the ordinate of the point on the graph having abscissa 12.64, it can be approximated by the ordinate of the point with abscissa 12.64 on the *line segment PQ*. Referring to Figure 5.12 we see that the latter ordinate is $1.1004 + d$. The number d can be approximated by using similar triangles. Referring to Figure 5.13, where the graph of $y = \log x$ has been deleted, we may form the following proportion:

$$\frac{d}{0.0034} = \frac{0.04}{0.1}.$$

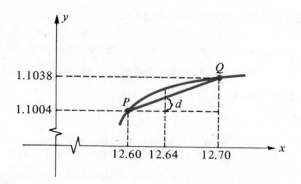

Figure 5.12

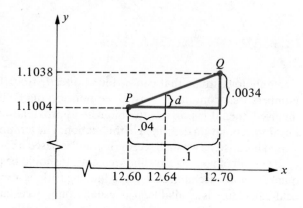

Figure 5.13

Hence

$$d = \frac{(0.04)(0.0034)}{0.1} = 0.00136.$$

When using this technique, we always round off decimals to the same number of places as appear in the body of the table. Consequently $d \approx 0.0014$ and

$$\log 12.64 \approx 1.1004 + 0.0014 = 1.1018.$$

Hereafter we shall not sketch a graph when interpolating. Instead we shall use the scheme illustrated in the next example.

Example 1   Approximate $\log 572.6$.

Solution   It is convenient to arrange our work as follows:

$$1.0\left\{0.6\left\{\begin{matrix} \log 572.0 \approx 2.7574 \\ \log 572.6 = \ ? \end{matrix}\right\}d\atop \log 573.0 \approx 2.7582\right\}0.0008$$

where we have indicated differences by appropriate symbols alongside of the braces. This leads to the proportion

$$\frac{d}{0.0008} = \frac{0.6}{1.0} = \frac{6}{10} \quad \text{or} \quad d = \left(\frac{6}{10}\right)(0.0008) = 0.00048 \approx 0.0005.$$

Hence

$$\log 572.6 \approx 2.7574 + 0.0005 = 2.7579.$$

Another way of working this type of problem is to reason that since 572.6 is $\frac{6}{10}$ of the way from 572.0 to 573.0, then $\log 572.6$ is (approximately) $\frac{6}{10}$ of the way from 2.7574 to 2.7582. Hence

$$\log 572.6 \approx 2.7574 + (\tfrac{6}{10})(0.0008) \approx 2.7574 + 0.0005 = 2.7579.$$

Example 2   Approximate $\log 0.003678$.

Solution   We begin by arranging our work as in the solution of Example 1. Thus,

$$10\left\{8\left\{\begin{array}{l}\log 0.003670 \approx 0.5647 + (-3)\}\!\!\}\,d \\ \log 0.003678 = ? \\ \log 0.003680 \approx 0.5658 + (-3)\end{array}\right.\right\}0.0011$$

Since we are only interested in ratios, we have used the numbers 8 and 10 on the left side because their ratio is the same as the ratio of 0.000008 to 0.000010. This leads to the proportion

$$\frac{d}{0.0011} = \frac{8}{10} = 0.8 \quad \text{or} \quad d = (0.0011)(0.8) = 0.00088 \approx 0.0009.$$

Hence

$$\log 0.003678 \approx [0.5647 + (-3)] + 0.0009$$
$$= 0.5656 + (-3).$$

If a number x is written in the form $x = c \cdot 10^k$, where $1 \le c < 10$, then before using Table 2 to find $\log x$ by interpolation, c should be rounded off to three decimal places. Another way of saying this is that x should be rounded off to four **significant figures**. Some examples will help to clarify the procedure. If $x = 36.4635$, we round off to 36.46 before approximating $\log x$. The number 684,279 should be rounded off to 684,300. For a decimal such as 0.096202 we write 0.09620, and so on. The reason for doing this is that Table 2 does not guarantee more than four-digit accuracy, since the mantissas which appear in it are approximations. This means that if *more* than four-digit accuracy is required in a problem, then Table 2 cannot be used. If, in more extensive tables, the logarithm of a number containing n digits can be found directly, then interpolation is allowed for numbers involving $n + 1$ digits and numbers should be rounded off accordingly.

The method of interpolation can also be used to find x when we are given $\log x$. If we use Table 2, then x may be found to four significant figures. In this case we are given the *ordinate* of a point on the graph of $y = \log x$ and are asked to find the *abscissa*. A geometric argument similar to the one given earlier can be used to justify the procedure illustrated in the next example.

Example 3   Find x to four significant figures if $\log x = 1.7949$.

Solution   The mantissa 0.7949 does not appear in Table 2, but it can be isolated between adjacent entries, namely the mantissas corresponding to 6.230 and 6.240. We shall arrange our work as follows:

$$0.1 \left\{ r \begin{cases} \log 62.30 \approx 1.7945 \\ \log x \quad = 1.7949 \end{cases} 0.0004 \\ \log 62.40 \approx 1.7952 \right\} 0.0007.$$

This leads to the proportion

$$\frac{r}{0.1} = \frac{0.0004}{0.0007} = \frac{4}{7} \quad \text{or} \quad r = (0.1)\left(\frac{4}{7}\right) \approx 0.06.$$

Hence,

$$x \approx 62.30 + 0.06 = 62.36.$$

If we are given $\log x$, then the number x is called the **antilogarithm** of $\log x$. In Example 3 the antilogarithm of $\log x = 1.7949$ is $x \approx 62.36$. Sometimes the notation antilog $(1.7949) \approx 62.36$ is used.

EXERCISES

Use linear interpolation to approximate the common logarithms of the numbers in Exercises 1–20.

1 25.48	2 421.6	3 5363
4 0.3817	5 0.001259	6 69,450
7 123,400	8 0.0212	9 0.7786
10 1.203	11 384.7	12 54.44
13 0.9462	14 7259	15 66,590
16 0.001428	17 0.04321	18 300,100
19 3.003	20 1.236	

In Exercises 21–40 use linear interpolation to approximate x.

21 $\log x = 1.4437$	22 $\log x = 3.7455$
23 $\log x = 4.6931$	24 $\log x = 0.5883$

25   $\log x = 9.1664 - 10$

26   $\log x = 8.3902 - 10$

27   $\log x = 3.8153 - 6$

28   $\log x = 5.9306 - 9$

29   $\log x = 2.3705$

30   $\log x = 4.2867$

31   $\log x = 0.1358$

32   $\log x = 0.0194$

33   $\log x = 8.9752 - 10$

34   $\log x = 2.4979 - 5$

35   $\log x = 5.0409$

36   $\log x = 1.3796$

37   $\log x = -2.8712$

38   $\log x = -1.8164$

39   $\log x = -0.6123$

40   $\log x = -3.1426$

7   COMPUTATIONS WITH LOGARITHMS

The importance of logarithms for numerical computations has diminished in recent years because of the development of computers and hand-held calculators. However, since mechanical devices are not always available, it is worthwhile to have some familiarity with the use of logarithms for solving arithmetic problems. At the same time practice in working numerical problems leads to a deeper understanding of logarithms and logarithmic functions. The following examples illustrate some computational techniques.

Example 1   Approximate $N = \dfrac{(59700)(0.0163)}{41.7}$.

Solution   Using (5.9) and Table 2 leads to

$$\begin{aligned}
\log N &= \log 59700 + \log 0.0163 - \log 41.7 \\
&\approx 4.7760 + (0.2122 - 2) - (1.6201) \\
&= 4.9882 - 3.6201 \\
&= 1.3681.
\end{aligned}$$

Referring to Table 2 for the antilogarithm we have, to three significant figures,

$$N \approx 23.3.$$

Example 2   Approximate $N = \sqrt[3]{56.11}$ to four significant figures.

Solution   Writing $N = (56.11)^{1/3}$ and using (iii) of (5.9) gives us

$$\log N = \tfrac{1}{3}\log 56.11.$$

To find $\log 56.11$ we interpolate from Table 2 as follows:

$$10\left\{{}^{1}\left\{\begin{array}{l} \log 56.10 \approx 1.7490 \\ \log 56.11 = ? \\ \log 56.20 \approx 1.7497 \end{array}\right\}{}^{d}\right\}0.0007$$

$$\frac{d}{0.0007} = \frac{1}{10}$$

$$d = 0.00007 \approx 0.0001.$$

Hence

$$\log 56.11 \approx 1.7490 + 0.0001 = 1.7491$$

and therefore

$$\log N \approx \tfrac{1}{3}(1.7491) \approx 0.5830.$$

The antilogarithm may be found by interpolation from Table 2 as follows

$$0.01\left\{{}^{r}\left\{\begin{array}{l} \log 3.820 \approx 0.5821 \\ \log N \quad \approx 0.5830 \\ \log 3.830 \approx 0.5832 \end{array}\right\}{}^{9}\right\}11$$

$$\frac{r}{0.01} = \frac{9}{11}$$

$$r = \frac{9}{11}(0.01) \approx 0.008.$$

Consequently,

$$N \approx 3.820 + 0.008 = 3.828.$$

If we were interested in only *three* significant figures, then the interpolations in Example 2 could have been avoided. In the remaining examples we shall, for simplicity, work only with three-digit numbers.

Example 3   Approximate $N = \dfrac{(1.32)^{10}}{\sqrt[5]{0.0268}}$.

Solution   Using (5.9) and Table 2,

$$\begin{aligned} \log N &= 10\log 1.32 - \tfrac{1}{5}\log 0.0268 \\ &\approx 10(0.1206) - \tfrac{1}{5}(3.4281 - 5) \\ &= 1.206 - 0.6856 + 1 \\ &= 1.5204. \end{aligned}$$

Finding the antilogarithm (to three significant figures), we obtain

$$N \approx 33.1.$$

Example 4 Find x if $x^{2.1} = 6.5$.

Solution Taking the common logarithm of both sides and using (iii) of (5.9) gives us

$$(2.1) \log x = \log 6.5.$$

Hence

$$\log x = \frac{\log 6.5}{2.1}$$

and by (5.5),

$$x = 10^{(\log 6.5)/(2.1)}.$$

If an approximation to x is desired, then from the second equation above we have

$$\log x = \frac{\log 6.5}{2.1} \approx \frac{0.8129}{2.1} \approx 0.3871.$$

Using Table 2, the antilogarithm (to three significant figures) is

$$x \approx 2.44.$$

Example 5 Approximate $N = \dfrac{69.3 + \sqrt[3]{56.1}}{\log 807}$.

Solution Since we have no formula for the logarithm of a sum, the two terms in the numerator must be *added* before the logarithm can be found. From Example 2, $\sqrt[3]{56.1} \approx 3.8$ and hence the numerator is approximately 73.1. Using Table 2 we see that

$$\log 807 \approx 2.9069 \approx 2.91.$$

Hence the given expression may be approximated by

$$N \approx \frac{73.1}{2.91}.$$

It is now an easy matter to find N, either by using logarithms or by long division. It is left to the reader to verify that $N \approx 25.1$.

EXERCISES

Use logarithms to approximate the numbers in Exercises 1–20 to three significant figures.

1 $(638)(57.2)$ 2 $(0.0178)(0.00729)$

3 $\dfrac{35,900}{8,430}$ 4 $\dfrac{3.14}{63.2}$

5 $(4.21)^{10}$ 6 $(0.712)^6$

7   $\sqrt[5]{0.517}$

8   $\sqrt[4]{2.23}$

9   $\dfrac{(26.7)^3(1.48)}{(67.4)^2}$

10   $\dfrac{(3.04)^3}{\sqrt{1.12}(86.6)}$

11   $\sqrt{1.65}\sqrt[3]{(70.8)^2}$

12   $\left[\dfrac{(11.1)^3}{\sqrt[5]{4.17}}\right]^{-1/2}$

13   $\sqrt{\dfrac{0.563}{(0.105)^3}}$

14   $\sqrt[5]{\dfrac{(124)^2}{(9.83)^3}}$

15   $10^{-3.14}$

16   $(100)^{0.523}$

17   $(4.12)^{0.220}$

18   $\sqrt{8.46\sqrt{3.07}}$

19   $\dfrac{\log 37.4}{\log 6.19}$

20   $\dfrac{56.8 + \log(7.13)}{\sqrt[10]{4.42}}$

Use interpolation in Table 2 to approximate the numbers in Exercises 21–26 to four significant figures.

21   $(2.461)^5$

22   $1/(33.89)^4$

23   $\sqrt[10]{0.5138}$

24   $\sqrt[5]{(17.04)^2}$

25   $(5.375)^{2/3}$

26   $(1776)^{11}$

27   The area A of a triangle with sides a, b, and c may be calculated from the formula $A = \sqrt{s(s-a)(s-b)(s-c)}$, where s is one-half the perimeter. Use logarithms to approximate the area of a triangle with sides 12.6, 18.2, and 14.1.

28   The volume V of a right circular cone of altitude h and radius of base r is $V = \frac{1}{3}\pi r^2 h$. Use logarithms to approximate the volume of a cone of radius 2.43 cm and altitude 7.28 cm.

29   The formula used in physics to approximate the period T (seconds) of a simple pendulum of length L (feet) is $T = 2\pi\sqrt{L/(32.2)}$. Approximate the period of a pendulum 33 inches long.

30   The pressure p (pounds per cubic foot) and volume v (cubic feet) of a certain gas are related by the formula $pv^{1.4} = 600$. Approximate the pressure if $v = 8.22$ cubic feet.

8   REVIEW

Concepts

Define or discuss the following.

1   The exponential function with base a

2   The natural exponential function

3   The logarithm of u with base a

4   The natural logarithmic function

5 The laws of logarithms

6 The logarithmic function with base a

7 Common logarithms

8 Mantissa

9 Characteristic

10 Linear interpolation

Exercises

Find the numbers in Exercises 1–6.

1 $\log_2 (1/16)$ 2 $\log_5 \sqrt[3]{5}$

3 $6^{\log_6 4}$ 4 $10^{3 \log 2}$

5 $\log 1{,}000{,}000$ 6 $\ln e$

In Exercises 7–16 sketch the graph of f.

7 $f(x) = 3^{x+2}$ 8 $f(x) = (3/5)^x$

9 $f(x) = (3/2)^{-x}$ 10 $f(x) = 3^{-2x}$

11 $f(x) = 3^{-x^2}$ 12 $f(x) = 1 - 3^{-x}$

13 $f(x) = \log_6 x$ 14 $f(x) = \log_3 (x^2)$

15 $f(x) = 2 \log_3 x$ 16 $f(x) = \log_2 (x + 4)$

Find the solutions of the equations in Exercises 17–24.

17 $\log_8 (x - 5) = 2/3$

18 $\log_4 (x + 1) = 2 + \log_4 (3x - 2)$

19 $2 \log_3 (x + 3) - \log_3 (x + 1) = 3 \log_3 2$

20 $\log \sqrt[4]{x + 1} = 1/2$

21 $2^{5-x} = 6$ 22 $3^{x^2} = 7$

23 $2^{5x+3} = 3^{2x+1}$ 24 $5^{\log_5 (x+1)} = 3$

25 Express $\log x^4 \sqrt[3]{y^2/z}$ in terms of logarithms of x, y, and z.

26 Express $\log (x^2/y^3) + 4 \log y - 6 \log \sqrt{xy}$ as one logarithm.

Solve the equations in Exercises 27 and 28 for x in terms of y.

27 $y = \dfrac{10^x + 10^{-x}}{10^x - 10^{-x}}$ 28 $y = \dfrac{1}{10^x + 10^{-x}}$

Use linear interpolation to approximate the logarithms in Exercises 29–32.

29 $\log 47.82$ 30 $\log 0.001347$

31 $\log 300{,}600$ 32 $\log 0.2143$

Use linear interpolation to approximate x in Exercises 33–36.

33 $\log x = 2.4995$

34 $\log x = 1.5045$

35 $\log x = 8.7970 - 10$

36 $\log x = -1.3146$

Use logarithms to approximate the numbers in Exercises 37–40.

37 $\dfrac{(38.2)^3}{\sqrt{4.67}}$

38 $\sqrt[5]{\dfrac{21.8}{62.2}}$

39 $(5.16)^{2.1}$

40 $(1.01)^{44}$

Systems of Equations and Inequalities

*In certain types of mathematical problems it is necessary to work simultaneously with more than one equation in several variables. We then refer to the given equations as a **system of equations**. It is usually desirable to find the solutions which are common to all equations in the system. In this chapter we shall investigate special types of systems of equations and develop methods for finding their common solutions. Of particular importance are the matrix techniques introduced for systems of linear equations. We shall also touch briefly on systems of inequalities and on linear programming.*

1 SYSTEMS OF EQUATIONS

If we are given an equation in two variables x and y, then as in Chapter Three, an ordered pair (a, b) is a **solution** of the equation if a true statement is obtained when a and b are substituted for x and y, respectively. Two equations in x and y are **equivalent** if they have exactly the same solutions. For example, the equation

$$x^2 - 4y - 2 = 3x + 2y + 6$$

is equivalent to

$$x^2 - 6y - 3x - 8 = 0.$$

As we know, the **graph** of an equation in x and y consists of all points in a coordinate plane which correspond to solutions.

By a **system** of two equations in x and y we mean any two equations in those variables. An ordered pair (a, b) is called a **solution of the system** if (a, b) is a solution of both equations. It follows that the points which correspond to the solutions are precisely the points at which the graphs of the two equations intersect.

As a concrete example, consider the system

$$x^2 - y = 0, \quad y - 2x - 3 = 0.$$

The following table exhibits some solutions of the equation $x^2 - y = 0$.

x	-2	-1	0	1	2	3	4
y	4	1	0	1	4	9	16

The graph is the parabola sketched in Figure 6.1. A similar table for $y - 2x - 3 = 0$ is

x	-2	-1	0	1	2	3	4
y	-1	1	3	5	7	9	11

The graph is the line in Figure 6.1. Note that the pairs $(3, 9)$ and $(-1, 1)$ are solutions of both equations and hence are solutions of the system. We shall see later that they are the *only* solutions. As pointed out before, the two pairs which are solutions of the system represent points of intersection of the two graphs.

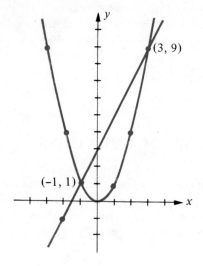

Figure 6.1. $\begin{cases} x^2 - y = 0 \\ y - 2x - 3 = 0 \end{cases}$

The graphical analysis used above is obviously of little value if the solutions of the system involve rational numbers such as 37/849, or irrational numbers, or if the equations are more complicated than those we have considered. Let us illustrate how the solutions of the above system can be found without reference to graphs. We shall begin by solving the first equation for y in terms of x, obtaining $y = x^2$. It follows that if an ordered pair (x, y) is a solution of the system, then it must be of the form (x, x^2). In particular, (x, x^2) must be a solution of the equation $y - 2x - 3 = 0$, that is,

$$x^2 - 2x - 3 = 0.$$

Factoring gives us

$$(x - 3)(x + 1) = 0$$

and hence either $x = 3$ or $x = -1$. The corresponding values for y (obtained from $y = x^2$) are 9 and 1, respectively. Thus the only possible solutions of the system are the ordered pairs $(3, 9)$ and $(-1, 1)$. That these are, indeed, solutions can be seen by checking each equation of the system.

To recapitulate, in order to find the solutions algebraically, we began by solving one equation of the system for y in terms of x, and then we substituted for y in the other equation, obtaining an equation in one variable, x. The solutions of the latter equation were the only possible x-values for the solutions of the system. The corresponding y-values were found by means of the equation which expressed y in terms of x. This algebraic technique is called the *method of substitution*. Sometimes it is more convenient to begin by solving one equation for x in terms of y, and then substituting for x in the other equation. In general the steps used in this process may be listed as follows.*

The Method of Substitution

(i) Solve one of the equations for one variable in terms of the other.

(ii) Substitute the expression obtained in step (i) in the other equation, obtaining an equation in one variable.

(iii) Find the solutions of the equation obtained in step (ii).

(iv) Use the solutions from step (iii) together with the expression obtained in step (i) to find the solutions of the system.

In the next example we find the solutions of the system considered at the beginning of this section by first solving one equation for x in terms of y.

Example 1 Find the solutions of the system

$$x^2 - y = 0, \quad y - 2x - 3 = 0.$$

Solution We may solve the second equation for x in terms of y as follows.

$$2x = y - 3, \quad \text{or} \quad x = \frac{y - 3}{2}$$

Substituting for x in the first equation gives us

$$\left(\frac{y - 3}{2}\right)^2 - y = 0$$

$$\frac{y^2 - 6y + 9}{4} - y = 0$$

$$y^2 - 6y + 9 - 4y = 0$$

$$y^2 - 10y + 9 = 0.$$

---

* For a proof that the method is valid in general, see E. W. Swokowski, *Functions and Graphs* (Boston: Prindle, Weber, & Schmidt, Inc., 1977) pp. 370–371.

Factoring we obtain

$$(y - 9)(y - 1) = 0$$

which has solutions $y = 9$ and $y = 1$. If we now substitute for y in the equation $x = (y - 3)/2$ we get the corresponding values $x = 3$ and $x = -1$. Hence, as before, the solutions of the system are $(3, 9)$ and $(-1, 1)$.

Example 2   Find the solutions of the system $x^2 + y^2 = 25$, $x^2 + y = 19$. Sketch the graph of each equation, showing the points of intersection.

Solution   Solving the second equation for y we obtain $y = 19 - x^2$. Substituting for y in the first equation leads to the following chain of equivalent equations:

$$x^2 + (19 - x^2)^2 = 25$$
$$x^4 - 37x^2 + 336 = 0$$
$$(x^2 - 16)(x^2 - 21) = 0.$$

The solutions of the last equation are $4, -4, \sqrt{21}$, and $-\sqrt{21}$. The corresponding y values are found by substituting for x in the equation $y = 19 - x^2$. Substitution of 4 or -4 for x gives us $y = 3$, whereas substitution of $\sqrt{21}$ or $-\sqrt{21}$ gives us $y = -2$. Hence the only possible solutions of the system are

$$(4, 3), (-4, 3), (\sqrt{21}, -2), \quad \text{and} \quad (-\sqrt{21}, -2).$$

It can be seen by direct substitution in each of the given equations that all four pairs are solutions. The graph of $x^2 + y^2 = 25$ is a circle of radius 5 with center at the origin, and the graph of $y = 19 - x^2$ is a parabola with a vertical axis. The graphs and their points of intersection are illustrated in Figure 6.2.

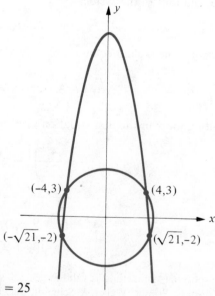

Figure 6.2. $\begin{cases} x^2 + y^2 = 25 \\ x^2 + y = 19 \end{cases}$

Example 3   Find the solutions of the system $y - \log(x + 3) = 1, 2 - y = \log x$.

Solution   From the second equation we obtain $y = 2 - \log x$. Substituting in the first equation and simplifying, we obtain

$$\log x + \log(x + 3) = 1.$$

This leads to the following chain of equivalent equations (supply the reasons):

$$\log x(x + 3) = 1$$
$$x(x + 3) = 10$$
$$x^2 + 3x - 10 = 0$$
$$(x + 5)(x - 2) = 0.$$

Hence the only possible solutions are $x = -5$ and $x = 2$. If we substitute -5 for x in the equation $y = 2 - \log x$, we arrive at the logarithm of a negative number. Since $\log x$ is defined only if $x > 0$, this value of x is extraneous. If we substitute 2 for x we get $y = 2 - \log 2$. Checking, we see that the system has one solution $(2, 2 - \log 2)$.

We can also consider equations in three variables x, y, and z, such as

$$x^2y + x \log z + 3^y = 4z^3.$$

We say that the equation has a **solution** (a, b, c) if substitution of a, b, and c for x, y, and z, respectively, produces a true statement. We refer to (a, b, c) as an **ordered triple** of real numbers. Equivalent equations are defined as before. A system of equations in three variables and the corresponding solutions are defined as in the two-variable case. In like manner, we can consider systems of *any* number of equations in *any* number of variables.

The method of substitution can be extended to these more complicated systems. For example, given three equations in three variables, suppose that it is possible to solve one of the equations for one variable in terms of the remaining two variables. By substituting that expression in each of the other equations, we obtain a system of two equations in two variables. The solutions of the latter system can then be used to find the solutions of the original system, as illustrated in the following example.

Example 4   Find the solutions of the system $x - y + z = 2$, $xyz = 0$, $2y + z = 1$.

Solution   Solving the third equation for z gives us

$$z = 1 - 2y.$$

Substituting for z in the first two equations of the system we obtain

$$\begin{cases} x - y + (1 - 2y) = 2 \\ \quad\quad xy(1 - 2y) = 0 \end{cases}$$

or equivalently,

$$\begin{cases} x - 3y - 1 = 0 \\ xy(1 - 2y) = 0. \end{cases}$$

We now find the solutions of the latter system. Solving the first equation for x in terms of y gives us

$$x = 3y + 1.$$

Substituting $3y + 1$ for x in the second equation $xy(1 - 2y) = 0$, we obtain

$$(3y + 1)y(1 - 2y) = 0$$

which has, for its solutions, the numbers $-\frac{1}{3}, 0, \frac{1}{2}$. These are the only possible y-values for the solutions of the system. To obtain the corresponding x-values, we use the equation $x = 3y + 1$, obtaining $x = 0, 1$, and $\frac{5}{2}$, respectively. Finally, returning to $z = 1 - 2y$, we obtain the z-values $\frac{5}{3}, 1$, and 0. It follows that the solutions of the original system consist of the ordered triples $(0, -\frac{1}{3}, \frac{5}{3})$, $(1, 0, 1)$, and $(\frac{5}{2}, \frac{1}{2}, 0)$.

EXERCISES

In Exercises 1–34 use the method of substitution to find the solutions of each system of equations.

1 $\begin{cases} y = x^2 - 4 \\ y = 2x - 1 \end{cases}$ 2 $\begin{cases} y = x^2 + 1 \\ x + y = 3 \end{cases}$

3 $\begin{cases} y^2 = 1 - x \\ x + 2y = 1 \end{cases}$ 4 $\begin{cases} y^2 = x \\ x + 2y + 3 = 0 \end{cases}$

5 $\begin{cases} 2y = x^2 \\ y = 4x^3 \end{cases}$ 6 $\begin{cases} x - y^3 = 1 \\ 2x = 9y^2 + 2 \end{cases}$

7 $\begin{cases} x + 2y = -1 \\ 2x - 3y = 12 \end{cases}$ 8 $\begin{cases} 3x - 4y + 20 = 0 \\ 3x + 2y + 8 = 0 \end{cases}$

9 $\begin{cases} 2x - 3y = 1 \\ -6x + 9y = 4 \end{cases}$ 10 $\begin{cases} 4x - 5y = 2 \\ 8x - 10y = -5 \end{cases}$

11 $\begin{cases} x + 3y = 5 \\ x^2 + y^2 = 25 \end{cases}$ 12 $\begin{cases} 3x - 4y = 25 \\ x^2 + y^2 = 25 \end{cases}$

13 $\begin{cases} x^2 + y^2 = 8 \\ y - x = 4 \end{cases}$ 14 $\begin{cases} x^2 + y^2 = 25 \\ 3x + 4y = -25 \end{cases}$

15 $\begin{cases} x^2 + y^2 = 9 \\ y - 3x = 2 \end{cases}$ 16 $\begin{cases} x^2 + y^2 = 16 \\ y + 2x = -1 \end{cases}$

17 $\begin{cases} x^2 + y^2 = 16 \\ 2y - x = 4 \end{cases}$ 18 $\begin{cases} x^2 + y^2 = 1 \\ y + 2x = -3 \end{cases}$

19 $\begin{cases} (x-1)^2 + (y+2)^2 = 10 \\ \qquad\qquad x + y = 1 \end{cases}$

20 $\begin{cases} \qquad\quad xy = 2 \\ 3x - y + 5 = 0 \end{cases}$

21 $\begin{cases} y = 20/x^2 \\ y = 9 - x^2 \end{cases}$

22 $\begin{cases} x = y^2 - 4y + 5 \\ x - y = 1 \end{cases}$

23 $\begin{cases} y^2 - 4x^2 = 4 \\ 9y^2 + 16x^2 = 140 \end{cases}$

24 $\begin{cases} 25y^2 - 16x^2 = 400 \\ 9y^2 - 4x^2 = 36 \end{cases}$

25 $\begin{cases} x^2 - y^2 = 4 \\ x^2 + y^2 = 12 \end{cases}$

26 $\begin{cases} 6x^3 - y^3 = 1 \\ 3x^3 + 4y^3 = 5 \end{cases}$

27 $\begin{cases} x + 2y - z = -1 \\ 2x - y + z = 9 \\ x + 3y + 3z = 6 \end{cases}$

28 $\begin{cases} 2x - 3y - z^2 = 0 \\ x - y - z^2 = -1 \\ x^2 - xy = 0 \end{cases}$

29 $\begin{cases} y = 5^x + 4 \\ y = 5^{2x} - 2 \end{cases}$

30 $\begin{cases} x + 3(2^y) = 2^{2y} \\ x - 2^{y+1} = 6 \end{cases}$

31 $\begin{cases} \log(3r + 1) = s + 5 \\ \log(r - 2) = s + 4 \end{cases}$

32 $\begin{cases} \log u = v - 3 \\ 5 = v + \log(u - 2) \end{cases}$

33 $\begin{cases} \log(3x + 5) - \log y = 1 \\ \qquad 5x - 2y = 6 \end{cases}$

34 $\begin{cases} y + 2 = \log(2x + 5) \\ y - 1 = \log x \end{cases}$

Solve Exercises 35–40 by introducing several variables and using a suitable system of equations.

35 Find two positive integers whose difference is 4 and whose squares differ by 88. (*Hint:* Let x and y denote the two integers.)

36 Find two real numbers whose difference and product both equal 4.

37 The perimeter of a rectangle is 40 inches and its area is 96 square inches. Find the length and width.

38 Generalize Exercise 37 to the case where the perimeter is P and area is A. Express the length and width in terms of P and A. What restrictions are necessary?

39 Find three numbers whose sum and product are 20 and 60 respectively, and such that one of the numbers equals the sum of the other two.

40 Find the values of b for which the system

$$\begin{cases} x^2 + y^2 = 4 \\ \qquad y = x + b \end{cases}$$

has solutions consisting of (a) one real number, (b) two real numbers, (c) no real numbers. Interpret the three cases geometrically.

2 SYSTEMS OF LINEAR EQUATIONS IN TWO VARIABLES

An equation of the form $ax + by + c = 0$ (or equivalently $ax + by = -c$) is called a linear equation in x and y. Similarly, a linear equation in three variables

x, y, and z is an equation of the form $ax + by + cz = d$, where the coefficients are real numbers. Linear equations in any number of variables may be defined in like manner. If a large number of variables occurs, a subscript notation is usually employed. If we let $x_1, x_2, \ldots, x_n$ denote variables, where n is any positive integer, then an expression of the form

(6.1)

$$a_1 x_1 + a_2 x_2 + \cdots + a_n x_n = a$$

where $a_1, a_2, \ldots, a_n$ and a are real numbers, is called a **linear equation in n variables** with real coefficients.

In modern applications of mathematics, perhaps the most common systems of equations are those in which all the equations are linear. In this section we shall only consider systems of two linear equations in two variables. Systems involving more than two variables are discussed in Section 3.

The method used to solve a system of equations in several variables is similar to that used for one equation in one variable, in the sense that the given system is replaced by a chain of equivalent systems until a stage is reached from which the solutions are easily obtained. Some general rules which allow us to transform a system of equations into an equivalent system are given in (6.2). Since we prefer not to specify the number or types of variables in (iii) of (6.2), we shall use the notation $p = 0$ and $q = 0$ for typical equations in a system. The reader should bear in mind that the symbols p and q represent expressions in the variables under consideration. For example, if $p = 3x - 2y + 5z - 1$ and $q = 6x + y - 7z - 4$, then the equations $p = 0$ and $q = 0$ have the form

$$3x - 2y + 5z - 1 = 0, \quad 6x + y - 7z - 4 = 0.$$

(6.2)   Transformations that Lead to Equivalent Systems

> The following transformations do not change the solutions of a system of equations.
> (i)   Interchanging the position of any two equations.
> (ii)   Multiplying both sides of an equation in the system by a nonzero real number.
> (iii)   Replacing an equation $q = 0$ of the system by $kp + q = 0$, where $p = 0$ is any other equation in the system and k is any real number.

Proof.   It is easy to show that (i) and (ii) do not change the solutions of the system and therefore we shall omit the proofs.

To prove (iii), we first note that a solution of the original system is a solution of both equations $p = 0$ and $q = 0$. Accordingly, each of the expressions p and q equals zero if the variables are replaced by appropriate real numbers, and hence the expression $kp + q$ will also equal zero. Since none of the other equations has been changed, this shows that any solution of the original system is also a solution of the transformed system obtained by replacing the equation $q = 0$ by $kp + q = 0$.

Conversely, given a solution of the transformed system, both of the expressions $kp + q$ and p equal zero if the variables are replaced by appropriate numbers. This implies, however, that $(kp + q) - kp$ equals 0. Since

$(kp + q) - kp = q$, we see that q must also equal zero when the substitution is made. Thus a solution of the transformed system is also a solution of the original system. This completes the proof.

For convenience we shall describe the process of using (iii) of (6.2) by the phrase "add to one equation of the system k times any other equation of the system." Of course, to *add* two equations means to add corresponding sides. To *multiply* an equation by k means to multiply both sides of the equation by k. The process may also be applied if 0 is not on one side of each equation, as illustrated in the next example.

Example 1   Find the solutions of the system

$$\begin{cases} x + 3y = -1 \\ 2x - y = 5. \end{cases}$$

Solution   By (ii) of (6.2) we may multiply the second equation by 3. This gives us the equivalent system

$$\begin{cases} x + 3y = -1 \\ 6x - 3y = 15. \end{cases}$$

Next, by (iii) of (6.2) we may add to the second equation 1 times the first. This gives us

$$\begin{cases} x + 3y = -1 \\ 7x = 14. \end{cases}$$

We see from the last equation that the only possible value for x is 2. The corresponding value for y may be found by substituting for x in the first equation. This gives us the following:

$$2 + 3y = -1, \quad 3y = -3, \quad y = -1.$$

Consequently, the given system has one solution, $(2, -1)$.

There are other methods of solution. For example, we could begin by multiplying the first equation by -2, obtaining

$$\begin{cases} -2x - 6y = 2 \\ 2x - y = 5. \end{cases}$$

If we next add the first equation to the second, we get

$$\begin{cases} -2x - 6y = 2 \\ - 7y = 7. \end{cases}$$

The last equation implies that $y = -1$. Substitution for y in the first equation gives us

$$-2x - 6(-1) = 2, \quad -2x = -4, \quad x = 2.$$

Again we see that the solution is $(2, -1)$.

The graphs of the two equations, showing the point of intersection $(2, -1)$, are sketched in Figure 6.3.

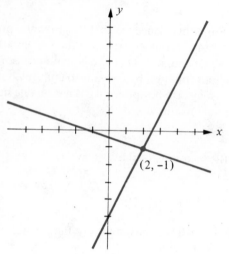

Figure 6.3. $\begin{cases} x + 3y = -1 \\ 2x - y = 5. \end{cases}$

The technique used in Example 1 is called the **method of elimination**, since it involves eliminating a variable from one of the equations by means of (6.2). The method of elimination usually leads to solutions in fewer steps than the method of substitution discussed in Section 1.

Example 2   Find the solutions of the system

$$\begin{cases} 3x + y = 6 \\ 6x + 2y = 12. \end{cases}$$

Solution   Multiplying the second equation by $\frac{1}{2}$ gives us

$$\begin{cases} 3x + y = 6 \\ 3x + y = 6. \end{cases}$$

Consequently (a, b) is a solution if and only if $3a + b = 6$, that is, $b = 6 - 3a$. It follows that the solutions consist of all ordered pairs of the form $(a, 6 - 3a)$, where a is a real number. If we wish to find particular solutions we may substitute various values for a. For example, a few solutions are $(0, 6), (1, 3), (3, -3), (-2, 12)$, $(\sqrt{2}, 6 - 3\sqrt{2})$.

Example 3   Find the solutions of the system

$$\begin{cases} 3x + y = 6 \\ 6x + 2y = 20. \end{cases}$$

Solution   If we add to the second equation -2 times the first equation we obtain the equivalent system

$$\begin{cases} 3x + y = 6 \\ 0 = 8. \end{cases}$$

Since the last equation is never true, the system has no solutions. Of course, this means that the graphs of the given equations do not intersect (see Figure 6.4).

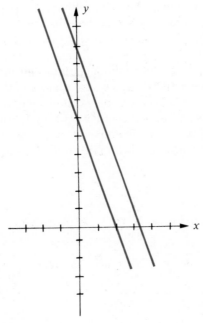

Figure 6.4. $\begin{cases} 3x + y = 6 \\ 6x + 2y = 20 \end{cases}$

Every system of two linear equations in two variables x and y with real coefficients has the form

(6.3)
$$\begin{cases} a_1x + a_2y = a \\ b_1x + b_2y = b \end{cases}$$

where the coefficients are real numbers. We know from (3.17) that the graph of each equation in the system is a straight line. It follows that precisely *one* of the following three possibilities must occur:

(i)   The lines are identical.
(ii)   The lines are parallel.
(iii)   The lines intersect in one and only one point.

Since the pairs which are solutions of the system (6.3) exhibit the coordinates of the points of intersection of the graphs of the two equations, we may interpret statements (i)–(iii) in the following way.

(6.4)

For each system of two linear equations in two variables, one and only one of the following statements is true:

(i)   The system has an infinite number of solutions.
(ii)   The system has no solutions.
(iii) The system has exactly one solution.

If (i) occurs, then every solution of one equation in (6.3) is also a solution of the other, and we say that the equations are **dependent**.

If (ii) occurs, there is no solution and the system is said to be **inconsistent**.

If (iii) occurs, there is a unique solution and the system is called **consistent**.

In practice, there should be little difficulty in determining which of the three cases occurs. In case (i) we have a solution similar to that for Example 2, where one of the equations can be transformed into the other by using (6.2). In case (ii) the lack of a solution is indicated by an absurdity such as the statement $0 = 8$, which appeared in Example 3. The case of the unique solution (iii) will be apparent when (6.2) is applied to the system, as illustrated in Example 1.

EXERCISES

Find the solutions of the systems in Exercises 1–20.

1 $\begin{cases} 2x + 3y = 2 \\ x - 2y = 8 \end{cases}$

2 $\begin{cases} 4x + 5y = 13 \\ 3x + y = -4 \end{cases}$

3 $\begin{cases} 2x + 5y = 16 \\ 3x - 7y = 24 \end{cases}$

4 $\begin{cases} 7x - 8y = 9 \\ 4x + 3y = -10 \end{cases}$

5 $\begin{cases} 3r + 4s = 3 \\ r - 2s = -4 \end{cases}$

6 $\begin{cases} 9u + 2v = 0 \\ 3u - 5v = 17 \end{cases}$

7 $\begin{cases} 5x - 6y = 4 \\ 3x + 7y = 8 \end{cases}$

8 $\begin{cases} 2x + 8y = 7 \\ 3x - 5y = 4 \end{cases}$

9 $\begin{cases} \frac{1}{3}c + \frac{1}{2}d = 5 \\ c - \frac{2}{3}d = -1 \end{cases}$

10 $\begin{cases} \frac{1}{2}t - \frac{1}{3}v = \frac{3}{2} \\ \frac{2}{3}t + \frac{1}{4}v = \frac{5}{12} \end{cases}$

11 $\begin{cases} \sqrt{3}x - \sqrt{2}y = 2\sqrt{3} \\ 2\sqrt{2}x + \sqrt{3}y = \sqrt{2} \end{cases}$

12 $\begin{cases} 0.11x - 0.03y = 0.25 \\ 0.12x + 0.05y = 0.70 \end{cases}$

13 $\begin{cases} 2x - 3y = 5 \\ -6x + 9y = 12 \end{cases}$

14 $\begin{cases} 3p - q = 7 \\ -12p + 4q = 3 \end{cases}$

15 $\begin{cases} 3m - 4n = 2 \\ -6m + 8n = -4 \end{cases}$

16 $\begin{cases} x - 5y = 2 \\ 3x - 15y = 6 \end{cases}$

17 $\begin{cases} 2y - 5x = 0 \\ 3y + 4x = 0 \end{cases}$

18 $\begin{cases} 3x + 7y = 9 \\ y = 5 \end{cases}$

19
$$\begin{cases} \dfrac{2}{x} + \dfrac{3}{y} = -2 \\ \dfrac{4}{x} - \dfrac{5}{y} = 1 \end{cases}$$
(*Hint:* Let $x' = 1/x$ and $y' = 1/y$.)

20
$$\begin{cases} \dfrac{3}{x-1} + \dfrac{4}{y+2} = 2 \\ \dfrac{6}{x-1} - \dfrac{7}{y+2} = -3 \end{cases}$$

21 Determine a and b so that the straight line with equation $ax + by = 10$ passes through the points $P_1(-3, 4)$ and $P_2(3, 1)$.

22 Generalize Exercise 21 to the case where the graph of $ax + by = 10$ passes through the points $P_1(x_1, y_1)$ and $P_2(x_2, y_2)$. (Express a and b in terms of x_1, y_1, x_2, and y_2.) Is it necessary to restrict P_1 or P_2 in any way?

Solve each of Exercises 23–30 by employing a system of linear equations.

23 The price of admission for a certain event was \$2.25 for adults and \$1.50 for children. If 450 tickets were sold for a total of \$777.75, how many of each kind were purchased?

24 A chemist wishes to obtain 80 cc of a 30 % acid solution by mixing a 50 % solution and a 25 % solution. How many cc of each solution should be used?

25 The sum of the digits of a certain two-digit number is 14. If the digits are reversed, the number is increased by 18. Find the number.

26 A man rows a boat 500 feet upstream against a constant current in 10 minutes. He then rows downstream (with the same current), covering 300 feet in 5 minutes. Find the speed of the current and the equivalent rate at which he can row in still water.

27 A collection of dimes and quarters amounts to \$7.00. If there are seven more dimes than quarters, find the number of quarters.

28 A man has \$10,000 invested in two funds which pay simple interest rates of 8 % and 7 %, respectively. If he receives yearly interest of \$772, how much is invested in each fund?

29 A man receives income from two investments at simple interest rates of $6\frac{1}{2}$ % and 8 %, respectively. He has twice as much invested at $6\frac{1}{2}$ % as at 8 %. If his annual income from the two investments is \$698.25, find how much is invested at each rate.

30 Two water pipes running at the same time can fill a swimming pool in 4 hours. If both pipes run for 2 hours and the first is then shut off, it takes 3 more hours for the second to fill the pool. How long does it take each pipe to fill the pool by itself?

3 SYSTEMS OF LINEAR EQUATIONS IN MORE THAN TWO VARIABLES

For systems of equations containing more than two variables we can use either the method of substitution explained in Section 1 or the techniques developed in Section 2. Let us illustrate this by solving a system in two different ways, first by the method of substitution and second by the method of elimination. You will see that the method of elimination is the shorter and more straightforward technique for finding solutions.

Example 1 Find the solutions of the system

$$\begin{cases} x - 2y + 3z = 4 \\ 2x + y - 4z = 3 \\ -3x + 4y - z = -2. \end{cases}$$

Solution 1 Using the method of substitution we solve the first equation for x, obtaining

$$x = 4 + 2y - 3z.$$

Substituting for x in the second and third equations gives us

$$\begin{cases} 2(4 + 2y - 3z) + y - 4z = 3 \\ -3(4 + 2y - 3z) + 4y - z = -2 \end{cases}$$

which simplifies to

$$\begin{cases} y - 2z = -1 \\ y - 4z = -5. \end{cases}$$

If we solve the first equation for y, obtaining $y = 2z - 1$, and then substitute for y in the second equation, we find that

$$(2z - 1) - 4z = -5 \quad \text{or} \quad -2z = -4.$$

The last equation has the solution $z = 2$. To find y we return to $y = 2z - 1$, obtaining $y = 2(2) - 1 = 3$. The corresponding value of x is found by substituting for y and z in the equation $x = 4 + 2y - 3z$. This gives us $x = 4 + 2(3) - 3(2) = 4$. Hence the given system has only one solution, the ordered triple (4, 3, 2). We shall leave it to the reader to show that this triple is a solution of *each* of the given equations. Of course, other approaches to the problem could be used, such as solving the third equation for z and substituting in the first and second equations, obtaining two equations in x and y, and so on.

Solution 2 We begin by eliminating x from the second and third equations. Adding to the second equation -2 times the first equation gives us the equivalent system

$$\begin{cases} x - 2y + 3z = 4 \\ 5y - 10z = -5 \\ -3x + 4y - z = -2. \end{cases}$$

Next, adding to the third equation 3 times the first equation, we obtain the equivalent system

$$\begin{cases} x - 2y + 3z = 4 \\ 5y - 10z = -5 \\ -2y + 8z = 10. \end{cases}$$

To simplify computations we multiply the second equation by $\frac{1}{5}$, obtaining

$$\begin{cases} x - 2y + 3z = 4 \\ y - 2z = -1 \\ -2y + 8z = 10. \end{cases}$$

We now eliminate y from the third equation by adding to it 2 times the second equation. This gives us the system

$$\begin{cases} x - 2y + 3z = 4 \\ y - 2z = -1 \\ 4z = 8. \end{cases}$$

The solutions of the last system are easy to obtain. From the third equation it is clear that the only possible z-value is 2. Substituting 2 for z in the second equation, $y - 2z = -1$, we get $y - 2(2) = -1$, or $y = 3$. Finally, the x-value is found by substituting for y and z in the first equation. This gives us $x - 2(3) + 3(2) = 4$ and hence $x = 4$. Thus there is one solution, $(4, 3, 2)$.

The general form for a system of three linear equations in three variables with real coefficients is

(6.5)
$$\begin{cases} a_1x + a_2y + a_3z = a \\ b_1x + b_2y + b_3z = b \\ c_1x + c_2y + c_3z = c \end{cases}$$

where the coefficients are real numbers. It can be shown that the system (6.5) has either a *unique solution*, an *infinite number of solutions*, or *no solutions*. As with the case of two equations in two variables, the terminology used to describe these cases is *consistent*, *dependent*, or *inconsistent*, respectively.

Similar remarks and methods of solution hold for systems of four linear equations in four variables or, for that matter, systems of n linear equations in n variables, where n is any positive integer. Using the method of substitution we can always solve one of the equations for one variable in terms of the remaining variables. By substituting in each of the remaining equations we obtain a system containing one less equation in one less variable. We keep repeating this process until we obtain a form from which the solution may be easily found. Of course, the method of elimination can also be used.

Sometimes it is necessary to work with systems in which the number of equations is not the same as the number of variables. Such systems can be attacked by using methods similar to those we have discussed.

Example 2   Find the solutions of the system

$$\begin{cases} 2x + 3y + 4z = 1 \\ 3x + 4y + 5z = 3. \end{cases}$$

Solution   We can eliminate x from the second equation by adding to it $-3/2$ times the first equation, obtaining the equivalent system

$$\begin{cases} 2x + 3y + 4z = 1 \\ \quad\;\; -\tfrac{1}{2}y - z = \tfrac{3}{2}. \end{cases}$$

To clear of fractions we multiply the second equation by -2, obtaining

$$\begin{cases} 2x + 3y + 4z = 1 \\ \quad\quad\;\; y + 2z = -3. \end{cases}$$

There are an infinite number of solutions (b, c) for the second equation. Indeed, if we write $y = -3 - 2z$, then by substituting any number c for z, the corresponding value b of y is given by $b = -3 - 2c$. For each such solution (b, c) of $y + 2z = -3$ there corresponds a number a such that the triple (a, b, c) is a solution of the first equation, that is,

$$2a + 3b + 4c = 1.$$

Solving the latter equation for a we obtain

$$a = \tfrac{1}{2}(1 - 3b - 4c) = \tfrac{1}{2} - \tfrac{3}{2}b - 2c.$$

Since $b = -3 - 2c$, we have

$$a = \tfrac{1}{2} - \tfrac{3}{2}(-3 - 2c) - 2c$$

which reduces to

$$a = 5 + c.$$

Consequently the solutions of the system consist of all ordered triples of the form

$$(5 + c, \; -3 - 2c, \; c)$$

where c is any real number. These solutions may be checked by substituting $5 + c$ for x, $-3 - 2c$ for y, and c for z in the two given equations. We can obtain any number of solutions for the system by substituting specific real numbers for c. For example, if $c = 0$, we obtain $(5, -3, 0)$; if $c = 2$, we have $(7, -7, 2)$: if $c = -1/2$, we get $(9/2, -2, -1/2)$; and so on.

Another method of solution for Example 2 is to rewrite the original system as

$$\begin{cases} 2x + 3y = 1 - 4z \\ 3x + 4y = 3 - 5z. \end{cases}$$

If a number c is substituted for z, we obtain a system of two equations in two variables x and y, and the solutions can be found as in Section 2. For a different approach we could solve for two *other* variables in terms of the third. This method can be extended. For example, given two equations in four variables, say x, y, z, and w, it may be possible to solve for two of the variables, say x and y, in terms of the remaining variables z and w. If we then substitute numbers c and d for z and w, respectively, and calculate the corresponding values for x and y, we obtain the solutions.

The system of linear equations (6.5) is called **homogeneous** if $a = b = c = 0$. The same terminology is used for more than three equations. A system of homogeneous equations always has the **trivial solution** obtained by substituting zero for all the variables. Nontrivial solutions sometimes exist. The procedure for finding solutions is the same as that used for nonhomogeneous systems.

Example 3   Find the solutions of the system

$$\begin{cases} x - y + 4z = 0 \\ 2x + y - z = 0 \\ -x - y + 2z = 0. \end{cases}$$

Solution   Adding to the second equation -2 times the first equation and then adding the first equation to the third gives us the equivalent system

$$\begin{cases} x - y + 4z = 0 \\ 3y - 9z = 0 \\ -2y + 6z = 0. \end{cases}$$

Multiplying the second equation by $\frac{1}{3}$ and the third equation by $-\frac{1}{2}$ leads to the same equation, $y - 3z = 0$. Hence the given system is equivalent to

$$\begin{cases} x - y + 4z = 0 \\ y - 3z = 0. \end{cases}$$

The solutions may now be found as in the previous example. If we substitute any value c for z, then the second equation gives us $y = 3c$. Substituting c for z and $3c$ for y in the first equation, we obtain $x - 3c + 4c = 0$, or $x = -c$. Thus the solutions consist of all ordered triples of the form $(-c, 3c, c)$, where c is any real number.

Example 4   Find the solutions of the system

$$\begin{cases} x + y + z = 0 \\ x - y + z = 0 \\ x - y - z = 0. \end{cases}$$

Solution   By the usual process we obtain the equivalent system

$$\begin{cases} x + y + \ z = 0 \\ \quad\ -2y \qquad = 0 \\ \quad\ -2y - 2z = 0. \end{cases}$$

Multiplying the second and third equations by $-\frac{1}{2}$ gives us

$$\begin{cases} x + y + z = 0 \\ \qquad y \quad\ = 0 \\ \qquad y + z = 0. \end{cases}$$

From the second equation of this system we see that the only possible y-value is 0. Substituting 0 for y in the last equation we obtain $0 + z = 0$, or $z = 0$. Finally, substituting 0 for both y and z in the first equation we see that x also must equal 0. Hence the only solution for the system is the trivial one, $(0, 0, 0)$.

The next example is an illustration of an applied problem which can be solved by means of a system of three linear equations.

Example 5   A merchant wishes to mix two grades of peanuts costing \$1.50 and \$2.50 per pound, respectively, with cashews costing \$4.00 per pound, to obtain 130 pounds of a mixture costing \$3.00 per pound. If the merchant also wants the amount of cheaper grade peanuts to be twice that of the better grade, how many pounds of each variety should be mixed?

Solution   Let us introduce three variables as follows:

$$x = \text{pounds of peanuts at \$1.50 per pound}$$
$$y = \text{pounds of peanuts at \$2.50 per pound}$$
$$z = \text{pounds of cashews at \$4.00 per pound.}$$

We refer to the statement of the problem and write the following system of equations

$$\begin{cases} x + y + z = 130 \\ 1.50x + 2.50y + 4.00z = 3.00(130) \\ \qquad\qquad\qquad\ x = 2y. \end{cases}$$

We shall use a combination of the methods of substitution and elimination to find the solution. Let us begin by substituting $2y$ for x in the first two equations, obtaining

$$\begin{cases} 2y + y + \ z = 130 \\ 1.5(2y) + 2.5y + 4z = 3(130) \end{cases}$$

or equivalently

$$\begin{cases} 3y + \ z = 130 \\ 5.5y + 4z = 390. \end{cases}$$

We next eliminate z from the second equation by adding to it -4 times the first equation. This gives us

$$\begin{cases} 3y + z = \quad 130 \\ -6.5y = -130. \end{cases}$$

It follows that

$$y = \frac{-130}{-6.5} = 20$$

$$x = 2y = 2(20) = 40$$

$$z = 130 - x - y = 130 - 40 - 20 = 70.$$

Thus the merchant should use 40 pounds of the $1.50 peanuts, 20 pounds of the $2.50 peanuts, and 70 pounds of cashews. The check is left to the reader.

EXERCISES

Solve the systems in Exercises 1–26.

1
$$\begin{cases} x - 2y - 3z = -1 \\ 2x + \ y + \ z = \quad 6 \\ x + 3y - 2z = \ 13 \end{cases}$$

2
$$\begin{cases} x + 3y - \ z = -3 \\ 3x - \ y + 2z = \quad 1 \\ 2x - \ y + \ z = -1 \end{cases}$$

3
$$\begin{cases} 5x + 2y - \ z = -7 \\ x - 2y + 2z = \quad 0 \\ 3y + \ z = \ 17 \end{cases}$$

4
$$\begin{cases} 4x - \ y + 3z = \quad 6 \\ -8x + 3y - 5z = -6 \\ 5x - 4y = -9 \end{cases}$$

5
$$\begin{cases} 2x + 6y - 4z = \quad 1 \\ x + 3y - 2z = \quad 4 \\ 2x + \ y - 3z = -7 \end{cases}$$

6
$$\begin{cases} 2x - \ y = \quad 5 \\ 5y + 3z = -2 \\ x - 7z = \quad 3 \end{cases}$$

7
$$\begin{cases} 2x - 3y + 2z = -3 \\ -3x + 2y + \ z = \quad 1 \\ 4x + \ y - 3z = \quad 4 \end{cases}$$

8
$$\begin{cases} 2x - 3y + z = \quad 2 \\ 3x + 2y - z = -5 \\ 5x - 2y + z = \quad 0 \end{cases}$$

9
$$\begin{cases} x + 3y + \ z = 0 \\ x + \ y - \ z = 0 \\ x - 2y - 4z = 0 \end{cases}$$

10
$$\begin{cases} 2x - \ y + \ z = 0 \\ x - \ y - 2z = 0 \\ 2x - 3y + \ z = 0 \end{cases}$$

11
$$\begin{cases} 2x + \ y + \ z = 0 \\ x - 2y - 2z = 0 \\ x + \ y + \ z = 0 \end{cases}$$

12
$$\begin{cases} x + y - 2z = 0 \\ x - y - 4z = 0 \\ z + \ y = 0 \end{cases}$$

13 $\begin{cases} 3x - 2y + 5z = 7 \\ x + 4y - z = -2 \end{cases}$

14 $\begin{cases} 2x - y + 4z = 8 \\ -3x + y - 2z = 5 \end{cases}$

15 $\begin{cases} 4x - 2y + z = 5 \\ 3x + y - 4z = 0 \end{cases}$

16 $\begin{cases} 5x + 2y - z = 10 \\ y + z = -3 \end{cases}$

17 $\begin{cases} x + 2y - z - 3w = 2 \\ 3x + y - 2z - w = 6 \\ x + y + 3z - 2w = -3 \\ 4x - 3y - z - 2w = -8 \end{cases}$

18 $\begin{cases} x - 2y - 5z + w = -1 \\ 2x - y + z + w = 1 \\ 3x - 2y - 4z - 2w = 1 \\ x + y + 3z - 2w = 2 \end{cases}$

19 $\begin{cases} 5x + 2z = 1 \\ y - 3z = 2 \\ 2x + y = 3 \end{cases}$

20 $\begin{cases} 2x - 5y = 4 \\ 3y + 2z = -3 \\ 7x - 3z = 1 \end{cases}$

21 $\begin{cases} 4x - 3y = 1 \\ 2x + y = -7 \\ -x + y = -1 \end{cases}$

22 $\begin{cases} 2x + 3y = -2 \\ x + y = 1 \\ x - 2y = 13 \end{cases}$

23 $\begin{cases} 2x + 3y = 5 \\ x - 3y = 4 \\ x + y = -2 \end{cases}$

24 $\begin{cases} 4x - y = 2 \\ 2x + 2y = 1 \\ 4x - 5y = 3 \end{cases}$

25 $\begin{cases} \dfrac{3}{x} - \dfrac{3}{y} + \dfrac{1}{z} = -1 \\ \dfrac{1}{x} + \dfrac{4}{y} + \dfrac{1}{z} = 5 \\ \dfrac{2}{x} + \dfrac{1}{y} - \dfrac{4}{z} = 0 \end{cases}$

26 $\begin{cases} \dfrac{1}{x} + \dfrac{1}{y} - \dfrac{1}{z} = 4 \\ \dfrac{2}{x} - \dfrac{1}{y} + \dfrac{1}{z} = 2 \\ \dfrac{4}{x} + \dfrac{2}{y} + \dfrac{3}{z} = -3 \end{cases}$

27 A collection of nickels, dimes, and quarters amounts to $8.00. If there are 72 coins in all and there are twice as many nickels as there are quarters, find the number of dimes.

28 A three-digit number is equal to 31 times the sum of its digits. If the three digits are reversed, the resulting number exceeds the given number by 99. The tens digit is three less than the sum of the hundreds and units digits. Find the number.

29 A chemist has three solutions containing a certain acid. The first contains 10% acid, the second 30%, and the third 50%. He wishes to use all three solutions to obtain a mixture of 50 liters containing 32% acid, using twice as much of the 50% solution as the 30% solution. How many liters of each solution should be used?

30 A swimming pool can be filled by three pipes A, B, and C. Pipe A can fill the pool by itself in 8 hours. If pipes A and C are used together, the pool can be filled in 6 hours. If B and C are used together, it takes 10 hours. How long does it take to fill the pool if all three pipes are used?

31 Find an equation of the circle which passes through the three points $P_1(2, 1)$, $P_2(-1, -4)$ and $P_3(3, 0)$. (*Hint:* An equation of the circle has the form $x^2 + y^2 + ax + by + c = 0$.)

32 Determine a, b, and c such that the graph of the equation

$$y = ax^2 + bx + c$$

passes through the points $P_1(3, -1)$, $P_2(1, -7)$, and $P_3(-2, 14)$.

4 MATRIX SOLUTIONS OF SYSTEMS OF EQUATIONS

In Example 1 of Section 3 we used two different methods to solve the system

$$\begin{cases} x - 2y + 3z = 4 \\ 2x + y - 4z = 3 \\ -3x + 4y - z = -2. \end{cases}$$

If we analyze the second method (the method of elimination), we see that the symbols used for the variables really have little effect on the solution of the problem. The *coefficients* of the variables are the important things to consider. Thus if different symbols such as r, s, and t are used for the variables in that example, we obtain the system

(6.6)
$$\begin{cases} r - 2s + 3t = 4 \\ 2r + s - 4t = 3 \\ -3r + 4s - t = -2. \end{cases}$$

The method of elimination could then proceed exactly as before. Since this is true, it is possible to use a process which reduces the amount of labor involved. Specifically, we introduce a notation for keeping track of the coefficients in such a way that the variables do not have to be written down. Referring to (6.6) and checking that corresponding variables appear underneath one another and terms not involving variables are to the right of the equals sign, we list the numbers which are involved in the equations in the following manner:

(6.7)
$$\begin{bmatrix} 1 & -2 & 3 & 4 \\ 2 & 1 & -4 & 3 \\ -3 & 4 & -1 & -2 \end{bmatrix}$$

If some variable had not appeared in one of the equations, we would have used a zero in the appropriate position. An array of numbers such as (6.7) is called a **matrix**. The **rows** of (6.7) are the numbers which appear next to one another *horizontally*. Thus the first row is 1 −2 3 4, the second row is 2 1 −4 3, and the third row is − 3 4 −1 − 2. The **columns** of (6.7) are the sets of numbers which form a *vertical* pattern. For example, the second column consists of the numbers − 2, 1, 4 (in that order); the fourth column consists of 4, 3, − 2; and so on.

Before discussing the method of solving (6.6) by means of the matrix (6.7), let us introduce a general definition of matrix. It will be convenient to use a **double subscript notation** for the symbols which represent the numbers in the matrix. Specifically, the symbol a_{ij} will denote the number which appears in row i and column j. We call i the **row subscript** and j the **column subscript** of a_{ij}.

(6.8)   Definition

> If m and n are positive integers, then an $m \times n$ **matrix** over the real number system is an array of the form
>
> $$\begin{bmatrix} a_{11} & a_{12} & a_{13} & \cdots & a_{1n} \\ a_{21} & a_{22} & a_{23} & \cdots & a_{2n} \\ a_{31} & a_{32} & a_{33} & \cdots & a_{3n} \\ \vdots & \vdots & \vdots & & \vdots \\ a_{m1} & a_{m2} & a_{m3} & \cdots & a_{mn} \end{bmatrix}$$
>
> where each a_{ij} is a real number.

The symbol $m \times n$ in Definition (6.8) is read "m by n." The matrix in (6.7) is a 3×4 matrix over the real numbers. It is also possible to consider matrices where the elements a_{ij} belong to other mathematical systems; however, we shall not do so in this book.

The rows and columns of a matrix are defined as before. Thus the matrix in (6.8) has m rows and n columns. The reader should carefully examine the notation for a matrix, observing, for example, that the symbol a_{23} appears in row 2 and column 3, whereas a_{32} appears in row 3 and column 2. Each a_{ij} is called an **element of the matrix**. The elements $a_{11}, a_{22}, a_{33}, \ldots$ are called the **main diagonal elements**. If $m \neq n$, the matrix is sometimes referred to as a **rectangular matrix**. If $m = n$, we refer to the matrix as a **square matrix of order n**.

It is possible to develop a comprehensive theory for matrices. In succeeding sections we shall define *equality* of matrices, the *sum* of two matrices, the *product* of two matrices, and the *inverse* of a matrix. These ideas are very important and useful in mathematics and applications. In particular, matrix techniques are well suited for the types of problems which are solved by computers. Our main purpose in introducing matrices at this time is for use as a tool in solving systems of linear equations.

Let us return to the system (6.6). The matrix (6.7) is called the **matrix of the system**. In certain cases we may wish to consider only the *coefficients* of the variables of a system of linear equations. The corresponding matrix is called the **coefficient matrix**. The coefficient matrix for the system (6.6) is

$$\begin{bmatrix} 1 & -2 & 3 \\ 2 & 1 & -4 \\ -3 & 4 & -1 \end{bmatrix}.$$

In order to distinguish the matrix of the system from the coefficient matrix, the former is sometimes called the **augmented matrix**. After forming the matrix of the system, we work with the rows of the matrix *just as though they were equations*. The only items missing are the symbols for the variables, the addition signs used between terms, and the equals signs. We simply keep in mind that the numbers in the first column are the coefficients of the first variable, the numbers in the second column are the coefficients of the second variable, and so on. Our rules for transforming a matrix are formulated in such a way that they always produce a matrix of an equivalent system of equations. The key to these rules is given by (6.2). As a matter of fact, we have the following theorem for matrices, which is a perfect analogue of Theorem (6.2) for equations.

(6.9) **Matrix Row Transformation Theorem**

> Given a matrix of a system of linear equations, each of the following transformations results in a matrix of an equivalent system of linear equations:
>
> (i)   Interchanging any two rows.
> (ii)   Multiplying all of the elements in a row by the same nonzero real number k.
> (iii) Adding to the elements in a row k times the corresponding elements of any other row, where k is any real number.

For convenience we shall refer to (ii) as the process of "multiplying a row by the number k" and (iii) will be described by saying "add k times any other row to a row." Rules (i)–(iii) of (6.9) are called the **elementary row transformations** of a matrix.

Since the rules given in (6.2) produce equivalent systems of equations, it is evident that each of the rules in (6.9) results in the matrix of an equivalent system. Let us find the solution of the system (6.6) by using (6.9). The reader should compare our work with the second solution of Example 1 of Section 3. We shall use arrows to indicate that the form of the matrices has been changed by means of elementary row transformations, and we shall justify each transformation by stating the reason for each step. Thus

$$\begin{bmatrix} 1 & -2 & 3 & 4 \\ 2 & 1 & -4 & 3 \\ -3 & 4 & -1 & -2 \end{bmatrix} \rightarrow \begin{bmatrix} 1 & -2 & 3 & 4 \\ 0 & 5 & -10 & -5 \\ -3 & 4 & -1 & -2 \end{bmatrix} \quad \text{Add to the second row } -2 \text{ times the first row.}$$

$$\rightarrow \begin{bmatrix} 1 & -2 & 3 & 4 \\ 0 & 5 & -10 & -5 \\ 0 & -2 & 8 & 10 \end{bmatrix} \quad \text{Add to the third row 3 times the first row.}$$

$$\rightarrow \begin{bmatrix} 1 & -2 & 3 & 4 \\ 0 & 1 & -2 & -1 \\ 0 & -2 & 8 & 10 \end{bmatrix} \quad \text{Multiply row 2 by } \tfrac{1}{5}.$$

$$\rightarrow \begin{bmatrix} 1 & -2 & 3 & 4 \\ 0 & 1 & -2 & -1 \\ 0 & 0 & 4 & 8 \end{bmatrix} \quad \text{Add to the third row 2 times the second row.}$$

The last matrix has zeros everywhere below the main diagonal and is said to be in **echelon form**. We now use that matrix to return to the system of equations

$$\begin{cases} r - 2s + 3t = 4 \\ s - 2t = -1 \\ 4t = 8 \end{cases}$$

which is equivalent to the given system. The solutions may now be found as was done in the previous section.

The method used above is completely general. Beginning with *any* system of linear equations, we can use (iii) of (6.9) many times and also interchange rows if necessary to introduce zeros everywhere in the first column after the first row; that is, we can obtain a matrix of an equivalent system which has the form

$$\begin{bmatrix} a_{11} & a_{12} & a_{13} & \cdots \\ 0 & b_{22} & b_{23} & \cdots \\ 0 & c_{32} & c_{33} & \cdots \\ 0 & \cdot & \cdot & \cdots \\ \cdot & \cdot & \cdot & \cdots \\ \cdot & \cdot & \cdot & \cdots \\ 0 & \cdot & \cdot & \cdots \end{bmatrix}$$

where the dots indicate that possibly more numbers appear. If the latter matrix has a nonzero element in the second column and after the first row, then again using (iii) of (6.9) we may transform the matrix into the form

$$\begin{bmatrix} a_{11} & a_{12} & a_{13} & \cdots \\ 0 & k_{22} & k_{23} & \cdots \\ 0 & 0 & l_{33} & \cdots \\ \cdot & \cdot & \cdot & \cdots \\ \cdot & \cdot & \cdot & \cdots \\ 0 & 0 & \cdot & \cdots \end{bmatrix}$$

where only zeros appear below k_{22} in the second column. This process is continued until we obtain the matrix of a system which is in echelon form. That will be true when all elements below the main diagonal are zero. The final matrix is then used to obtain a system of equations which is equivalent to the original system. The solutions are then found as in previous sections. Let us illustrate the technique by solving a system of four linear equations.

Example   Find the solutions of the system

$$\begin{cases} x - 2z + 2w = 1 \\ -2x + 3y + 4z = -1 \\ y + z - w = 0 \\ 3x + y - 2z - w = 3. \end{cases}$$

Solution   Notice that we have arranged the equations so that the same variables appear in vertical columns. We shall begin with the matrix of the system and proceed to the solution.

$$
\begin{bmatrix}
1 & 0 & -2 & 2 & 1 \\
-2 & 3 & 4 & 0 & -1 \\
0 & 1 & 1 & -1 & 0 \\
3 & 1 & -2 & -1 & 3
\end{bmatrix}
\rightarrow
\begin{bmatrix}
1 & 0 & -2 & 2 & 1 \\
0 & 3 & 0 & 4 & 1 \\
0 & 1 & 1 & -1 & 0 \\
0 & 1 & 4 & -7 & 0
\end{bmatrix}
$$

Add to the second row 2 times the first row and then add to the fourth row -3 times the first row.

$$
\rightarrow
\begin{bmatrix}
1 & 0 & -2 & 2 & 1 \\
0 & 1 & 1 & -1 & 0 \\
0 & 3 & 0 & 4 & 1 \\
0 & 1 & 4 & -7 & 0
\end{bmatrix}
$$

Interchange rows 2 and 3.

$$
\rightarrow
\begin{bmatrix}
1 & 0 & -2 & 2 & 1 \\
0 & 1 & 1 & -1 & 0 \\
0 & 0 & -3 & 7 & 1 \\
0 & 0 & 3 & -6 & 0
\end{bmatrix}
$$

Add to the third row -3 times the second row and then add to the fourth row -1 times the second row.

$$
\rightarrow
\begin{bmatrix}
1 & 0 & -2 & 2 & 1 \\
0 & 1 & 1 & -1 & 0 \\
0 & 0 & -3 & 7 & 1 \\
0 & 0 & 0 & 1 & 1
\end{bmatrix}
$$

Add row 3 to row 4.

The final matrix corresponds to the system of equations

$$
\begin{cases}
x \quad\ \ - 2z + 2w = 1 \\
\quad y + \ z - \ w = 0 \\
\qquad\ \ - 3z + 7w = 1 \\
\qquad\qquad\quad w = 1.
\end{cases}
$$

From the last equation we see that $w = 1$. Substituting in the third equation, we get $-3z + 7(1) = 1$, or $z = 2$. Using the second equation, we get $y + 2 - 1 = 0$, or $y = -1$. Finally, from the first equation, we have $x - 2(2) + 2(1) = 1$, or $x = 3$. Hence the system has only one solution, $x = 3$, $y = -1$, $z = 2$, and $w = 1$.

EXERCISES

1–20   Use matrices to solve Exercises 1–20 of Section 3.
21–28   Use matrices to solve Exercises 1–8 of Section 2.

Solve Exercises 29 and 30 by means of matrices.

29
$$\begin{cases} 2x - y - 2z + 2s - 5t = 2 \\ x + 3y - 2z + s - 2t = -5 \\ -x + 4y + 2z - 3s + 8t = -4 \\ 3x - 2y - 4z + s - 3t = -3 \\ 4x - 6y + z - 2s + t = 10 \end{cases}$$

30
$$\begin{cases} 3x + 2y + z + 3u + v + w = 1 \\ 2x + y - 2z + 3u - v + 4w = 6 \\ 6x + 3y + 4z - u + 2v + w = -6 \\ x + y + z + u - v - w = 8 \\ -2x - 2y + z - 3u + 2v - 3w = -10 \\ x - 3y + 2z + u + 3v + w = -1 \end{cases}$$

5   THE ALGEBRA OF MATRICES

It is possible to develop a comprehensive theory for matrices which has many mathematical and scientific applications. In this section we shall discuss several basic algebraic properties of matrices which serve as the starting point for such a theory. Our work will be restricted to matrices whose elements are real numbers.

In order to conserve space it is sometimes convenient to denote an $m \times n$ matrix A of the type displayed in (6.8) by the symbol (a_{ij}). If (b_{ij}) denotes another $m \times n$ matrix B, then we say that A and B are **equal** and we write

$$\boxed{A = B \quad \text{if and only if} \quad a_{ij} = b_{ij}}$$

for every i and j. For example,

$$\begin{bmatrix} 1 & 0 & 5 \\ \sqrt[3]{8} & 3^2 & -2 \end{bmatrix} = \begin{bmatrix} (-1)^2 & 0 & \sqrt{25} \\ 2 & 9 & -2 \end{bmatrix}.$$

If $A = (a_{ij})$ and $B = (b_{ij})$ are $m \times n$ matrices, then their **sum** $A + B$ is defined as the $m \times n$ matrix $C = (c_{ij})$, where $c_{ij} = a_{ij} + b_{ij}$ for all i and j. Thus, to add two matrices we add the elements which appear in corresponding positions in each matrix. Note that two matrices can be added only if they have the same number of rows and the same number of columns. Using the parentheses notation, we have

$$\boxed{(a_{ij}) + (b_{ij}) = (a_{ij} + b_{ij})}$$

Although we have used the symbol $+$ in two different ways, there is little chance for confusion, since whenever $+$ appears between symbols for matrices it refers to matrix addition, and when $+$ is used between real numbers it denotes their sum.

An example of the sum of two 3×2 matrices is

$$\begin{bmatrix} 4 & -5 \\ 0 & 4 \\ -6 & 1 \end{bmatrix} + \begin{bmatrix} 3 & 2 \\ 7 & -4 \\ -2 & 1 \end{bmatrix} = \begin{bmatrix} 7 & -3 \\ 7 & 0 \\ -8 & 2 \end{bmatrix}.$$

It is not difficult to prove that addition of matrices is both commutative and associative, that is,

$$A + B = B + A, \quad A + (B + C) = (A + B) + C$$

for $m \times n$ matrices A, B, and C.

The **$m \times n$ zero matrix**, denoted by O, is the matrix with m rows and n columns in which every element is 0. It is an identity element relative to addition, since

$$A + O = A$$

for every $m \times n$ matrix A. For example,

$$\begin{bmatrix} a_{11} & a_{12} \\ a_{21} & a_{22} \\ a_{31} & a_{32} \end{bmatrix} + \begin{bmatrix} 0 & 0 \\ 0 & 0 \\ 0 & 0 \end{bmatrix} = \begin{bmatrix} a_{11} & a_{12} \\ a_{21} & a_{22} \\ a_{31} & a_{32} \end{bmatrix}.$$

The **additive inverse** $-A$ of the matrix $A = (a_{ij})$ is, by definition, the matrix $(-a_{ij})$ obtained by changing the sign of each element of A. For example,

$$-\begin{bmatrix} 2 & -3 & 4 \\ -1 & 0 & 5 \end{bmatrix} = \begin{bmatrix} -2 & 3 & -4 \\ 1 & 0 & -5 \end{bmatrix}.$$

It follows that for every $m \times n$ matrix A,

$$A + (-A) = O$$

Subtraction of two $m \times n$ matrices is defined by

$$A - B = A + (-B)$$

Using the parentheses notation for matrices, this implies that

$$(a_{ij}) - (b_{ij}) = (a_{ij}) + (-b_{ij}) = (a_{ij} - b_{ij}).$$

Thus, to subtract two matrices we merely subtract the elements in the same positions.

The **product** of a real number c and an $m \times n$ matrix $A = (a_{ij})$ is defined by

$$cA = (ca_{ij})$$

Thus, to find cA we multiply each element of A by c. For example,

$$3\begin{bmatrix} 4 & -1 \\ 2 & 3 \end{bmatrix} = \begin{bmatrix} 12 & -3 \\ 6 & 9 \end{bmatrix}.$$

The following results may be established, where A and B are $m \times n$ matrices, and c and d are any real numbers.

(6.10)

$$c(A + B) = cA + cB$$
$$(c + d)A = cA + dA$$
$$(cd)A = c(dA)$$

The following definition of the product of two matrices may appear unusual to the beginning student; however, there are many applications which justify the form of the definition. In order to define the product AB of two matrices A and B, *the number of columns of A must be the same as the number of rows of B.* Suppose that $A = (a_{ij})$ is $m \times n$ and $B = (b_{ij})$ is $n \times p$. To determine the element c_{ij} of the product, we single out row i of A and column j of B as illustrated in (6.11).

(6.11)

$$\begin{bmatrix} a_{11} & a_{12} & \cdots & a_{1n} \\ \vdots & & & \vdots \\ \boxed{a_{i1} \quad a_{i2} \quad \cdots \quad a_{in}} \\ \vdots & \vdots & & \vdots \\ a_{m1} & a_{m2} & \cdots & a_{mn} \end{bmatrix} \begin{bmatrix} b_{11} & \cdots & b_{1j} & \cdots & b_{1p} \\ b_{21} & \cdots & b_{2j} & \cdots & b_{2p} \\ \vdots & & \vdots & & \vdots \\ b_{n1} & \cdots & b_{nj} & \cdots & b_{np} \end{bmatrix}$$

Next we multiply pairs of elements and then add them, according to the following formula:

(6.12)

$$c_{ij} = a_{i1}b_{1j} + a_{i2}b_{2j} + \cdots + a_{in}n_{nj}$$

For example, the element c_{11} in the first row and the first column of AB is given by

$$c_{11} = a_{11}b_{11} + a_{12}b_{21} + \cdots + a_{1n}b_{n1}.$$

The element c_{12} in the first row and second column of AB is

$$c_{12} = a_{11}b_{12} + a_{12}b_{22} + \cdots + a_{1n}b_{n2}$$

and so on. By definition, the product AB has the same number of rows as A and the same number of columns as B. In particular, if A is $m \times n$ and B is $n \times p$, then AB is $m \times p$. This is illustrated by the following product of a 2×3 matrix and a 3×4 matrix.

$$\begin{bmatrix} 1 & 2 & -3 \\ 4 & 0 & -2 \end{bmatrix} \begin{bmatrix} 5 & -4 & 2 & 0 \\ -1 & 6 & 3 & 1 \\ 7 & 0 & 4 & 8 \end{bmatrix} = \begin{bmatrix} -18 & 8 & -4 & -22 \\ 6 & -16 & 0 & -16 \end{bmatrix}.$$

Here are some typical computations of the elements c_{ij} in the product:

$$c_{11} = (1)(5) + (2)(-1) + (-3)(7) = 5 - 2 - 21 = -18$$
$$c_{13} = (1)(2) + (2)(3) + (-3)(4) = 2 + 6 - 12 = -4$$
$$c_{23} = (4)(2) + (0)(3) + (-2)(4) = 8 + 0 - 8 = 0$$
$$c_{24} = (4)(0) + (0)(1) + (-2)(8) = 0 + 0 - 16 = -16.$$

The reader should check the remaining elements.

The product operation for matrices is not commutative. Indeed, if A is 2×3 and B is 3×4, then AB may be found, but BA is undefined since the number of columns of B is different from the number of rows of A. Even if AB and BA are both defined, it is often true that these products are different. This is illustrated in the next example, along with the fact that the product of two nonzero matrices may equal a zero matrix.

Example   If $A = \begin{bmatrix} 2 & 2 \\ -1 & -1 \end{bmatrix}$ and $B = \begin{bmatrix} 1 & 2 \\ 1 & 2 \end{bmatrix}$, show that $AB \neq BA$.

Solution   Using the definition of product we obtain

$$AB = \begin{bmatrix} 4 & 8 \\ -2 & -4 \end{bmatrix} \quad \text{and} \quad BA = \begin{bmatrix} 0 & 0 \\ 0 & 0 \end{bmatrix}.$$

Hence $AB \neq BA$. Note that BA is a zero matrix.

It is possible to show that matrix multiplication is associative in the sense that

$$\boxed{A(BC) = (AB)C}$$

provided that the indicated products are defined. That will be the case when A is $m \times n$, B is $n \times p$, and C is $p \times q$. The Distributive Properties also hold if the matrices involved have the proper number of rows and columns. Thus if A_1 and A_2 are $m \times n$ matrices and if B_1 and B_2 are $n \times p$ matrices, then

$$\boxed{A_1(B_1 + B_2) = A_1 B_1 + A_1 B_2}$$

and

$$\boxed{(A_1 + A_2)B_1 = A_1 B_1 + A_2 B_1}$$

As a special case, if all matrices are square, of order n, then the Associative and Distributive Properties are true. We shall not give proofs for these theorems.

EXERCISES

In Exercises 1–8 find $A + B$, $A - B$, $2A$, and $-3B$.

1   $A = \begin{bmatrix} 5 & -2 \\ 1 & 3 \end{bmatrix}$, $B = \begin{bmatrix} 4 & 1 \\ -3 & 2 \end{bmatrix}$ 2 $A = \begin{bmatrix} 3 & 0 \\ -1 & 2 \end{bmatrix}$, $B = \begin{bmatrix} 3 & -4 \\ 1 & 1 \end{bmatrix}$

3   $A = \begin{bmatrix} 6 & -1 \\ 2 & 0 \\ -3 & 4 \end{bmatrix}$, $B = \begin{bmatrix} 3 & 1 \\ -1 & 5 \\ 6 & 0 \end{bmatrix}$

4   $A = \begin{bmatrix} 0 & -2 & 7 \\ 5 & 4 & -3 \end{bmatrix}$, $B = \begin{bmatrix} 8 & 4 & 0 \\ 0 & 1 & 4 \end{bmatrix}$

5   $A = \begin{bmatrix} 4 & -3 & 2 \end{bmatrix}$, $B = \begin{bmatrix} 7 & 0 & -5 \end{bmatrix}$

6   $A = \begin{bmatrix} 7 \\ -16 \end{bmatrix}$, $B = \begin{bmatrix} -11 \\ 9 \end{bmatrix}$

7   $A = \begin{bmatrix} 0 & 4 & 0 & 3 \\ 1 & 2 & 0 & -5 \end{bmatrix}$, $B = \begin{bmatrix} -3 & 0 & 1 & 3 \\ 2 & 0 & 7 & -2 \end{bmatrix}$

8   $A = \begin{bmatrix} -7 \end{bmatrix}$, $B = \begin{bmatrix} 9 \end{bmatrix}$

In Exercises 9–18 find AB and BA.

9   $A = \begin{bmatrix} 2 & 6 \\ 3 & -4 \end{bmatrix}$, $B = \begin{bmatrix} 5 & -2 \\ 1 & 7 \end{bmatrix}$ 10 $A = \begin{bmatrix} 4 & -2 \\ -2 & 1 \end{bmatrix}$, $B = \begin{bmatrix} 2 & 1 \\ 4 & 2 \end{bmatrix}$

11   $A = \begin{bmatrix} 3 & 0 & -1 \\ 0 & 4 & 2 \\ 5 & -3 & 1 \end{bmatrix}$, $B = \begin{bmatrix} 1 & -5 & 0 \\ 4 & 1 & -2 \\ 0 & -1 & 3 \end{bmatrix}$

12   $A = \begin{bmatrix} 5 & 0 & 0 \\ 0 & -3 & 0 \\ 0 & 0 & 2 \end{bmatrix}$, $B = \begin{bmatrix} 3 & 0 & 0 \\ 0 & 4 & 0 \\ 0 & 0 & -2 \end{bmatrix}$

13   $A = \begin{bmatrix} 4 & -3 & 1 \\ -5 & 2 & 2 \end{bmatrix}$, $B = \begin{bmatrix} 2 & 1 \\ 0 & 1 \\ -4 & 7 \end{bmatrix}$

14   $A = \begin{bmatrix} 2 & 1 & -1 & 0 \\ 3 & -2 & 0 & 5 \\ -2 & 1 & 4 & 2 \end{bmatrix}$, $B = \begin{bmatrix} 5 & -3 & 1 \\ 1 & 2 & 0 \\ -1 & 0 & 4 \\ 0 & -2 & 3 \end{bmatrix}$

15   $A = \begin{bmatrix} 1 & 2 & 3 \\ 4 & 5 & 6 \\ 7 & 8 & 9 \end{bmatrix}$, $B = \begin{bmatrix} 1 & 0 & 0 \\ 0 & 1 & 0 \\ 0 & 0 & 1 \end{bmatrix}$

16   $A = \begin{bmatrix} 1 & 2 & 3 \\ 2 & 3 & 1 \\ 3 & 1 & 2 \end{bmatrix}$, $B = \begin{bmatrix} 2 & 0 & 0 \\ 0 & 2 & 0 \\ 0 & 0 & 2 \end{bmatrix}$

17   $A = \begin{bmatrix} -3 & 7 & 2 \end{bmatrix}$, $B = \begin{bmatrix} 1 \\ 4 \\ -5 \end{bmatrix}$ 18 $A = \begin{bmatrix} 4 & 8 \end{bmatrix}$, $B = \begin{bmatrix} -3 \\ 2 \end{bmatrix}$

In Exercises 19–22 find AB.

19   $A = \begin{bmatrix} 4 & -2 \\ 0 & 3 \\ -7 & 5 \end{bmatrix}$, $B = \begin{bmatrix} 3 \\ 4 \end{bmatrix}$ 20 $A = \begin{bmatrix} 4 \\ -3 \\ 2 \end{bmatrix}$, $B = \begin{bmatrix} 5 & 1 \end{bmatrix}$

21 $A = \begin{bmatrix} 2 & 1 & 0 & -3 \\ -7 & 0 & -2 & 4 \end{bmatrix}$, $B = \begin{bmatrix} 4 & -2 & 0 \\ 1 & 1 & -2 \\ 0 & 0 & 5 \\ -3 & -1 & 0 \end{bmatrix}$

22 $A = \begin{bmatrix} 1 & 2 & -3 \\ 4 & -5 & 6 \end{bmatrix}$, $B = \begin{bmatrix} 1 & -1 & 0 & 2 \\ -2 & 3 & 1 & 0 \\ 0 & 4 & 0 & -3 \end{bmatrix}$

23 If $A = \begin{bmatrix} 1 & 2 \\ 0 & -3 \end{bmatrix}$, and $B = \begin{bmatrix} 2 & -1 \\ 3 & 1 \end{bmatrix}$, show that $(A + B)(A - B) \neq A^2 - B^2$, where $A^2 = AA$ and $B^2 = BB$.

24 If A and B are the matrices of Exercise 23, show that $(A + B)(A + B) \neq A^2 + 2AB + B^2$.

25 If A and B are the matrices of Exercise 23 and $C = \begin{bmatrix} 3 & 1 \\ -2 & 0 \end{bmatrix}$, prove that

$A(B + C) = AB + AC$.

26 If A, B, and C are the matrices of Exercise 25 prove that $A(BC) = (AB)C$.

Prove the identities given in Exercises 27–30, where $A = \begin{bmatrix} a_{11} & a_{12} \\ a_{21} & a_{22} \end{bmatrix}$,

$B = \begin{bmatrix} b_{11} & b_{12} \\ b_{21} & b_{22} \end{bmatrix}$, $C = \begin{bmatrix} c_{11} & c_{12} \\ c_{21} & c_{22} \end{bmatrix}$, and c, d are real numbers.

27 $c(A + B) = cA + cB$ 28 $(c + d)A = cA + dA$

29 $A(B + C) = AB + AC$ 30 $A(BC) = (AB)C$

6   INVERSES OF MATRICES

Throughout this section we shall concentrate on square matrices. The symbol I_n will be used to denote the square matrix of order n which has 1 in each position on the main diagonal and 0 elsewhere. For example,

$$I_2 = \begin{bmatrix} 1 & 0 \\ 0 & 1 \end{bmatrix}, \quad I_3 = \begin{bmatrix} 1 & 0 & 0 \\ 0 & 1 & 0 \\ 0 & 0 & 1 \end{bmatrix}$$

and so on. It can be shown that if A is any square matrix of order n, then

$$AI_n = A = I_n A$$

For that reason I_n is called an **identity matrix of order** n. To illustrate, if $A = (a_{ij})$ is of order 2, then a direct calculation shows that

$$\begin{bmatrix} a_{11} & a_{12} \\ a_{21} & a_{22} \end{bmatrix}\begin{bmatrix} 1 & 0 \\ 0 & 1 \end{bmatrix} = \begin{bmatrix} a_{11} & a_{12} \\ a_{21} & a_{22} \end{bmatrix} = \begin{bmatrix} 1 & 0 \\ 0 & 1 \end{bmatrix}\begin{bmatrix} a_{11} & a_{12} \\ a_{21} & a_{22} \end{bmatrix}.$$

Some, but not all, $n \times n$ matrices A have an **inverse** in the sense that there is a matrix B such that $AB = I_n = BA$. If A has an inverse we denote it by A^{-1} and write

$$AA^{-1} = I_n = A^{-1}A.$$

The symbol A^{-1} is read "A inverse." In matrix theory it is *not* acceptable to use the symbol $1/A$ in place of A^{-1}.

Let us describe a technique for finding the inverse of a square matrix A, whenever it exists. We shall not attempt to justify the following procedure, since that would require concepts which are beyond the scope of this text. Given the $n \times n$ matrix $A = (a_{ij})$, we begin by forming the $n \times (2n)$ matrix:

(6.13)

$$\begin{bmatrix} a_{11} & a_{12} & \cdots & a_{1n} & 1 & 0 & 0 & \cdots & 0 \\ a_{21} & a_{22} & \cdots & a_{2n} & 0 & 1 & 0 & \cdots & 0 \\ \vdots & \vdots & \cdots & \vdots & \vdots & \vdots & \vdots & \cdots & \vdots \\ a_{n1} & a_{n2} & \cdots & a_{nn} & 0 & 0 & 0 & \cdots & 1 \end{bmatrix}$$

where the $n \times n$ identity matrix I_n appears "to the right" of the matrix A, as indicated. We next apply a succession of elementary row transformations to (6.13) until we arrive at a matrix of the form

(6.14)

$$\begin{bmatrix} 1 & 0 & 0 & \cdots & 0 & b_{11} & b_{12} & \cdots & b_{1n} \\ 0 & 1 & 0 & \cdots & 0 & b_{21} & b_{22} & \cdots & b_{2n} \\ \vdots & \vdots & \vdots & \cdots & \vdots & \vdots & \vdots & \cdots & \vdots \\ 0 & 0 & 0 & \cdots & 1 & b_{n1} & b_{n2} & \cdots & b_{nn} \end{bmatrix}$$

where the identity matrix I_n appears "to the left" of the $n \times n$ matrix (b_{ij}). It can be shown that (b_{ij}) is the desired inverse A^{-1}. If A does not have an inverse, then it is impossible to change (6.13) into the form (6.14) by means of elementary row transformations.

Example 1   Find A^{-1} if $A = \begin{bmatrix} 3 & 5 \\ 1 & 4 \end{bmatrix}$.

Solution   As in (6.13), we begin with the matrix

$$\begin{bmatrix} 3 & 5 & 1 & 0 \\ 1 & 4 & 0 & 1 \end{bmatrix}.$$

We next perform elementary row transformations until we reach the form (8.14). Thus,

$$\begin{bmatrix} 3 & 5 & 1 & 0 \\ 1 & 4 & 0 & 1 \end{bmatrix} \rightarrow \begin{bmatrix} 1 & 4 & 0 & 1 \\ 3 & 5 & 1 & 0 \end{bmatrix} \qquad \text{Interchange rows 1 and 2.}$$

$$\rightarrow \begin{bmatrix} 1 & 4 & 0 & 1 \\ 0 & -7 & 1 & -3 \end{bmatrix}$$ Add to the second row -3 times the first row.

$$\rightarrow \begin{bmatrix} 1 & 4 & 0 & 1 \\ 0 & 1 & -\frac{1}{7} & \frac{3}{7} \end{bmatrix}$$ Multiply the second row by $-1/7$.

$$\rightarrow \begin{bmatrix} 1 & 0 & \frac{4}{7} & -\frac{5}{7} \\ 0 & 1 & -\frac{1}{7} & \frac{3}{7} \end{bmatrix}$$ Add to the first row -4 times the second row.

According to the previous discussion,

$$A^{-1} = \begin{bmatrix} \frac{4}{7} & -\frac{5}{7} \\ -\frac{1}{7} & \frac{3}{7} \end{bmatrix}.$$

To check our work the reader should verify that

$$\begin{bmatrix} 3 & 5 \\ 1 & 4 \end{bmatrix} \begin{bmatrix} \frac{4}{7} & -\frac{5}{7} \\ -\frac{1}{7} & \frac{3}{7} \end{bmatrix} = \begin{bmatrix} 1 & 0 \\ 0 & 1 \end{bmatrix} = \begin{bmatrix} \frac{4}{7} & -\frac{5}{7} \\ -\frac{1}{7} & \frac{3}{7} \end{bmatrix} \begin{bmatrix} 3 & 5 \\ 1 & 4 \end{bmatrix}.$$

Example 2 Find A^{-1} if $A = \begin{bmatrix} -1 & 3 & 1 \\ 2 & 5 & 0 \\ 3 & 1 & -2 \end{bmatrix}$.

Solution As in Example 1, we begin with form (6.13)

$$\begin{bmatrix} -1 & 3 & 1 & 1 & 0 & 0 \\ 2 & 5 & 0 & 0 & 1 & 0 \\ 3 & 1 & -2 & 0 & 0 & 1 \end{bmatrix}$$

and apply elementary row operations until we arrive at form (6.14). Thus,

$$\begin{bmatrix} -1 & 3 & 1 & 1 & 0 & 0 \\ 2 & 5 & 0 & 0 & 1 & 0 \\ 3 & 1 & -2 & 0 & 0 & 1 \end{bmatrix}$$

$$\rightarrow \begin{bmatrix} 1 & -3 & -1 & -1 & 0 & 0 \\ 2 & 5 & 0 & 0 & 1 & 0 \\ 3 & 1 & -2 & 0 & 0 & 1 \end{bmatrix}$$ Multiply the first row by -1.

$$\rightarrow \begin{bmatrix} 1 & -3 & -1 & -1 & 0 & 0 \\ 0 & 11 & 2 & 2 & 1 & 0 \\ 0 & 10 & 1 & 3 & 0 & 1 \end{bmatrix}$$ Add to the second row -2 times the first row and then add to the third row -3 times the first row.

$$\rightarrow \begin{bmatrix} 1 & -3 & -1 & -1 & 0 & 0 \\ 0 & 1 & 1 & -1 & 1 & -1 \\ 0 & 10 & 1 & 3 & 0 & 1 \end{bmatrix}$$

Add to the second row -1 times the third row.

$$\rightarrow \begin{bmatrix} 1 & 0 & 2 & -4 & 3 & -3 \\ 0 & 1 & 1 & -1 & 1 & -1 \\ 0 & 0 & -9 & 13 & -10 & 11 \end{bmatrix}$$

Add to the first row 3 times the second row and then add to the third row -10 times the second row.

$$\rightarrow \begin{bmatrix} 1 & 0 & 2 & -4 & 3 & -3 \\ 0 & 1 & 1 & -1 & 1 & -1 \\ 0 & 0 & 1 & -\frac{13}{9} & \frac{10}{9} & -\frac{11}{9} \end{bmatrix}$$

Multiply the third row by $-1/9$.

$$\rightarrow \begin{bmatrix} 1 & 0 & 0 & -\frac{10}{9} & \frac{7}{9} & -\frac{5}{9} \\ 0 & 1 & 0 & \frac{4}{9} & -\frac{1}{9} & \frac{2}{9} \\ 0 & 0 & 1 & -\frac{13}{9} & \frac{10}{9} & -\frac{11}{9} \end{bmatrix}$$

Add to the first row -2 times the third row and then add to the second row -1 times the third row.

Consequently,

$$A^{-1} = \begin{bmatrix} -\frac{10}{9} & \frac{7}{9} & -\frac{5}{9} \\ \frac{4}{9} & -\frac{1}{9} & \frac{2}{9} \\ -\frac{13}{9} & \frac{10}{9} & -\frac{11}{9} \end{bmatrix} = \frac{1}{9}\begin{bmatrix} -10 & 7 & -5 \\ 4 & -1 & 2 \\ -13 & 10 & -11 \end{bmatrix}.$$

The reader should check the fact that

$$AA^{-1} = I_3 = A^{-1}A.$$

There are many uses for inverses of matrices. One application concerns solutions of systems of linear equations. To illustrate, let us consider the case of two linear equations in two unknowns:

$$\begin{cases} a_{11}x + a_{12}y = k_1 \\ a_{21}x + a_{22}y = k_2. \end{cases}$$

We may express the system by means of matrices as follows:

(6.15)
$$\begin{bmatrix} a_{11}x + a_{12}y \\ a_{21}x + a_{22}y \end{bmatrix} = \begin{bmatrix} k_1 \\ k_2 \end{bmatrix}.$$

If we let

$$A = \begin{bmatrix} a_{11} & a_{12} \\ a_{21} & a_{22} \end{bmatrix}, \quad X = \begin{bmatrix} x \\ y \end{bmatrix}, \quad \text{and} \quad B = \begin{bmatrix} k_1 \\ k_2 \end{bmatrix},$$

then (6.15) may be written in the form

$$AX = B.$$

(6.16)

If A^{-1} exists, then multiplying both sides of (6.16) by A^{-1} gives us $A^{-1}AX = A^{-1}B$. Since $A^{-1}A = I_n$ and $I_nX = X$, this leads to

$$X = A^{-1}B$$

from which the solution (x, y) may be found. The above technique may be extended to systems of n linear equations in n unknowns.

Example 3 Solve the following system of equations:

$$\begin{cases} -x + 3y + z = 1 \\ 2x + 5y \quad\quad = 3 \\ 3x + y - 2z = -2. \end{cases}$$

Solution If we let

$$A = \begin{bmatrix} -1 & 3 & 1 \\ 2 & 5 & 0 \\ 3 & 1 & -2 \end{bmatrix}, \quad X = \begin{bmatrix} x \\ y \\ z \end{bmatrix}, \quad \text{and} \quad B = \begin{bmatrix} 1 \\ 3 \\ -2 \end{bmatrix},$$

then as in (6.15) and (6.16), the given system may be written in terms of matrices as $AX = B$. From the preceding discussion this implies that $X = A^{-1}B$. The matrix A^{-1} was found in Example 2. Substituting for X, A^{-1}, and B in the last equation gives us

$$\begin{bmatrix} x \\ y \\ z \end{bmatrix} = \frac{1}{9} \begin{bmatrix} -10 & 7 & -5 \\ 4 & -1 & 2 \\ -13 & 10 & -11 \end{bmatrix} \begin{bmatrix} 1 \\ 3 \\ -2 \end{bmatrix} = \frac{1}{9} \begin{bmatrix} 21 \\ -3 \\ 39 \end{bmatrix} = \begin{bmatrix} \frac{7}{3} \\ -\frac{1}{3} \\ \frac{13}{3} \end{bmatrix}.$$

It follows that $x = \frac{7}{3}$, $y = -\frac{1}{3}$, and $z = \frac{13}{3}$. Hence the ordered triple $(\frac{7}{3}, -\frac{1}{3}, \frac{13}{3})$ is the solution of the given system.

The method of solution employed in Example 3 is beneficial only if A^{-1} is known, or if many systems with the same coefficient matrix are to be considered. The preferred technique for solving a system of linear equations is still either the method of elimination or the matrix method discussed in Section 4.

EXERCISES

In Exercises 1–12 find the inverse of the given matrix, if it exists.

1 $\begin{bmatrix} 2 & -4 \\ 1 & 3 \end{bmatrix}$

2 $\begin{bmatrix} 3 & 2 \\ 4 & 5 \end{bmatrix}$

3 $\begin{bmatrix} 2 & 4 \\ 4 & 8 \end{bmatrix}$

4 $\begin{bmatrix} 3 & -1 \\ 6 & -2 \end{bmatrix}$

5 $\begin{bmatrix} 3 & -1 & 0 \\ 2 & 2 & 0 \\ 0 & 0 & 4 \end{bmatrix}$

6 $\begin{bmatrix} 3 & 0 & 2 \\ 0 & 1 & 0 \\ -4 & 0 & 2 \end{bmatrix}$

7 $\begin{bmatrix} -2 & 2 & 3 \\ 1 & -1 & 0 \\ 0 & 1 & 4 \end{bmatrix}$

8 $\begin{bmatrix} 1 & 2 & 3 \\ -2 & 1 & 0 \\ 3 & -1 & 1 \end{bmatrix}$

9 $\begin{bmatrix} 2 & 0 & 0 \\ 0 & 4 & 0 \\ 0 & 0 & 6 \end{bmatrix}$

10 $\begin{bmatrix} 1 & 1 & 1 \\ 2 & 2 & 2 \\ 3 & 3 & 3 \end{bmatrix}$

11 $\begin{bmatrix} 1 & -1 & 0 & 1 \\ 0 & 1 & -2 & 0 \\ -1 & 2 & 1 & 2 \\ -2 & 1 & 2 & 0 \end{bmatrix}$

12 $\begin{bmatrix} 1 & 2 & 0 & 1 \\ 0 & -1 & 1 & -2 \\ 0 & 0 & 2 & 0 \\ 0 & 0 & 0 & 1 \end{bmatrix}$

13 State conditions on a and b which guarantee that the matrix $\begin{bmatrix} a & 0 \\ 0 & b \end{bmatrix}$ has an inverse, and find a formula for the inverse when it exists.

14 If $abc \neq 0$, find the inverse of $\begin{bmatrix} a & 0 & 0 \\ 0 & b & 0 \\ 0 & 0 & c \end{bmatrix}$.

15 If $A = \begin{bmatrix} a_{11} & a_{12} & a_{13} \\ a_{21} & a_{22} & a_{23} \\ a_{31} & a_{32} & a_{33} \end{bmatrix}$, prove that $AI_3 = A = I_3A$.

16 Prove that $AI_4 = A = I_4A$ for every square matrix A of order 4.

Solve the systems in Exercises 17–20 by the method of Example 3. (Refer to inverses of matrices found in Exercises 1–8.)

17 $\begin{cases} 2x - 4y = 3 \\ x + 3y = 1 \end{cases}$

18 $\begin{cases} 3x + 2y = -1 \\ 4x + 5y = 1 \end{cases}$

19 $\begin{cases} -2x + 2y + 3z = 1 \\ x - y = 3 \\ y + 4z = -2 \end{cases}$

20 $\begin{cases} x + 2y + 3z = -1 \\ -2x + y = 4 \\ 3x - y + z = 2 \end{cases}$

7   DETERMINANTS

Throughout this section and the next it is assumed that all matrices under discussion are *square* matrices. Associated with each such matrix A is a number called the **determinant** of A. Determinants can be used to solve systems of linear equations if the number of equations is the same as the number of variables. In this section we shall state the definition and give some basic properties of determinants. It may be difficult to see exactly why the definitions given below are used. One reason will be pointed out in Section 9, where it is shown that our definitions arise naturally when solving systems of linear equations.

The determinant of a square matrix A will be denoted by $|A|$. This notation should not be confused with the symbol used for the absolute value of a real number. To avoid any misunderstanding, the expression det A is used in some mathematics texts instead of $|A|$. We shall define $|A|$ by beginning with the case in which A has order 1 and then by increasing the order a step at a time.

If A is a square matrix of order 1, then A has only one element. Thus, $A = [a_{11}]$ and we define $|A| = a_{11}$. If A is a square matrix of order 2, then we may write

$$A = \begin{bmatrix} a_{11} & a_{12} \\ a_{21} & a_{22} \end{bmatrix}$$

and the determinant of A is defined by

$$|A| = a_{11}a_{22} - a_{21}a_{12}.$$

Another notation for $|A|$ is obtained by replacing the brackets in the symbol for A with vertical bars as follows:

(6.17)
$$|A| = \begin{vmatrix} a_{11} & a_{12} \\ a_{21} & a_{22} \end{vmatrix} = a_{11}a_{22} - a_{21}a_{12}$$

Example 1   Find $|A|$ if $A = \begin{bmatrix} 2 & -1 \\ 4 & -3 \end{bmatrix}$.

Solution   Using (6.17),

$$|A| = \begin{vmatrix} 2 & -1 \\ 4 & -3 \end{vmatrix} = (2)(-3) - (4)(-1) = -6 + 4 = -2.$$

For matrices of higher order it is convenient to introduce additional terminology as follows.

(6.18)   If A is a matrix of order 3, then the **minor** M_{ij} of an element a_{ij} is the determinant of the matrix of order 2 obtained by deleting row i and column j of A.

Thus, to determine the minor of an element we discard the row and column in which the element appears and then find the determinant of the resulting matrix. To illustrate, if

(6.19)
$$A = \begin{bmatrix} a_{11} & a_{12} & a_{13} \\ a_{21} & a_{22} & a_{23} \\ a_{31} & a_{32} & a_{33} \end{bmatrix}$$

then using (6.18) we obtain

(6.20)

$$M_{11} = \begin{vmatrix} a_{22} & a_{23} \\ a_{32} & a_{33} \end{vmatrix} = a_{22}a_{33} - a_{32}a_{23}$$

$$M_{12} = \begin{vmatrix} a_{21} & a_{23} \\ a_{31} & a_{33} \end{vmatrix} = a_{21}a_{33} - a_{31}a_{23}$$

$$M_{13} = \begin{vmatrix} a_{21} & a_{22} \\ a_{31} & a_{32} \end{vmatrix} = a_{21}a_{32} - a_{31}a_{22}$$

$$M_{23} = \begin{vmatrix} a_{11} & a_{12} \\ a_{31} & a_{32} \end{vmatrix} = a_{11}a_{32} - a_{31}a_{12}$$

and likewise for the other minors M_{21}, M_{22}, M_{31}, M_{32}, and M_{33}.
We shall also make use of the following concept.

(6.21)

> The **cofactor** A_{ij} of the element a_{ij} is defined by
>
> $$A_{ij} = (-1)^{i+j}M_{ij}.$$

Thus to obtain the cofactor of a_{ij} we find the minor and multiply it by 1 or -1 depending on whether the sum of i and j is even or odd, respectively. An easy way to remember the sign $(-1)^{i+j}$ associated with the cofactor A_{ij} is to consider the following "checkerboard "

(6.22)

$$\begin{bmatrix} + & - & + \\ - & + & - \\ + & - & + \end{bmatrix}$$

where we may regard the + signs as occurring on red squares and the − signs on black squares.

Example 2   If

$$A = \begin{bmatrix} 1 & -3 & 3 \\ 4 & 2 & 0 \\ -2 & -7 & 5 \end{bmatrix}$$

find M_{11}, M_{21}, M_{22}, A_{11}, A_{21}, and A_{22}.

Solution   By definition

$$M_{11} = \begin{vmatrix} 2 & 0 \\ -7 & 5 \end{vmatrix} = (2)(5) - (-7)(0) = 10$$

$$M_{21} = \begin{vmatrix} -3 & 3 \\ -7 & 5 \end{vmatrix} = (-3)(5) - (-7)(3) = 6$$

$$M_{22} = \begin{vmatrix} 1 & 3 \\ -2 & 5 \end{vmatrix} = (1)(5) - (-2)(3) = 11.$$

All that is necessary to obtain the cofactors is to prefix the corresponding minors with the proper signs. Thus, using (6.21)

$$A_{11} = (-1)^{1+1}M_{11} = (1)(10) = 10$$
$$A_{21} = (-1)^{2+1}M_{21} = (-1)(6) = -6$$
$$A_{22} = (-1)^{2+2}M_{22} = (1)(11) = 11.$$

The checkerboard (6.22) could also be used to determine the proper signs.

The determinant $|A|$ of a square matrix of order 3 is defined by

(6.23)
$$|A| = \begin{vmatrix} a_{11} & a_{12} & a_{13} \\ a_{21} & a_{22} & a_{23} \\ a_{31} & a_{32} & a_{33} \end{vmatrix} = a_{11}A_{11} + a_{12}A_{12} + a_{13}A_{13}$$

Since $A_{11} = (-1)^{1+1}M_{11} = M_{11}$, $A_{12} = (-1)^{1+2}M_{12} = -M_{12}$, and $A_{13} = (-1)^{1+3}M_{13} = M_{13}$, Definition (6.23) may also be written

(6.24)
$$|A| = a_{11}M_{11} - a_{12}M_{12} + a_{13}M_{13}$$

If we substitute for M_{11}, M_{12}, and M_{13}, from (6.20) we obtain the following formula for $|A|$ in terms of the elements of A:

(6.25)
$$|A| = a_{11}a_{22}a_{33} - a_{11}a_{32}a_{23} - a_{12}a_{21}a_{33} + a_{12}a_{31}a_{23} + a_{13}a_{21}a_{32} \\ - a_{13}a_{31}a_{22}$$

Formula (6.23) displays a pattern of multiplying each element in row 1 by its cofactor and then adding to find $|A|$. This is referred to as *expanding* $|A|$ *by the first row*. By actually carrying out the computations, it is not difficult to show that $|A|$ *can be expanded in similar fashion by using any row or column*. As an illustration, the expansion by the second column in (6.19) is:

$$|A| = a_{12}A_{12} + a_{22}A_{22} + a_{32}A_{32}$$
$$= a_{12}\left(-\begin{vmatrix} a_{21} & a_{23} \\ a_{31} & a_{32} \end{vmatrix}\right) + a_{22}\left(+\begin{vmatrix} a_{11} & a_{13} \\ a_{31} & a_{33} \end{vmatrix}\right) + a_{32}\left(-\begin{vmatrix} a_{11} & a_{13} \\ a_{21} & a_{23} \end{vmatrix}\right).$$

If we use (6.17) for the determinants in parentheses, multiply as indicated, and rearrange the terms in the sum, then the expression for $|A|$ given in (6.25) will result. Similarly, the expansion of (6.19) by the third row is

$$|A| = a_{31}A_{31} + a_{32}A_{32} + a_{33}A_{33}$$
$$= a_{31}\left(+\begin{vmatrix} a_{12} & a_{13} \\ a_{22} & a_{23} \end{vmatrix}\right) + a_{32}\left(-\begin{vmatrix} a_{11} & a_{13} \\ a_{21} & a_{23} \end{vmatrix}\right) + a_{33}\left(+\begin{vmatrix} a_{11} & a_{12} \\ a_{21} & a_{22} \end{vmatrix}\right).$$

Once again it can be shown that this equals (6.25).

Example 3   Find $|A|$ if

$$A = \begin{bmatrix} -1 & 3 & 1 \\ 2 & 5 & 0 \\ 3 & 1 & -2 \end{bmatrix}.$$

Solution   Expanding $|A|$ by the second row gives us

$$|A| = (2)A_{21} + (5)A_{22} + (0)A_{23}.$$

Using the definition of cofactor we have

$$A_{21} = (-1)^3 M_{21} = -\begin{vmatrix} 3 & 1 \\ 1 & -2 \end{vmatrix} = -[(3)(-2) - (1)(1)] = 7$$

$$A_{22} = (-1)^4 M_{22} = \begin{vmatrix} -1 & 1 \\ 3 & -2 \end{vmatrix} = [(-1)(-2) - (3)(1)] = -1.$$

Consequently

$$|A| = (2)(7) + (5)(-1) + (0)A_{23} = 14 - 5 + 0 = 9.$$

If A is a matrix of order 4, we define the minor M_{ij} of the element a_{ij} as the determinant of the matrix of order 3 obtained by deleting row i and column j of A. The cofactor A_{ij} is again given by $(-1)^{i+j} M_{ij}$. In a manner analogous to (6.23) we define

$$|A| = a_{11}A_{11} + a_{12}A_{12} + a_{13}A_{13} + a_{14}A_{14}.$$

The last formula is called *the expansion of $|A|$ by the first row*. In terms of minors, the formula may be written as

$$|A| = a_{11}M_{11} - a_{12}M_{12} + a_{13}M_{13} - a_{14}M_{14}.$$

It can be shown that the same number is obtained if $|A|$ is expanded by any other row or column.

The method of defining determinants of matrices of arbitrary order n should now be apparent. If a_{ij} is an element of a matrix of order $n > 1$, then the **minor** M_{ij} is defined as the determinant of the matrix of order $n - 1$ obtained by deleting row i and column j. The **cofactor** A_{ij} is defined as $(-1)^{i+j} M_{ij}$. The sign $(-1)^{i+j}$ associated with A_{ij} can be remembered by using a checkerboard similar to (6.22), extending the rows and columns as far as necessary. We then define the determinant $|A|$ of a matrix A of order n as the expansion by the first row, that is

(6.26)

$$\boxed{|A| = a_{11}A_{11} + a_{12}A_{12} + \cdots + a_{1n}A_{1n}}$$

or, in terms of minors,

(6.27)
$$|A| = a_{11}M_{11} - a_{12}M_{12} + \cdots + a_{1n}(-1)^{1+n}M_{1n}$$

As was the case with matrices of small order, the number $|A|$ may be found by using *any* row or column. Specifically, we have the following theorem.

(6.28) **Expansion Theorem for Determinants**

> If A is a square matrix of order $n > 1$, then the determinant $|A|$ may be found by multiplying the elements of any row (or column) by their respective cofactors and adding the resulting products.

The proof of (6.28) is difficult and may be found in texts on matrix theory. The theorem is quite useful if many zeros appear in a row or column, as illustrated in the following example.

Example 4 Find $|A|$ if

$$A = \begin{bmatrix} 1 & 0 & 2 & 5 \\ -2 & 1 & 5 & 0 \\ 0 & 0 & -3 & 0 \\ 0 & -1 & 0 & 3 \end{bmatrix}.$$

Solution Note that all but one of the elements in the third row is zero. Hence if we expand $|A|$ by the third row there will be at most one nonzero term. Specifically,

$$|A| = (0)A_{31} + (0)A_{32} + (-3)A_{33} + (0)A_{34} = -3A_{33}$$

where

$$A_{33} = \begin{vmatrix} 1 & 0 & 5 \\ -2 & 1 & 0 \\ 0 & -1 & 3 \end{vmatrix}.$$

Expanding A_{33} by column 1, we obtain

$$A_{33} = (1)\begin{vmatrix} 1 & 0 \\ -1 & 3 \end{vmatrix} + (-2)\left(-\begin{vmatrix} 0 & 5 \\ -1 & 3 \end{vmatrix}\right) + 0\begin{vmatrix} 0 & 5 \\ 1 & 0 \end{vmatrix}.$$

$$= 3 + 10 + 0 = 13.$$

Therefore

$$|A| = -3A_{33} = (-3)(13) = -39.$$

In general, if all but one element a in some row (or column) of A is zero and if the determinant $|A|$ is expanded by that row (or column), then all terms drop out except the product of the element a with its cofactor. We will make important use of this fact in the next section.

If *every* element in a row (or column) of a matrix A is zero, then upon expanding $|A|$ by that row (or column) we obtain the number 0. This gives us the following result.

(6.29)   Theorem

> If every element of a row (or column) of a square matrix A is zero, then $|A| = 0$.

EXERCISES

In each of Exercises 1–4 find all the minors and cofactors of the elements in the given matrix.

1. $\begin{bmatrix} 2 & 4 & -1 \\ 0 & 3 & 2 \\ -5 & 7 & 0 \end{bmatrix}$

2. $\begin{bmatrix} 5 & -2 & 1 \\ 4 & 7 & 0 \\ -3 & 4 & -1 \end{bmatrix}$

3. $\begin{bmatrix} 7 & -1 \\ 5 & 0 \end{bmatrix}$

4. $\begin{bmatrix} -6 & 4 \\ 3 & 2 \end{bmatrix}$

5–8   Find the determinants of the matrices given in Exercises 1–4.

Find the determinants of the matrices in Exercises 9–20.

9. $\begin{bmatrix} -5 & 4 \\ -3 & 2 \end{bmatrix}$

10. $\begin{bmatrix} 6 & 4 \\ -3 & 2 \end{bmatrix}$

11. $\begin{bmatrix} a & -a \\ b & -b \end{bmatrix}$

12. $\begin{bmatrix} c & d \\ -d & c \end{bmatrix}$

13. $\begin{bmatrix} 3 & 1 & -2 \\ 4 & 2 & 5 \\ -6 & 3 & -1 \end{bmatrix}$

14. $\begin{bmatrix} 2 & -5 & 1 \\ -3 & 1 & 6 \\ 4 & -2 & 3 \end{bmatrix}$

15. $\begin{bmatrix} -5 & 4 & 1 \\ 3 & -2 & 7 \\ 2 & 0 & 6 \end{bmatrix}$

16. $\begin{bmatrix} 2 & 7 & -3 \\ 1 & 0 & 4 \\ 4 & -1 & -2 \end{bmatrix}$

17. $\begin{bmatrix} 3 & -1 & 2 & 0 \\ 4 & 0 & -3 & 5 \\ 0 & 6 & 0 & 0 \\ 1 & 3 & -4 & 2 \end{bmatrix}$

18. $\begin{bmatrix} 2 & 5 & 1 & 0 \\ -4 & 0 & -3 & 0 \\ 3 & -2 & 1 & 6 \\ -1 & 4 & 2 & 0 \end{bmatrix}$

$$
19 \quad \begin{bmatrix} 0 & b & 0 & 0 \\ 0 & 0 & c & 0 \\ a & 0 & 0 & 0 \\ 0 & 0 & 0 & d \end{bmatrix}
\qquad\qquad
20 \quad \begin{bmatrix} a & u & v & w \\ 0 & b & x & y \\ 0 & 0 & c & z \\ 0 & 0 & 0 & d \end{bmatrix}
$$

Verify the identities in Exercises 21–28 by expanding each determinant.

21 $\quad \begin{vmatrix} a & b \\ c & d \end{vmatrix} = - \begin{vmatrix} c & d \\ a & b \end{vmatrix}$
$\qquad\qquad$
22 $\quad \begin{vmatrix} a & b \\ c & d \end{vmatrix} = - \begin{vmatrix} b & a \\ d & c \end{vmatrix}$

23 $\quad \begin{vmatrix} a & kb \\ c & kd \end{vmatrix} = k \begin{vmatrix} a & b \\ c & d \end{vmatrix}$
$\qquad\qquad$
24 $\quad \begin{vmatrix} a & b \\ kc & kd \end{vmatrix} = k \begin{vmatrix} a & b \\ c & d \end{vmatrix}$

25 $\quad \begin{vmatrix} a & b \\ c & d \end{vmatrix} = \begin{vmatrix} a & b \\ ka+c & kb+d \end{vmatrix}$
$\qquad\qquad$
26 $\quad \begin{vmatrix} a & b \\ c & d \end{vmatrix} = \begin{vmatrix} a & ka+b \\ c & kc+d \end{vmatrix}$

27 $\quad \begin{vmatrix} a & b \\ c & d \end{vmatrix} + \begin{vmatrix} a & e \\ c & f \end{vmatrix} = \begin{vmatrix} a & b+e \\ c & d+f \end{vmatrix}$
$\qquad\qquad$
28 $\quad \begin{vmatrix} a & b \\ c & d \end{vmatrix} + \begin{vmatrix} a & b \\ e & f \end{vmatrix} = \begin{vmatrix} a & b \\ c+e & d+f \end{vmatrix}$

29   Prove that if a matrix A of order 2 has two identical rows or columns, then $|A| = 0$.

30   Verify (6.25) for the matrix

$$
A = \begin{bmatrix} a_{11} & a_{12} & a_{13} \\ a_{21} & a_{22} & a_{23} \\ a_{31} & a_{32} & a_{33} \end{bmatrix} .
$$

8   PROPERTIES OF DETERMINANTS

The method of evaluating a determinant by means of the Expansion Theorem (6.28) is not very efficient for matrices of high order. For example, if a determinant of a matrix of order 10 is expanded by any row, a sum of 10 terms is obtained, where each term contains the determinant of a matrix of order 9 (that is, a cofactor of the original matrix). If any of the latter determinants is expanded by a row (or column), a sum of 9 terms is obtained, each containing the determinant of a matrix of order 8. Hence, at this stage there are 90 determinants of matrices of order 8 to evaluate! The process could be continued until only determinants of matrices of order 2 remain. Unless many elements of the original matrix are zero, it is an enormous task to carry out all of the computations.

We shall now consider some rules which make the process of evaluating determinants simpler. These rules are used mainly for introducing zeros into the determinant. They may also be used to change the determinant to echelon form, that is, a form in which the elements below the main diagonal elements a_{ii} are all zero. The transformations on rows given in (6.30) are the same as the elementary row transformations of a matrix given in (6.9). However, for determinants we may also employ similar transformations on columns.

(6.30)   Row and Column Transformations of a Determinant

> Let A be a matrix of order n.
>
> (i)   If a matrix B is obtained from A by interchanging two rows (or columns), then $|B| = -|A|$.
>
> (ii)   If B is obtained from A by multiplying every element of one row (or column) of A by a real number k, then $|B| = k|A|$.
>
> (iii)   If B is obtained from A by adding to any row (or column) of A, k times another row (or column), where k is any real number, then $|B| = |A|$.

When using (6.30) to justify manipulations with determinants, we shall refer to the rows (or columns) of the *determinant* in the obvious way. For example, property (iii) may be phrased: "Adding the product of k times another row (or column) to any row (or column) of a determinant does not affect the value of the determinant."

We shall not give the general proof of (6.30). For the case of matrices of orders 2 or 3 the theorem can be proved by evaluating $|B|$. For example, given a matrix of order 3 as in (6.19), suppose that B is the matrix obtained by adding to row 2 the product of k times row 1, that is,

$$B = \begin{bmatrix} a_{11} & a_{12} & a_{13} \\ ka_{11} + a_{21} & ka_{12} + a_{22} & ka_{13} + a_{23} \\ a_{31} & a_{32} & a_{33} \end{bmatrix}.$$

To evaluate $|B|$ we expand by row 2, obtaining

$$|B| = (ka_{11} + a_{21})\left(-\begin{vmatrix} a_{12} & a_{13} \\ a_{32} & a_{33} \end{vmatrix}\right) + (ka_{12} + a_{22})\begin{vmatrix} a_{11} & a_{13} \\ a_{31} & a_{33} \end{vmatrix}$$

$$+ (ka_{13} + a_{23})\left(-\begin{vmatrix} a_{11} & a_{12} \\ a_{31} & a_{32} \end{vmatrix}\right).$$

The determinants which appear in this equation are those associated with the cofactors A_{21}, A_{22}, and A_{23} of the original matrix. Hence

$$|B| = (ka_{11} + a_{21})A_{21} + (ka_{12} + a_{22})A_{22} + (ka_{13} + a_{23})A_{23}$$

which may also be written in the form

$$|B| = k(a_{11}A_{21} + a_{12}A_{22} + a_{13}A_{23}) + (a_{21}A_{21} + a_{22}A_{22} + a_{23}A_{23}).$$

By (6.28) the second expression in parentheses equals $|A|$. Actually carrying out the computations shows that the first expression in parentheses is zero. Consequently

$$|B| = 0 + |A| = |A|.$$

Statement (i) of (6.30) is often phrased: "Interchanging two rows (or columns) changes the sign of a determinant." As illustrations we have

$$\begin{vmatrix} 2 & 0 & 1 \\ 6 & 4 & 3 \\ 0 & 3 & 5 \end{vmatrix} = -\begin{vmatrix} 6 & 4 & 3 \\ 2 & 0 & 1 \\ 0 & 3 & 5 \end{vmatrix} \quad \text{and} \quad \begin{vmatrix} 2 & 0 & 1 \\ 6 & 4 & 3 \\ 0 & 3 & 5 \end{vmatrix} = -\begin{vmatrix} 1 & 0 & 2 \\ 3 & 4 & 6 \\ 5 & 3 & 0 \end{vmatrix}$$

where in the first case rows 1 and 2 were interchanged and in the second case we interchanged columns 1 and 3.

As an illustration of (iii) of (6.30),

$$\begin{vmatrix} 1 & -3 & 4 \\ 2 & -1 & 0 \\ 3 & 1 & 6 \end{vmatrix} = \begin{vmatrix} 1 & -3 & 4 \\ 0 & 5 & -8 \\ 3 & 1 & 6 \end{vmatrix}$$

where we have added to the second row the product of -2 times the first row. This type of manipulation is very important for evaluating determinants of large order as will be shown below in Example 1. In like manner,

$$\begin{vmatrix} 1 & -3 & 4 \\ 2 & -1 & 0 \\ 3 & 1 & 6 \end{vmatrix} = \begin{vmatrix} -5 & -3 & 4 \\ 0 & -1 & 0 \\ 5 & 1 & 6 \end{vmatrix}$$

where we have added to the first column the product of 2 times the second column.

Rule (ii) of (6.30) will be illustrated in Examples 2 and 3.

(6.31)   Theorem

> If two rows (or columns) of a square matrix A are identical, then $|A| = 0$.

Proof. If B is the matrix obtained from A by interchanging the two identical rows (or columns), then B and A are the same and consequently $|B| = |A|$. However, by (i) of (6.30), $|B| = -|A|$ and hence $-|A| = |A|$, which implies that $|A| = 0$.

We shall illustrate the use of the previous theorems by means of several examples.

Example 1   Find $|A|$ if

$$A = \begin{bmatrix} 2 & 3 & 0 & 4 \\ 0 & 5 & -1 & 6 \\ 1 & 0 & -2 & 3 \\ -3 & 2 & 0 & -5 \end{bmatrix}.$$

Solution   We plan to use (iii) of (6.30) to introduce many zeros in some row or column. To do this, it is convenient to work with an element of the matrix which equals 1 or -1, since this enables us to avoid the use of fractions. If no such element appears in the original matrix, it is always possible to introduce the number 1 by using (iii) or (ii) of (6.30). In this example there is no such problem, since 1 appears in row 3 and -1 in row 2. Let us work with the element 1 and introduce zero everywhere else in the first column, as shown below:

$$|A| = \begin{vmatrix} 0 & 3 & 4 & -2 \\ 0 & 5 & -1 & 6 \\ 1 & 0 & -2 & 3 \\ 0 & 2 & -6 & 4 \end{vmatrix}$$

Add to the first row -2 times the third row and then add to the fourth row 3 times the third row.

$$= (1) \begin{vmatrix} 3 & 4 & -2 \\ 5 & -1 & 6 \\ 2 & -6 & 4 \end{vmatrix}$$

Expand by column 1.

$$= \begin{vmatrix} 23 & 4 & 22 \\ 0 & -1 & 0 \\ -28 & -6 & -32 \end{vmatrix}$$

Add to the first column 5 times the second column and then add to the third column 6 times the second column.

$$= (-1) \begin{vmatrix} 23 & 22 \\ -28 & -32 \end{vmatrix}$$

Expand by row 2.

$$= (-1)[(23)(-32) - (-28)(22)]$$ By (6.17)

$$= 120.$$

Part (ii) of (6.30) is useful for finding factors of determinants. To illustrate, for a determinant of a matrix of order 3, we have the following:

$$\begin{vmatrix} a_{11} & a_{12} & a_{13} \\ ka_{21} & ka_{22} & ka_{23} \\ a_{31} & a_{32} & a_{33} \end{vmatrix} = k \begin{vmatrix} a_{11} & a_{12} & a_{13} \\ a_{21} & a_{22} & a_{23} \\ a_{31} & a_{32} & a_{33} \end{vmatrix}.$$

Similar formulas hold if k is a common factor of the elements of some other row or column. When (6.30) is used in this way, we often use the phrase "k is a common factor in the row (or column)."

Example 2   Find $|A|$ if

$$A = \begin{bmatrix} 14 & -6 & 4 \\ 4 & -5 & 12 \\ -21 & 9 & -6 \end{bmatrix}.$$

Solution

$$|A| = 2 \begin{vmatrix} 7 & -3 & 2 \\ 4 & -5 & 12 \\ -21 & 9 & -6 \end{vmatrix}$$

2 is a common factor in row 1.

$$= (2)(-3) \begin{vmatrix} 7 & -3 & 2 \\ 4 & -5 & 12 \\ 7 & -3 & 2 \end{vmatrix}$$

-3 is a common factor in row 3.

$$= 0$$

By (6.31).

Example 3 Without expanding, show that $a - b$ is a factor of

$$\begin{vmatrix} 1 & 1 & 1 \\ a & b & c \\ a^2 & b^2 & c^2 \end{vmatrix}.$$

Solution

$$\begin{vmatrix} 1 & 1 & 1 \\ a & b & c \\ a^2 & b^2 & c^2 \end{vmatrix} = \begin{vmatrix} 0 & 1 & 1 \\ a-b & b & c \\ a^2-b^2 & b^2 & c^2 \end{vmatrix}$$

Add to the first column -1 times the second column.

$$= (a-b) \begin{vmatrix} 0 & 1 & 1 \\ 1 & b & c \\ a+b & b^2 & c^2 \end{vmatrix}$$

$a - b$ is a common factor of column 1.

EXERCISES

Without expanding, explain why the statements in Exercises 1–14 are true.

1 $\begin{vmatrix} 1 & 0 & 1 \\ 0 & 1 & 1 \\ 1 & 1 & 0 \end{vmatrix} = - \begin{vmatrix} 1 & 0 & 1 \\ 1 & 1 & 0 \\ 0 & 1 & 1 \end{vmatrix}$

2 $\begin{vmatrix} 1 & 0 & 1 \\ 0 & 1 & 1 \\ 1 & 1 & 0 \end{vmatrix} = - \begin{vmatrix} 1 & 1 & 0 \\ 0 & 1 & 1 \\ 1 & 0 & 1 \end{vmatrix}$

3 $\begin{vmatrix} 1 & 0 & 1 \\ 2 & 1 & 0 \\ 1 & 1 & 2 \end{vmatrix} = \begin{vmatrix} 1 & 0 & 1 \\ 2 & 1 & 0 \\ 0 & 1 & 1 \end{vmatrix}$

4 $\begin{vmatrix} 1 & 1 & 2 \\ 1 & 0 & 1 \\ 2 & 1 & 1 \end{vmatrix} = \begin{vmatrix} 0 & 1 & 1 \\ 1 & 0 & 1 \\ 2 & 1 & 1 \end{vmatrix}$

5 $\begin{vmatrix} 2 & 4 & 2 \\ 1 & 2 & 4 \\ 2 & 6 & 4 \end{vmatrix} = 4 \begin{vmatrix} 1 & 2 & 1 \\ 1 & 2 & 4 \\ 1 & 3 & 2 \end{vmatrix}$

6 $\begin{vmatrix} 2 & 1 & 6 \\ 4 & 3 & 3 \\ 2 & 1 & 3 \end{vmatrix} = 6 \begin{vmatrix} 1 & 1 & 2 \\ 2 & 3 & 1 \\ 1 & 1 & 1 \end{vmatrix}$

7 $\begin{vmatrix} 1 & -1 & 2 \\ 1 & 2 & -1 \\ 1 & -1 & 2 \end{vmatrix} = 0$

8 $\begin{vmatrix} 1 & -1 & 1 \\ 0 & 1 & 0 \\ -1 & 0 & -1 \end{vmatrix} = 0$

9 $\begin{vmatrix} 1 & 5 \\ -3 & 2 \end{vmatrix} = - \begin{vmatrix} 1 & 5 \\ 3 & -2 \end{vmatrix}$

10 $\begin{vmatrix} 2 & -2 \\ 1 & 1 \end{vmatrix} = - \begin{vmatrix} -2 & 2 \\ 1 & 1 \end{vmatrix}$

11 $\begin{vmatrix} 0 & 0 & 1 \\ 1 & 0 & 0 \\ 0 & 0 & 1 \end{vmatrix} = 0$
12 $\begin{vmatrix} 1 & 0 & 1 \\ 0 & 0 & 0 \\ 1 & 1 & 0 \end{vmatrix} = 0$

13 $\begin{vmatrix} 1 & -1 & -2 \\ -1 & 2 & 1 \\ 0 & 1 & 1 \end{vmatrix} = \begin{vmatrix} 1 & -1 & 0 \\ -1 & 2 & -1 \\ 0 & 1 & 1 \end{vmatrix}$

14 $\begin{vmatrix} a & 0 & 0 \\ 0 & b & 0 \\ 0 & 0 & c \end{vmatrix} = - \begin{vmatrix} 0 & 0 & a \\ 0 & b & 0 \\ c & 0 & 0 \end{vmatrix}$

In each of Exercises 15–24 find the determinant of the matrix after introducing zeros as in Example 1.

15 $\begin{bmatrix} 3 & 1 & 0 \\ -2 & 0 & 1 \\ 1 & 3 & -1 \end{bmatrix}$
16 $\begin{bmatrix} -3 & 0 & 4 \\ 1 & 2 & 0 \\ 4 & 1 & -1 \end{bmatrix}$

17 $\begin{bmatrix} 5 & 4 & 3 \\ -3 & 2 & 1 \\ 0 & 7 & -2 \end{bmatrix}$
18 $\begin{bmatrix} 0 & 2 & -6 \\ 5 & 1 & -3 \\ 6 & -2 & 5 \end{bmatrix}$

19 $\begin{bmatrix} 2 & 2 & -3 \\ 3 & 6 & 9 \\ -2 & 5 & 4 \end{bmatrix}$
20 $\begin{bmatrix} 3 & 8 & 5 \\ 5 & 3 & -6 \\ 2 & 4 & -2 \end{bmatrix}$

21 $\begin{bmatrix} 3 & 1 & -2 & 2 \\ 2 & 0 & 1 & 4 \\ 0 & 1 & 3 & 5 \\ -1 & 2 & 0 & -3 \end{bmatrix}$
22 $\begin{bmatrix} 3 & 2 & 0 & 4 \\ -2 & 0 & 5 & 0 \\ 4 & -3 & 1 & 6 \\ 2 & -1 & 2 & 0 \end{bmatrix}$

23 $\begin{bmatrix} 2 & -2 & 0 & 0 & -3 \\ 3 & 0 & 3 & 2 & -1 \\ 0 & 1 & -2 & 0 & 2 \\ -1 & 2 & 0 & 3 & 0 \\ 0 & 4 & 1 & 0 & 0 \end{bmatrix}$
24 $\begin{bmatrix} 2 & 0 & -1 & 0 & 2 \\ 1 & 3 & 0 & 0 & 1 \\ 0 & 4 & 3 & 0 & -1 \\ -1 & 2 & 0 & -2 & 0 \\ 0 & 1 & 5 & 0 & -4 \end{bmatrix}$

25 Prove that

$$\begin{vmatrix} 1 & 1 & 1 \\ a & b & c \\ a^2 & b^2 & c^2 \end{vmatrix} = (a - b)(b - c)(c - a).$$

(Hint: See Example 3.)

26 Prove that

$$\begin{vmatrix} 1 & 1 & 1 \\ a & b & c \\ a^3 & b^3 & c^3 \end{vmatrix} = (a - b)(b - c)(c - a)(a + b + c).$$

27   If A is a matrix of order 4 of the form

$$A = \begin{bmatrix} a_{11} & a_{12} & a_{13} & a_{14} \\ 0 & a_{22} & a_{23} & a_{24} \\ 0 & 0 & a_{33} & a_{34} \\ 0 & 0 & 0 & a_{44} \end{bmatrix}$$

show that $|A| = a_{11}a_{22}a_{33}a_{44}$.

28   If

$$A = \begin{bmatrix} a & b & 0 & 0 \\ c & d & 0 & 0 \\ 0 & 0 & e & f \\ 0 & 0 & g & h \end{bmatrix}$$

prove that

$$|A| = \begin{vmatrix} a & b \\ c & d \end{vmatrix} \begin{vmatrix} e & f \\ g & h \end{vmatrix}.$$

29   If $A = (a_{ij})$ and $B = (b_{ij})$ are arbitrary square matrices of order 2, prove that $|AB| = |A||B|$.

30   If $A = (a_{ij})$ is a square matrix of order n and k is any real number, prove that $|kA| = k^n|A|$. (*Hint:* Use (ii) of (8.30).)

9   CRAMER'S RULE

Determinants arise naturally in the study of solutions of systems of linear equations. To illustrate, let us consider the case of two linear equations in two unknowns:

(6.32)
$$\begin{cases} a_{11}x + a_{12}y = k_1 \\ a_{21}x + a_{22}y = k_2 \end{cases}$$

where at least one nonzero coefficient appears in each equation. We may as well assume that $a_{11} \neq 0$, for otherwise $a_{12} \neq 0$ and we could regard y as the "first" variable instead of x. We shall use the matrix method to obtain the matrix of an equivalent system. Transforming the matrix of the system by means of (6.9), we obtain

$$\begin{bmatrix} a_{11} & a_{12} & k_1 \\ a_{21} & a_{22} & k_2 \end{bmatrix}$$

$$\rightarrow \begin{bmatrix} a_{11} & a_{12} & k_1 \\ 0 & a_{22} - \left(\dfrac{a_{12}a_{21}}{a_{11}}\right) & k_2 - \left(\dfrac{a_{21}k_1}{a_{11}}\right) \end{bmatrix} \quad \begin{array}{l}\text{Add to the second row} \\ -a_{21}/a_{11} \text{ times the first} \\ \text{row.}\end{array}$$

$$\rightarrow \begin{bmatrix} a_{11} & a_{12} & k_1 \\ 0 & (a_{11}a_{22} - a_{12}a_{21}) & (a_{11}k_2 - a_{21}k_1) \end{bmatrix} \quad \text{Multiply row 2 by } a_{11}.$$

Thus the given system is equivalent to

$$\begin{cases} a_{11}x + a_{12}y = k_1 \\ (a_{11}a_{22} - a_{12}a_{21})y = a_{11}k_2 - a_{21}k_1. \end{cases}$$

Notice that the numbers in the second equation may be written in determinant form as follows:

(6.33)

$$\begin{cases} a_{11}x + a_{12}y = k_1 \\ \begin{vmatrix} a_{11} & a_{12} \\ a_{21} & a_{22} \end{vmatrix} y = \begin{vmatrix} a_{11} & k_1 \\ a_{21} & k_2 \end{vmatrix}. \end{cases}$$

The following results now follow from the discussion in Section 2.

(a)   If

$$\begin{vmatrix} a_{11} & a_{12} \\ a_{21} & a_{22} \end{vmatrix} = 0 \quad \text{and} \quad \begin{vmatrix} a_{11} & k_1 \\ a_{21} & k_2 \end{vmatrix} = 0$$

then the equations are dependent.

(b)   If

$$\begin{vmatrix} a_{11} & a_{12} \\ a_{21} & a_{22} \end{vmatrix} = 0 \quad \text{and} \quad \begin{vmatrix} a_{11} & k_1 \\ a_{21} & k_2 \end{vmatrix} \neq 0$$

then the equations are inconsistent.

(c)   If

$$\begin{vmatrix} a_{11} & a_{12} \\ a_{21} & a_{22} \end{vmatrix} \neq 0$$

then the equations are consistent.

If (c) occurs, then we can solve the second equation of (6.33) for y, obtaining

(6.34)

$$y = \frac{\begin{vmatrix} a_{11} & k_1 \\ a_{21} & k_2 \end{vmatrix}}{\begin{vmatrix} a_{11} & a_{12} \\ a_{21} & a_{22} \end{vmatrix}}.$$

The corresponding value for x may be found by substituting for y in the first equation. It can be shown that this leads to

(6.35)

$$x = \frac{\begin{vmatrix} k_1 & a_{12} \\ k_2 & a_{22} \end{vmatrix}}{\begin{vmatrix} a_{11} & a_{12} \\ a_{21} & a_{22} \end{vmatrix}}.$$

This proves that *if the determinant of the coefficient matrix of* (6.32) *is not zero, then the system has a unique solution given by* (6.34) *and* (6.35). The last two formulas constitute what is known as **Cramer's Rule** for the solution of a system of two linear equations in two variables.

There is an easy way to remember Cramer's Rule. Let

$$D = \begin{bmatrix} a_{11} & a_{12} \\ a_{21} & a_{22} \end{bmatrix}$$

be the coefficient matrix of the system and let D_x denote the matrix obtained from D by replacing the coefficients a_{11}, a_{21} of x by the numbers k_1, k_2, respectively. Similarly, let D_y denote the matrix obtained from D by replacing the coefficients a_{12}, a_{22} of y by the numbers k_1, k_2, respectively. Thus

$$D_x = \begin{bmatrix} k_1 & a_{12} \\ k_2 & a_{22} \end{bmatrix}, \quad D_y = \begin{bmatrix} a_{11} & k_1 \\ a_{21} & k_2 \end{bmatrix}.$$

According to (6.35) and (6.34), if $|D| \neq 0$, the solution (x, y) is given by

(6.36)

$$\boxed{x = \frac{|D_x|}{|D|}, \quad y = \frac{|D_y|}{|D|}}$$

Example 1   Use Cramer's Rule to solve the system

$$\begin{cases} 2x - 3y = -4 \\ 5x + 7y = 1. \end{cases}$$

Solution   The determinant of the coefficient matrix is

$$|D| = \begin{vmatrix} 2 & -3 \\ 5 & 7 \end{vmatrix} = 29.$$

Using the notation introduced above,

$$|D_x| = \begin{vmatrix} -4 & -3 \\ 1 & 7 \end{vmatrix} = -25, \quad |D_y| = \begin{vmatrix} 2 & -4 \\ 5 & 1 \end{vmatrix} = 22.$$

Hence by (6.36)

$$x = \frac{|D_x|}{|D|} = \frac{-25}{29}, \quad y = \frac{|D_y|}{|D|} = \frac{22}{29}.$$

Thus the system has the unique solution $(-25/29, 22/29)$.

Let us briefly consider the case in which each of the equations in (6.32) is homogeneous, that is, $k_1 = 0$ and $k_2 = 0$. In this event the determinants $|D_x|$ and $|D_y|$ in (6.36) are both zero. (Why?) Consequently, if

$$|D| = \begin{vmatrix} a_{11} & a_{12} \\ a_{21} & a_{22} \end{vmatrix} \neq 0$$

then $x = 0/|D| = 0$ and $y = 0/|D| = 0$; that is, the only solution is the trivial one $(0, 0)$. This proves that *if a system of homogeneous equations*

$$\begin{cases} a_{11}x + a_{12}y = 0 \\ a_{21}x + a_{22}y = 0 \end{cases}$$

has a nontrivial solution, then

$$\begin{vmatrix} a_{11} & a_{12} \\ a_{21} & a_{22} \end{vmatrix} = 0.$$

It can be shown, conversely, that if the determinant of the coefficient matrix is zero, then a homogeneous system has a nontrivial solution.

The previous discussion can be extended to systems of n linear equations in n variables. It is possible to show that such a system has a unique solution if and only if the determinant of the coefficient matrix is different from zero. If the system is homogeneous, then nontrivial solutions exist if and only if the determinant of the coefficient matrix is zero.

Cramer's Rule can be extended to systems of n linear equations in n variables $x_1, x_2, \ldots, x_n$, where each equation is written in the form

$$a_1 x_1 + a_2 x_2 + \cdots + a_n x_n = a.$$

To solve such a system, let D denote the coefficient matrix and let D_{x_i} denote the matrix obtained by replacing the coefficients of x_i in D by the column of numbers $k_1, \ldots, k_n$, which appears to the right of the equals signs in the system. It can be shown that if $|D| \neq 0$, then the system has a unique solution given by

(6.37)

$$x_1 = \frac{|D_{x_1}|}{|D|}, \quad x_2 = \frac{|D_{x_2}|}{|D|}, \ldots, \quad x_n = \frac{|D_{x_n}|}{|D|}$$

It can be shown that if $|D| = 0$, the equations are dependent or inconsistent, depending on whether all the D_{x_i} are zero or at least one of them is not zero.

Example 2   Use Cramer's Rule to solve the system

$$\begin{cases} x & - 2z = 3 \\ & - y + 3z = 1 \\ 2x & + 5z = 0. \end{cases}$$

Solution   We shall merely list the various determinants which are used, leaving the reader to check the answers:

$$|D| = \begin{vmatrix} 1 & 0 & -2 \\ 0 & -1 & 3 \\ 2 & 0 & 5 \end{vmatrix} = -9, \quad |D_x| = \begin{vmatrix} 3 & 0 & -2 \\ 1 & -1 & 3 \\ 0 & 0 & 5 \end{vmatrix} = -15,$$

$$|D_y| = \begin{vmatrix} 1 & 3 & -2 \\ 0 & 1 & 3 \\ 2 & 0 & 5 \end{vmatrix} = 27, \quad |D_z| = \begin{vmatrix} 1 & 0 & 3 \\ 0 & -1 & 1 \\ 2 & 0 & 0 \end{vmatrix} = 6.$$

By (6.37) the solution is

$$x = \frac{|D_x|}{|D|} = \frac{-15}{-9} = \frac{5}{3}, \quad y = \frac{|D_y|}{|D|} = \frac{27}{-9} = -3, \quad z = \frac{|D_z|}{|D|} = \frac{6}{-9} = -\frac{2}{3}.$$

Cramer's Rule is an inefficient method to apply if there are a large number of equations, since many determinants of matrices of high order must be evaluated. Note also that Cramer's Rule cannot be used directly if $|D| = 0$ or if the number of equations is not the same as the number of variables. In general, the method of elimination or the matrix method is far superior to Cramer's Rule.

EXERCISES

1–18   Use Cramer's Rule to solve the systems in Exercises 1–18 of Section 2.

19–26   Use Cramer's Rule to solve the systems in Exercises 1–8 of Section 3.

27–30   Use Cramer's Rule to solve the systems in Exercises 17–20 of Section 3.

10   SYSTEMS OF INEQUALITIES

Our previous work with inequalities was restricted to inequalities in one variable. The notion of inequalities in several variables can be developed in a manner similar to our work with equations in several variables. For example, expressions of the form

$$3x + y < 5y^2 + 1$$
$$2x^2 \geq 4 - 3y$$

and so on, are called **inequalities in x and y.** As with equations, a **solution** of an inequality in x and y is defined as an ordered pair (a, b) which produces a true statement when a and b are substituted for x and y, respectively. The **graph of an inequality** is the graph of the totality of solutions. Two inequalities are **equivalent** if they have exactly the same solutions. It is possible to prove the analogues of (2.3) and (2.4) in the present situation; that is, an inequality in x and y can be simplified by adding an expression in x and y to both sides or by multiplying both sides by such an expression, provided we are careful about signs. Similar definitions and remarks hold for inequalities in more than two variables. We shall restrict our discussion, however, to the case of inequalities in two variables.

Example 1   Find the solutions and sketch the graph of the inequality $3x - 3 < 5x - y.$

Solution   Adding the expression $y + 3 - 3x$ to both sides, we obtain the equivalent inequality $y < 2x + 3.$ Hence the solutions consist of all ordered pairs (x, y) such that $y < 2x + 3$. It is convenient to denote the set of solutions as follows:

$$\{(x, y): y < 2x + 3\}.$$

There is a close relationship between the graph of the inequality $y < 2x + 3$ and the graph of the equation $y = 2x + 3$. The graph of the equation is the straight line sketched in Figure 6.5. For each real number a, the point on the line

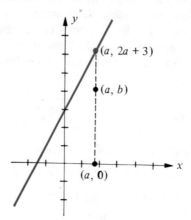

Figure 6.5

with abscissa a has coordinates $(a, 2a + 3)$. A point $P(a, b)$ belongs to the graph of the *inequality* if and only if $b < 2a + 3$; that is, if and only if the point $P(a, b)$ lies directly below the point with coordinates $(a, 2a + 3)$ as shown in Figure 6.5. It follows that the graph of the inequality $y < 2x + 3$ consists of all points in the plane which lie below the line $y = 2x + 3$. In Figure 6.6 we have shaded a portion of the graph. Dashes used for the line indicate that it is not part of the graph of the inequality.

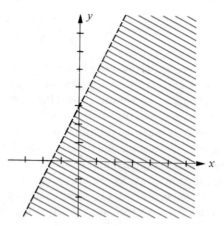

Figure 6.6. $y < 2x + 3$

A region of the type shown in Figure 6.6 is called a **half-plane**. More precisely, when the line is *not* included, we refer to such a region as an **open** half-plane. If the line *is* included, as would be the case for the graph of the inequality $y \leq 2x + 3$, then the region is called a **closed half-plane**.

By an argument similar to that used in Example 1, it can be shown that the graph of the inequality $y > 2x + 3$ is the open half-plane which lies *above* the line with equation $y = 2x + 3$.

If an inequality involves only polynomials of the first degree in x and y, as was the case in Example 1, it is called a **linear inequality**.

The procedure used in Example 1 can be generalized to inequalities of the form $y < f(x)$, where f is a function. Specifically, the following can be established.

(6.38)

> If f is a function, then the graph of the inequality $y < f(x)$ is the set of points which lie *below* the graph of the equation $y = f(x)$. Similarly, the graph of $y > f(x)$ is the set of points which lie *above* the graph of $y = f(x)$.

Example 2 Find the solutions and sketch the graph of the inequality
$x(x + 1) - 2y > 3(x - y)$.

Solution The given inequality is equivalent to

$$x^2 + x - 2y > 3x - 3y.$$

Adding $3y - x^2 - x$ to both sides we obtain

$$y > 2x - x^2.$$

Hence the solutions are given by $\{(x, y): y > 2x - x^2\}$. To find the graph, we begin by sketching the graph of $y = 2x - x^2$ (a parabola) with dashes as illustrated in Figure 6.7. Using (6.38), the graph is the region above the parabola, as indicated by the shaded portion of the figure.

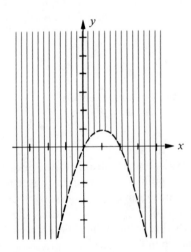

Figure 6.7. $y > 2x - x^2$

The following result is analogous to (6.38).

(6.39)   If g is a function, then the graph of the inequality $x < g(y)$ is the set of points to the *left* of the graph of the equation $x = g(y)$. Similarly, the graph of $x > g(y)$ is the set of points to the *right* of the graph of $x = g(y)$.

Example 3   Sketch the graph of $x \geq y^2$.

Solution   The graph of the equation $x = y^2$ is a parabola. By (6.39), the graph of the inequality consists of all points on the parabola together with the points in the region to the right of the parabola (see Figure 6.8).

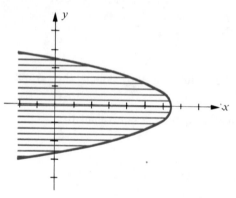

Figure 6.8. $x \geq y^2$

It is sometimes necessary to work simultaneously with several inequalities in two variables. In this case we refer to the given inequalities as a **system of inequalities**. The **solutions of a system** of inequalities are, by definition, the common solutions of all the inequalities in the system. It should be clear how to define **equivalent systems** and the **graph of a system** of inequalities. The following examples illustrate a method for solving systems of inequalities.

Example 4   Find the solutions and sketch the graph of the system $\begin{cases} x + y \leq 4 \\ 2x - y \leq 4. \end{cases}$

Solution   The given system is equivalent to

$$\begin{cases} y \leq 4 - x \\ y \geq 2x - 4. \end{cases}$$

We begin by sketching the graphs of the lines $y = 4 - x$ and $y = 2x - 4$. The lines intersect at the point $(\frac{8}{3}, \frac{4}{3})$ shown in Figure 6.9. The graph of $y \leq 4 - x$ includes the points on the graph of $y = 4 - x$ together with the points which lie below this line. The graph of $y \geq 2x - 4$ includes the points on the graph of $y = 2x - 4$ together with the points which lie above this line. A portion of each of these regions is shown in Figure 6.9. The graph of the system consists of the points that are in *both* regions. This corresponds to the cross-hatched region shown in Figure 6.9.

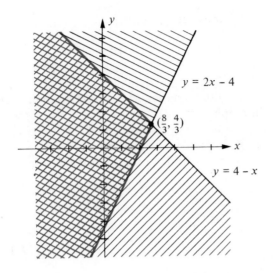

Figure 6.9

Example 5   Sketch the graph of the system

$$\begin{cases} x + y \le 4 \\ 2x - y \le 4 \\ x \ge 0 \\ y \ge 0. \end{cases}$$

Solution   The first two inequalities are the same as those considered in Example 4 and hence the points on the graph of the present system must lie within the region shown in Figure 6.10. In addition, the third and fourth inequalities in the system tell us that the points must lie in the first quadrant. This gives us the region shown in Figure 6.10.

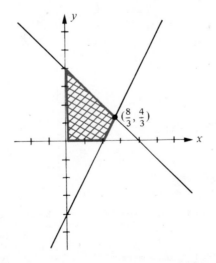

Figure 6.10

Example 6   Sketch the graph of the system

$$\begin{cases} x^2 + y^2 \le 1 \\ (x-1)^2 + y^2 \le 1. \end{cases}$$

Solution   The graph of the equation $x^2 + y^2 = 1$ is a unit circle with center at the origin, and the graph of $(x-1)^2 + y^2 = 1$ is a unit circle with center at the point $C(1,0)$. To find the points of intersection of the two circles, let us solve the equation $x^2 + y^2 = 1$ for y^2, obtaining $y^2 = 1 - x^2$. Substituting for y^2 in $(x-1)^2 + y^2 = 1$ leads to the following equations:

$$(x-1)^2 + (1 - x^2) = 1$$
$$x^2 - 2x + 1 + 1 - x^2 = 1$$
$$-2x = -1$$
$$x = 1/2.$$

The corresponding values for y are given by

$$y^2 = 1 - x^2 = 1 - (1/2)^2 = 3/4$$

and hence $y = \pm\sqrt{3}/2$. Thus the points of intersection are $(1/2, \sqrt{3}/2)$ and $(1/2, -\sqrt{3}/2)$ as shown in Figure 6.11. By the distance formula, the graphs of the given inequalities are the regions within and on the two circles. The graph of the system consists of the points common to both regions, as indicated by the shaded portion of the figure.

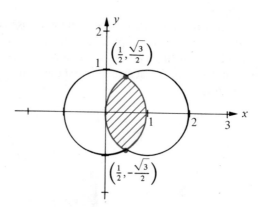

Figure 6.11

EXERCISES

In each of Exercises 1–10 find the solutions and sketch the graph of the given inequality.

1   $3x - 2y < 6$ 2 $4x + 3y < 12$

3   $2x + 3y \ge 2y + 1$ 4 $2x - y > 3$

5   $y + 2 < x^2$ 6 $y^2 - x \le 0$

7 $x^2 + 1 \le y$

8 $y - x^3 < 1$

9 $yx^2 \ge 1$

10 $x^2 + 4 \ge y$

In each of Exercises 11–24 sketch the graph of the given system.

11 $\begin{cases} 3x + y < 3 \\ 4 - y < 2x \end{cases}$

12 $\begin{cases} y + 2 < 2x \\ y - x > 4 \end{cases}$

13 $\begin{cases} y - x < 0 \\ 2x + 5y < 10 \end{cases}$

14 $\begin{cases} 2y - x \le 4 \\ 3y + 2x < 6 \end{cases}$

15 $\begin{cases} 3x + y \le 6 \\ y - 2x \ge 1 \\ x \ge -2 \\ y \le 4 \end{cases}$

16 $\begin{cases} 3x - 4y \ge 12 \\ x - 2y \le 2 \\ x \ge 9 \\ y \le 5 \end{cases}$

17 $\begin{cases} x^2 + y^2 \le 4 \\ x + y \ge 1 \end{cases}$

18 $\begin{cases} x^2 + y^2 > 1 \\ x^2 + y^2 < 4 \end{cases}$

19 $\begin{cases} x^2 \le 1 - y \\ x \ge 1 + y \end{cases}$

20 $\begin{cases} x - y^2 < 0 \\ x + y^2 > 0 \end{cases}$

21 $\begin{cases} y < 3^x \\ y > 2^x \\ x \ge 0 \end{cases}$

22 $\begin{cases} y \ge \log x \\ y - x \le 1 \\ x \ge 1 \end{cases}$

23 $\begin{cases} y \le \log x \\ y + x \ge 1 \\ x \le 10 \end{cases}$

24 $\begin{cases} y \le 3^{-x} \\ y \ge 2^{-x} \\ y < 9 \end{cases}$

25 The manager of a baseball team wishes to buy bats and balls, costing $3.50 and $2.50, respectively. If the maximum amount she can spend is $40 and she wants to buy at least two balls and three bats, find a system of inequalities describing all the possibilities and sketch the graph.

26 An office worker wishes to purchase some 40-cent postage stamps and also some 50-cent stamps, totaling not more than $35. Moreover, he wants at least twice as many 40-cent stamps as 50-cent stamps and more than ten 40-cent stamps. Find a system of inequalities describing all the possibilities and sketch the graph.

27 A store sells two brands of television sets. Customer demand indicates that it is necessary to stock at least twice as many sets of brand A as of brand B. It is also necessary to have on hand at least 20 of brand A and 10 of brand B. If there is room for not more than 100 sets in the store, find a system of inequalities describing all possibilities and sketch the graph.

28 An auditorium contains 600 seats. For a certain event it is planned to charge $4.00 for certain seats and $3.00 for others. At least 225 tickets are to be sold for $3.00, and total sales of more than $2,000 is desired. Find a system of inequalities describing all possibilities and sketch the graph.

29 A woman wishes to invest $15,000 in two different savings accounts. She also wants to have at least $2,000 in each account, with the amount in one account being at least three times that in the other. Find a system of inequalities describing all possibilities and sketch the graph.

30 The manager of a college book store stocks two types of notebooks, the first wholesaling for 55 cents and the second for 85 cents. If the maximum amount he may spend is $600 and he wants an inventory of at least 300 of the 85-cent variety and 400 of the 55-cent variety, find a system of inequalities describing all possibilities and sketch the graph.

11   LINEAR PROGRAMMING

In applications, problems sometimes arise which require finding solutions of systems of inequalities. A typical problem is that of finding maximum and minimum values of certain expressions involving variables which are subject to various constraints. If all the expressions and inequalities are linear in the variables, then a technique called **linear programming** may be used to help solve such problems. This technique has become very important in businesses where decisions must be made concerning the best use of stock, parts, manufacturing processes, and so on. Usually the objective of management is to maximize profit or to minimize cost. Since there are often many choices, it may be extremely difficult to arrive at a correct decision. A mathematical theory such as that afforded by linear programming can simplify the task considerably. The logical development of the theorems and techniques which are needed would take us beyond the objectives of this text. We shall, therefore, limit ourselves to several examples.

Example 1   A manufacturer of a certain product has two warehouses W_1 and W_2. There are 80 units of his product stored at W_1 and 70 units at W_2. Two customers A and B order 35 units and 60 units, respectively. The shipping cost from each warehouse to A and B is determined according to the following table. How should the order be filled so as to minimize the total shipping cost?

Warehouse	Customer	Shipping cost per unit
W_1	A	$ 8
W_1	B	12
W_2	A	10
W_2	B	13

Solution   If we let x denote the number of units to be sent to A from W_1, then $35 - x$ units must be sent from W_2 to A. Similarly, if y denotes the number of units to be sent from W_1 to B, then $60 - y$ units must be sent from W_2 to B. We wish to determine values for x and y which make the total shipping costs minimal. We first note that since x and y are between 35 and 60, respectively, the pair (x, y) must be a solution of the following system of inequalities:

(6.40) $$0 \leq x \leq 35, \quad 0 \leq y \leq 60.$$

The graph of this system is the rectangular region shown in Figure 6.12.

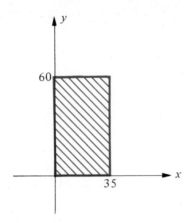

Figure 6.12

There are further constraints on x and y which make it possible to reduce the size of the region above. Since the total number of units shipped from W_1 cannot exceed 80 and the total shipped from W_2 cannot exceed 70, the pair (x, y) must also be a solution of the system

$$\begin{cases} x + y \le 80 \\ (35 - x) + (60 - y) \le 70. \end{cases}$$

This system is equivalent to

(6.41)
$$\begin{cases} x + y \le 80 \\ x + y \ge 25. \end{cases}$$

The graph of system (6.41) is the region between the parallel lines $x + y = 80$ and $x + y = 25$ (see Figure 6.13). The pair (x, y) which we seek must be a solution of

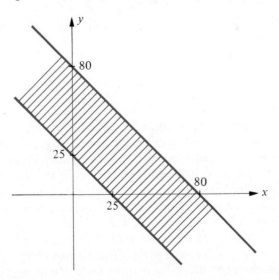

Figure 6.13

both (6.40) and (6.41) and hence the corresponding point must lie in the region shown in Figure 6.14.

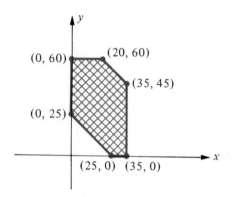

Figure 6.14

Now let C denote the total cost (in dollars) of shipping the merchandise to A and B. We see from the table that the cost of shipping the 35 units to A is $8x + 10(35 - x)$, whereas the cost of shipping the 60 units to B is $12y + 13(60 - y)$. Hence the total cost is

$$C = (8x + 350 - 10x) + (12y + 780 - 13y)$$

or

$$C = 1130 - 2x - y.$$

For each point (x, y) of the region shown in Figure 6.14 we obtain a value for C. For example, at $(20, 40)$

$$C = 1130 - 40 - 40 = 1050;$$

at $(10, 50)$,

$$C = 1130 - 20 - 50 = 1060;$$

and so on.

Since x and y must be integers, there are only a finite number of possible values for C. By checking each possibility, we could find the pair (x, y) which produces the smallest cost. However, since there are a very large number of pairs, the task of checking each one would be very tedious. This is where the theory developed in linear programming is helpful. It can be shown that if we are interested in the value C of a linear expression $ax + by + c$, where each pair (x, y) is a solution of a system of linear inequalities and hence corresponds to a point which is in the intersection of half-planes, then C takes on its maximum and minimum values at a point of intersection of the lines which determine the half-planes. This means that in order to determine the minimum (or maximum) value of C we need only check the points $(0, 25)$, $(0, 60)$, $(20, 60)$, $(35, 45)$, $(35, 0)$, and $(25, 0)$ shown in Figure 6.14. The values are arranged in tabular form below.

Point	$1130 - 2x - y = C$
$(0, 25)$	$1130 - 2(0) - 25 = 1105$
$(0, 60)$	$1130 - 2(0) - 60 = 1070$
$(20, 60)$	$1130 - 2(20) - 60 = 1030$
$(35, 45)$	$1130 - 2(35) - 45 = 1015$
$(35, 0)$	$1130 - 2(35) - 0 = 1060$
$(25, 0)$	$1130 - 2(25) - 0 = 1080$

According to our remarks, the minimal shipping cost $1,015 occurs if $x = 35$ and $y = 45$. This means that the manufacturer should ship all of the units to A from W_1. In addition, the manufacturer should ship 45 units to B from W_1 and 15 units to B from W_2. Note that the *maximum* shipping cost will occur when $x = 0$ and $y = 25$, that is, when all 35 units are shipped to A from W_2 and when B receives 25 units from W_1 and 35 units from W_2.

The preceding example illustrates how linear programming can be used to minimize the cost in a certain situation. The next example has to do with maximization of profit.

Example 2 A firm manufactures two products X and Y. For each product it is necessary to use three different machines A, B, and C. In order to manufacture one unit of product X, machine A must be used for 3 hours, machine B for 1 hour, and machine C for 1 hour. To manufacture one unit of product Y requires 2 hours on A, 2 hours on B, and 1 hour on C. The profit on product X is $500 per unit and the profit on product Y is $350 per unit. Machine A is available for a total of 24 hours per day; however, B can only be used for 16 hours and C for 9 hours. If the machines are available when needed (subject to the noted total hour restrictions), determine the number of units of each product that should be manufactured each day in order to maximize the profit.

Solution The following table summarizes the data given in the statement of the problem.

Machine	Hours required for 1 unit of X	Hours required for 1 unit of Y	Hours available
A	3	2	24
B	1	2	16
C	1	1	9

Let x and y denote the number of units of products X and Y, respectively, to be produced per day. Since each unit of product X requires 3 hours on machine A, x units require $3x$ hours. Similarly, since each unit of product Y requires 2 hours on A, y units require $2y$ hours. Hence the total number of hours per day that machine A must be used is $3x + 2y$. Since A can be used for at most 24 hours per day, we have

$$3x + 2y \leq 24.$$

Using the same type of reasoning on rows two and three of the table we see that

$$x + 2y \le 16$$
$$x + y \le 9.$$

This system of three linear inequalities, together with the obvious inequalities

$$x \ge 0, \quad y \ge 0$$

states, in mathematical form, the restraints which occur in the manufacturing process. The graph of the above system of five linear inequalities is sketched in Figure 6.15. The points shown in the figure are found by solving systems of linear equations. Specifically, $(6, 3)$ is a solution of the system $3x + 2y = 24$, $x + y = 9$, and $(2, 7)$ is a solution of the system $x + 2y = 16$, $x + y = 9$.

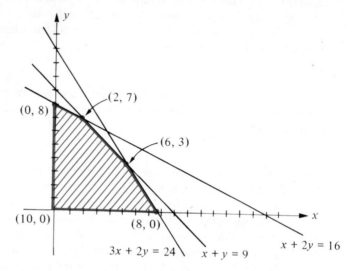

Figure 6.15

Since the production of each unit of product X results in a profit of $500, and each unit of product Y yields a profit of $350, the profit P obtained by producing x units of X together with y units of Y is given by

$$P = 500x + 350y.$$

As pointed out in the solution of Example 1, the maximum and minimum values of P occur at certain of the points shown in Figure 6.15. The values of P at all the points are given in the following table.

(x, y)	$500x + 350y = P$
$(0, 8)$	$500(0) + 350(8) = 2800$
$(2, 7)$	$500(2) + 350(7) = 3450$
$(6, 3)$	$500(6) + 350(3) = 4050$
$(8, 0)$	$500(8) + 350(0) = 4000$
$(0, 0)$	$500(0) + 350(0) = 0$

We see from the table that a maximum profit of $4,050 occurs if there is a daily production of 6 units of product X and 3 units of product Y.

The two illustrations we have given in this section are elementary problems in linear programming which can be solved by rather crude methods. The much more complicated problems that occur in practice are usually solved by employing matrix techniques which are adapted for solutions by computers.

EXERCISES

1   A manufacturer of tennis rackets makes a profit of $15 on each Set Point racket and $8 on each Double Fault racket. To meet dealer demand, daily production of Double Faults should be between 30 and 80, whereas the number of Set Points should be between 10 and 30. In order to maintain high quality, the total number of rackets produced should not exceed 80 per day. How many of each type should be manufactured daily to maximize the profit?

2   A manufacturer of CB radios makes a profit of $25 on a deluxe model and $30 on a standard model. The company wishes to produce at least 80 deluxe models and at least 100 standard models per day. To maintain high quality, the daily production should not exceed 200 radios. How many of each type should be produced daily in order to maximize the profit?

3   Two substances S and T each contain two types of ingredients I and G. One pound of S contains 2 ounces of I and 4 ounces of G. One pound of T contains 2 ounces of I and 6 ounces of G. It is desired to combine quantities of the two substances to obtain a mixture which contains at least 9 ounces of I and 20 ounces of G. If the cost of S is $3.00 per pound and the cost of T is $4.00 per pound, how much of each substance should be used to keep the cost to a minimum?

4   A stationery company makes two types of notebooks. Type M sells for $1.25 and type N sells for $0.90. It costs the company $1.00 to produce one type M notebook and $0.75 to produce one of type N. The company has the facilities to manufacture between 2,000 and 3,000 of type M and between 3,000 and 6,000 of type N, but not more than 7,000 altogether. How many notebooks of each type should be manufactured to maximize the difference between the selling prices and the costs of production?

5   In Example 1 of this section, if the shipping costs are $12 per unit from W_1 to X, $10 per unit from W_2 to X, $16 per unit from W_1 to Y, and $12 per unit from W_2 to Y, determine how the order should be filled so as to minimize shipping costs.

6   A coffee company purchases mixed lots of coffee beans and then grades them into premium, regular and unusable beans. The company needs at least 280 tons of premium-grade and 200 tons of regular-grade coffee beans. The company can purchase ungraded coffee from two suppliers A and B in any amount desired. Samples from the two suppliers contain the following percentages of premium, regular, and unusable beans:

Supplier	Premium	Regular	Unusable
A	20%	50%	30%
B	40%	20%	40%

If A charges $125 per ton and B charges $200 per ton, how much should the company purchase from each supplier to fulfill its needs at minimum cost?

7   A farmer has 100 acres available for planting two crops A and B. The seed for crop A costs $4 per acre and the seed for crop B costs $6 per acre. The total cost of labor will amount to $20 per acre for crop A and $10 per acre for crop B. The expected income from crop A is $110 per acre, whereas from crop B it is $150 per acre. If the farmer does not wish to spend more than $480 for seed and $1,400 for labor, how many acres of each crop should be planted in order to obtain the maximum profit?

8   A firm manufactures two products A and B. For each product it is necessary to employ two different machines X and Y. To manufacture product A, machine X must be used for $\frac{1}{2}$ hour and machine Y for 1 hour. To manufacture product B, machine X must be used for 2 hours and machine Y for 2 hours. The profit on product A is $20 per unit and the profit on B is $50 per unit. If machine X can be used for 8 hours per day and machine Y for 12 hours per day, determine how many units of each product should be manufactured each day in order to maximize the profit.

9   Three substances X, Y, and Z each contain four ingredients A, B, C, and D. The percentage of each ingredient and the cost in cents per ounce of the substances are given in the following table.

Substance	A	B	C	D	Cost/ Ounce
X	20%	10%	25%	45%	25¢
Y	20%	40%	15%	25%	35¢
Z	10%	20%	25%	45%	50¢

If the cost is to be minimum, how many ounces of each substance should be combined in order to obtain a mixture of 20 ounces containing at least 14% A, 16% B, and 20% C? What combination would make the cost greatest?

10   A man plans to operate a stand at a one-day fair in which he will sell bags of peanuts and bags of candy. He has $100 available to purchase his stock, which will cost 10¢ per bag of peanuts and 20¢ per bag of candy. He intends to sell the peanuts at 15¢ and the candy at 26¢ per bag. His stand can accommodate up to 500 bags of peanuts and 400 bags of candy. From past experience he knows that he will sell no more than a total of 700 bags. Find the number of bags of each that he should have available in order to maximize his profit. What is the maximum profit?

12   REVIEW

Concepts

Define or discuss each of the following.

1   System of equations

2   Solution of a system of equations

3   Equivalent systems of equations

4 System of linear equations

5 Homogeneous system of linear equations

6 An $m \times n$ matrix

7 A square matrix of order n

8 The coefficient matrix of a system of linear equations; the augmented matrix

9 Elementary row transformations

10 The sum and product of two matrices

11 Zero matrix

12 Identity matrix

13 Inverse of a matrix

14 Minor

15 Cofactor

16 Determinant

17 Properties of determinants

18 Cramer's Rule

19 System of inequalities

20 Linear programming

Exercises

Find the solutions of the systems of equations in Exercises 1–16.

1 $\begin{cases} 2x - 3y = 4 \\ 5x + 4y = 1 \end{cases}$

2 $\begin{cases} x - 3y = 4 \\ -2x + 6y = 2 \end{cases}$

3 $\begin{cases} y + 4 = x^2 \\ 2x + y = -1 \end{cases}$

4 $\begin{cases} x^2 + y^2 = 25 \\ x - y = 7 \end{cases}$

5 $\begin{cases} 9x^2 + 16y^2 = 140 \\ x^2 - 4y^2 = 4 \end{cases}$

6 $\begin{cases} 2x = y^2 + 3z \\ x = y^2 + z - 1 \\ x^2 = xz \end{cases}$

7 $\begin{cases} \dfrac{1}{x} + \dfrac{3}{y} = 7 \\ \dfrac{4}{x} - \dfrac{2}{y} = 1 \end{cases}$

8 $\begin{cases} 2^x + 3^{y+1} = 10 \\ 2^{x+1} - 3^y = 5 \end{cases}$

9 $\begin{cases} 3x + y - 2z = -1 \\ 2x - 3y + z = 4 \\ 4x + 5y - z = -2 \end{cases}$

10 $\begin{cases} x + 3y = 0 \\ y - 5z = 3 \\ 2x + z = -1 \end{cases}$

11 $\begin{cases} 4x - 3y - z = 0 \\ x - y - z = 0 \\ 3x - y + 3z = 0 \end{cases}$

12 $\begin{cases} 2x + y - z = 0 \\ x - 2y + z = 0 \\ 3x + 3y + 2z = 0 \end{cases}$

13 $\begin{cases} 4x + 2y - z = 1 \\ 3x + 2y + 4z = 2 \end{cases}$

14 $\begin{cases} 2x + y = 6 \\ x - 3y = 17 \\ 3x + 2y = 7 \end{cases}$

15 $\begin{cases} \dfrac{4}{x} + \dfrac{1}{y} + \dfrac{2}{z} = 4 \\[2mm] \dfrac{2}{x} + \dfrac{3}{y} - \dfrac{1}{z} = 1 \\[2mm] \dfrac{1}{x} + \dfrac{1}{y} + \dfrac{1}{z} = 4 \end{cases}$

16 $\begin{cases} 2x - y + 3z - w = -3 \\ 3x + 2y - z + w = 13 \\ x - 3y + z - 2w = -4 \\ -x + y + 4z + 3w = 0 \end{cases}$

Find the solutions and sketch the graphs of the systems in Exercises 17–20.

17 $\begin{cases} x^2 + y^2 < 16 \\ y - x^2 > 0 \end{cases}$

18 $\begin{cases} y - x \le 0 \\ y + x \ge 2 \\ x \le 5 \end{cases}$

19 $\begin{cases} x - 2y \le 2 \\ y - 3x \le 4 \\ 2x + y \le 4 \end{cases}$

20 $\begin{cases} x^2 - y < 0 \\ y - 2x < 5 \\ xy < 0 \end{cases}$

Find the determinants of the matrices in Exercises 21–30.

21 $[-6]$

22 $\begin{bmatrix} 3 & 4 \\ -6 & -5 \end{bmatrix}$

23 $\begin{bmatrix} 3 & -4 \\ 6 & 8 \end{bmatrix}$

24 $\begin{bmatrix} 0 & 4 & -3 \\ 2 & 0 & 4 \\ -5 & 1 & 0 \end{bmatrix}$

25 $\begin{bmatrix} 2 & -3 & 5 \\ -4 & 1 & 3 \\ 3 & 2 & -1 \end{bmatrix}$

26 $\begin{bmatrix} 3 & 1 & -2 \\ -5 & 2 & -4 \\ 7 & 3 & -6 \end{bmatrix}$

27 $\begin{bmatrix} 5 & 0 & 0 & 0 \\ 6 & -3 & 0 & 0 \\ 1 & 4 & -4 & 0 \\ 7 & 2 & 3 & 2 \end{bmatrix}$

28 $\begin{bmatrix} 1 & 2 & 0 & 3 & 1 \\ -2 & -1 & 4 & 1 & 2 \\ 3 & 0 & -1 & 0 & -1 \\ 2 & -3 & 2 & -4 & 2 \\ -1 & 1 & 0 & 1 & 3 \end{bmatrix}$

29 $\begin{bmatrix} 2 & 0 & 1 & 0 & -1 \\ 0 & 1 & 0 & 1 & 2 \\ 2 & -2 & 1 & -2 & 0 \\ 0 & 0 & -2 & 0 & 1 \\ 1 & -1 & 0 & -1 & 0 \end{bmatrix}$

30 $\begin{bmatrix} 1 & 2 & 0 & 0 & 0 \\ 3 & 4 & 0 & 0 & 0 \\ 0 & 0 & 1 & 2 & 3 \\ 0 & 0 & 2 & -1 & 1 \\ 0 & 0 & 1 & 3 & -1 \end{bmatrix}$

31 Find the determinant of the $n \times n$ matrix (a_{ij}), where $a_{ij} = 0$ if $i \ne j$.

32 Without expanding show that

$$\begin{vmatrix} 1 & a & b + c \\ 1 & b & a + c \\ 1 & c & a + b \end{vmatrix} = 0.$$

Find the inverses of the matrices in Exercises 33–36.

33 $\begin{bmatrix} 5 & -4 \\ -3 & 2 \end{bmatrix}$

34 $\begin{bmatrix} 2 & -1 & 0 \\ 1 & 4 & 2 \\ 3 & -2 & 1 \end{bmatrix}$

35 $\begin{bmatrix} 3 & -1 & 0 & 0 \\ 1 & 2 & 0 & 0 \\ 0 & 0 & -1 & -2 \\ 0 & 0 & 5 & 3 \end{bmatrix}$

36 $\begin{bmatrix} 2 & 0 & 0 & 0 \\ 0 & 3 & 0 & 0 \\ 0 & 0 & 4 & 0 \\ 0 & 0 & 0 & 5 \end{bmatrix}$

In Exercises 37–46 express as a single matrix.

37 $\begin{bmatrix} 2 & -1 & 0 \\ 3 & 0 & -2 \end{bmatrix} \begin{bmatrix} 2 & -1 & 3 \\ 0 & 3 & 0 \\ 1 & 4 & 2 \end{bmatrix}$

38 $\begin{bmatrix} 4 & 2 \\ 5 & -3 \end{bmatrix} \begin{bmatrix} 3 \\ 7 \end{bmatrix}$

39 $\begin{bmatrix} 2 & 0 \\ 1 & 4 \\ -2 & 3 \end{bmatrix} \begin{bmatrix} 0 & 2 & -3 \\ 4 & 5 & 1 \end{bmatrix}$

40 $\begin{bmatrix} 0 & -2 & 3 \\ 4 & 1 & 2 \end{bmatrix} \begin{bmatrix} 2 & 0 \\ 3 & 8 \\ 2 & -7 \end{bmatrix}$

41 $2\begin{bmatrix} 0 & -1 & -4 \\ 3 & 2 & 1 \end{bmatrix} - 3\begin{bmatrix} 4 & -2 & 1 \\ 0 & 5 & -1 \end{bmatrix}$

42 $\begin{bmatrix} 1 & 3 \\ 2 & 4 \end{bmatrix} \begin{bmatrix} a & 0 \\ 0 & a \end{bmatrix}$

43 $\begin{bmatrix} a & 0 \\ 0 & b \end{bmatrix} \begin{bmatrix} 1 & 3 \\ 2 & 4 \end{bmatrix}$

44 $\begin{bmatrix} 3 & 2 \\ 0 & 0 \end{bmatrix} \begin{bmatrix} -2 & 0 \\ 3 & 0 \end{bmatrix}$

45 $\begin{bmatrix} 1 & 2 \\ 3 & 4 \end{bmatrix} \left\{ \begin{bmatrix} 2 & -4 \\ 3 & 7 \end{bmatrix} + \begin{bmatrix} 1 & 5 \\ -2 & -3 \end{bmatrix} \right\}$

46 $\begin{bmatrix} 3 & 2 & 5 \\ -3 & 4 & 7 \\ 6 & 5 & 1 \end{bmatrix} \begin{bmatrix} 3 & 2 & 5 \\ -3 & 4 & 7 \\ 6 & 5 & 1 \end{bmatrix}^{-1}$

Verify Exercises 47 and 48 without expanding the determinants.

47 $\begin{vmatrix} 2 & 4 & -6 \\ 1 & 4 & 3 \\ 2 & 2 & 0 \end{vmatrix} = 12\begin{vmatrix} 1 & 1 & -1 \\ 1 & 2 & 1 \\ 2 & 1 & 0 \end{vmatrix}$

48 $\begin{vmatrix} a & b & c \\ d & e & f \\ g & h & k \end{vmatrix} = \begin{vmatrix} d & e & f \\ g & h & k \\ a & b & c \end{vmatrix}$

49 Suppose that $A = (a_{ij})$ is a square matrix of order n such that $a_{ij} = 0$ if $i < j$. Prove that

$$|A| = a_{11}a_{22}\ldots a_{nn}.$$

50 If $A = (a_{ij})$ is any 2×2 matrix such that $|A| \neq 0$, prove that A has an inverse and find a general formula for A^{-1}.

CHAPTER 7

Complex
Numbers

Although real numbers are adequate for many mathematical and scientific problems, there is a serious defect in the system when it comes to solving some equations. Indeed, since the square of a real number cannot be negative, an equation such as $x^2 = -5$ has no solutions if x is restricted to $\mathbb{R}$. *For many applications we need a mathematical system which* contains *the real numbers and which has the additional property that equations such as $x^2 = -5$ do have solutions. Fortunately, it is possible to construct such a system:* the *system of complex numbers* discussed in this chapter.

1 DEFINITION OF COMPLEX NUMBERS

Let us consider the problem of inventing a new mathematical system $\mathbb{C}$ which contains the real number system $\mathbb{R}$ and which has certain other properties. If $\mathbb{C}$ is to be used to find solutions of equations, then it must possess *operations*, that is, rules which may be applied to every pair of its elements to obtain another element. As a matter of fact, we would like to define operations called *addition* and *multiplication* in such a way that if we restrict the elements to the subset $\mathbb{R}$, then the operations behave in the same way as addition and multiplication of real numbers. Since we are extending the notions of addition and multiplication to the set $\mathbb{C}$, we shall continue to use the symbols $+$ and $\cdot$ for those operations.

In order to gain some insight into the construction of $\mathbb{C}$, let us begin by taking an intuitive approach. We first note that if we want equations of the form $x^2 = -k$, where k is a positive real number, to have solutions, then in particular when $k = 1$ it is necessary for $\mathbb{C}$ to contain some element i such that $i^2 = -1$. If b is in $\mathbb{R}$, then b is also in $\mathbb{C}$, and since $\mathbb{C}$ is to be closed relative to multiplication, bi must be in $\mathbb{C}$. Moreover, if a is in $\mathbb{R}$ and if $\mathbb{C}$ is to be closed relative to addition, then $a + bi$ is in $\mathbb{C}$. Thus, $\mathbb{C}$ contains elements of the form $a + bi, c + di$, and so on, where a, b, c, and d are real numbers and $i^2 = -1$. If we want the Commutative, Associative, and Distributive Properties (1.1)–(1.3) to be valid, then these elements must add as follows:

$$(a + bi) + (c + di) = (a + c) + (bi + di)$$

or

(7.1)

$$\boxed{(a + bi) + (c + di) = (a + c) + (b + d)i}$$

Similarly, if $i^2 = -1$ and (1.1)–(1.3) are to hold in $\mathbb{C}$, then

$$(a + bi)(c + di) = (a + bi)c + (a + bi)(di)$$
$$= ac + (bi)c + a(di) + (bi)(di)$$
$$= ac + (bc)i + (ad)i + (bd)(i^2)$$
$$= ac + (bd)(-1) + (ad)i + (bc)i$$
$$= (ac - bd) + (ad + bc)i.$$

To summarize, the following rule must hold in $\mathbb{C}$:

(7.2)

$$\boxed{(a + bi)(c + di) = (ac - bd) + (ad + bc)i}$$

The preceding discussion indicates the manner in which elements behave if a system of the required type is to exist. Moreover, our discussion provides a key to the actual construction of $\mathbb{C}$. Thus, we begin by *defining* a **complex number** as any symbol of the form $a + bi$, where a and b are real numbers. The real number a is called the **real part** of the complex number and bi is called the **imaginary part**. At the outset the letter i is given no specific meaning and the $+$ sign which appears in $a + bi$ is not to be interpreted as the symbol for addition, but only as part of the notation for a complex number. As above, $\mathbb{C}$ will denote the set of all complex numbers. Two complex numbers $a + bi$ and $c + di$ are said to be **equal**, and we write

(7.3)

$$\boxed{a + bi = c + di \quad \text{if and only if} \quad a = c \text{ and } b = d}$$

Next we *define* addition and multiplication of complex numbers by means of (7.1) and (7.2). It should be observed that the $+$ sign is used in three different ways in (7.1). First, by our previous remarks, it is part of the symbol for a complex number. Second, it is used to denote addition of the complex numbers $a + bi$ and $c + di$. Third, it is the addition sign for real numbers, as in the expressions $a + c$ and $b + d$ on the right-hand side of (7.1). The need for remembering this threefold use of $+$ will disappear after we agree on the notational conventions which follow.

Let us consider the subset $\mathbb{R}'$ of $\mathbb{C}$ consisting of all complex numbers of the form $a + 0i$, where a is a real number. By associating a with $a + 0i$, we obtain a one-to-one correspondence between the sets $\mathbb{R}$ and $\mathbb{R}'$. Applying (7.1) and (7.2) to the elements of $\mathbb{R}'$ (by letting $b = d = 0$), we obtain

$$(a + 0i) + (c + 0i) = (a + c) + 0i$$
$$(a + 0i)(c + 0i) = ac + 0i.$$

This shows that in order to add (or multiply) two elements of $\mathbb{R}'$, we merely add (or multiply) the real parts, *disregarding* the imaginary parts. Hence, as far as properties of addition and multiplication are concerned, the only difference between $\mathbb{R}$ and $\mathbb{R}'$ is the notation for the elements. Accordingly, we shall use the symbol a in place of $a + 0i$. For example, an element such as 3 of $\mathbb{R}$ (or $\mathbb{C}$) is considered the same as the element $3 + 0i$ of $\mathbb{C}$. It is also convenient to abbreviate the complex number $0 + bi$ by the symbol bi. Applying (7.1) gives us

$$(a + 0i) + (0 + bi) = (a + 0) + (0 + b)i = a + bi.$$

This indicates that $a + bi$ may be thought of as the sum of two complex numbers a and bi (that is, $a + 0i$ and $0 + bi$). With these agreements on notation, all the $+$ signs in (7.1) may be regarded as addition of complex numbers.

Example 1   Express each of the following in the form $a + bi$, where a and b are real numbers.

(a) $(3 + 4i) + (2 + 5i)$

(b) $(3 + 4i)(2 + 5i)$

(c) $(3 + 4i)^2$

Solutions   (a) Applying (7.1),

$$(3 + 4i) + (2 + 5i) = (3 + 2) + (4 + 5)i = 5 + 9i.$$

(b) Using (7.2),

$$(3 + 4i)(2 + 5i) = (3 \cdot 2 - 4 \cdot 5) + (3 \cdot 5 + 4 \cdot 2)i = -14 + 23i.$$

(c) Exponents are defined in $\mathbb{C}$ exactly as they are in $\mathbb{R}$. Thus,

$$
\begin{aligned}
(3 + 4i)^2 &= (3 + 4i)(3 + 4i) \\
&= (3 \cdot 3 - 4 \cdot 4) + (3 \cdot 4 + 4 \cdot 3)i \\
&= -7 + 24i.
\end{aligned}
$$

It is not difficult to show that complex numbers obey Properties (1.1)–(1.5). In particular, the Commutative and Associative Properties for addition and multiplication are true, as are the Distributive Properties. The identity element relative to addition is 0 (or equivalently $0 + 0i$), since

$$
\begin{aligned}
(a + bi) + 0 &= (a + bi) + (0 + 0i) \\
&= (a + 0) + (b + 0)i \\
&= a + bi.
\end{aligned}
$$

As usual, we refer to 0 as **zero** or the **zero element**. It follows from (7.2) that the *product* of zero and any complex number is zero. We may also use (7.2) to prove that 1 (that is, $1 + 0i$) is the identity element relative to multiplication.

If $(-a) + (-bi)$ is added to $a + bi$, we obtain 0. This implies that $(-a) + (-b)i$ is the additive inverse of $a + bi$, that is,

$$\boxed{-(a + bi) = (-a) + (-b)i}$$

We shall postpone the discussion of multiplicative inverses until the next section.

Subtraction of complex numbers is defined using additive inverses as follows:

$$(a + bi) - (c + di) = (a + bi) + [-(c + di)].$$

Since $-(c + di) = (-c) + (-d)i$, it follows that

$$\boxed{(a + bi) - (c + di) = (a - c) + (b - d)i}$$

The special case with $b = c = 0$ gives us

$$(a + 0i) - (0 + di) = (a - 0) + (0 - d)i$$

which may be written in the form

$$a - di = a + (-d)i.$$

The preceding formula is useful when the real number associated with the imaginary part of a complex number is negative.

If c, d, and k are real numbers, then by (7.2) and our agreement on notation,

$$k(c + di) = (k + 0i)(c + di) = (kc - 0d) + (kd + 0c)i$$

that is,

(7.4)

$$\boxed{k(c + di) = kc + (kd)i}$$

One illustration of that formula is

$$3(5 + 2i) = 15 + 6i.$$

The special case of (7.4) with $k = -1$ gives us

$$(-1)(c + di) = (-c) + (-d)i = -(c + di).$$

Hence, as in $\mathbb{R}$, the additive inverse of a complex number may be found by multiplying it by -1.

The complex number $0 + 1i$ (or equivalently $1i$) will be denoted by i. Applying (7.4) with $k = b, c = 0$, and $d = 1$ gives us

$$b(0 + 1i) = (b \cdot 0) + (b \cdot 1)i = 0 + bi = bi.$$

Thus the symbol bi which has been used throughout this section may be regarded as the *product* of b and i.

Finally, using (7.2) with $a = c = 0$ and $b = d = 1$, we obtain

$$\begin{aligned} i^2 &= (0 + 1i)(0 + 1i) \\ &= (0 \cdot 0 - 1 \cdot 1) + (0 \cdot 1 + 1 \cdot 0)i \\ &= -1 + 0i. \end{aligned}$$

This gives us the following important rule:

$$\boxed{i^2 = -1}$$

If we collect all of the formulas and remarks made in this section it becomes evident that when working with complex numbers *we may treat all symbols just as though they represented real numbers with exactly one exception: wherever the symbol i^2 appears it may be replaced by* -1. Consequently, manipulations can be carried out without referring to (7.1) or (7.2), which is what we had in mind from the very beginning of our discussion! We shall use this technique in the solution of the next example. If, as in Example 2, we are asked to write an expression in the form

$a + bi$, we shall also accept the form $a - di$ since we have seen that it equals $a + (-d)i$.

Example 2   Write each of the following in the form $a + bi$.

(a) $4(2 + 5i) - (3 - 4i)$ (b) $(4 - 3i)(2 + i)$

(c) $i(3 - 2i)^2$ (d) i^{51}

Solutions   (a) $4(2 + 5i) - (3 - 4i) = 8 + 20i - 3 + 4i = 5 + 24i$

(b) $(4 - 3i)(2 + i) = 8 - 6i + 4i - 3i^2 = 11 - 2i$

(c) $i(3 - 2i)^2 = i(9 - 12i + 4i^2) = i(5 - 12i) = 5i - 12i^2 = 12 + 5i$

(d)   Taking successive powers of i, we obtain $i^1 = i, i^2 = -1, i^3 = -i, i^4 = 1$, and then the cycle starts over: $i^5 = i, i^6 = i^2 = -1$, and so on. In particular, $i^{51} = i^{48}i^3 = (i^4)^{12}i^3 = (1)^{12}i^3 = i^3 = -i$.

Finally, it should be pointed out that there is another way to define complex numbers. Observe that each symbol $a + bi$ determines a unique ordered pair (a, b) of real numbers. Conversely, every ordered pair (a, b) can be used to obtain a symbol $a + bi$. In this way we obtain a one-to-one correspondence between the symbols $a + bi$ and ordered pairs (a, b). The correspondence suggests using ordered pairs of real numbers to define the system $\mathbb{C}$. Formulas (7.1) and (7.2) can then be used to motivate definitions for addition and multiplication. Specifically, we define $\mathbb{C}$ as the set of all ordered pairs of real numbers subject to the following two laws:

$$(a, b) + (c, d) = (a + c, b + d)$$
$$(a, b)(c, d) = (ac - bd, ad + bc).$$

Notice the manner in which (7.1) and (7.2) were used to help formulate the definition. We merely replaced symbols such as $a + bi$ by (a, b) and translated the rules accordingly. It may then be shown that properties (1.1)–(1.5) are true for ordered pairs. By a suitable change in notation we can obtain the $a + bi$ form for complex numbers introduced in this section.

EXERCISES

In each of Exercises 1–36 write the expression in the form $a + bi$.

1   $(3 + 2i) + (-5 + 4i)$ 2 $(8 - 5i) + (2 - 3i)$

3   $(-4 + 5i) + (2 - i)$ 4 $(5 + 7i) + (-8 - 4i)$

5   $(16 + 10i) - (9 + 15i)$ 6 $(2 - 6i) - (7 + 2i)$

7   $-(-2 + 7i) + (-6 + 6i)$ 8 $-(5 - 3i) - (-3 - 4i)$

9   $7 - (3 - 7i)$ 10 $-9 + (5 + 9i)$

11   $5i - (6 + 2i)$ 12 $(10 + 7i) - 12i$

13   $(4 + 3i)(-1 + 2i)$

14   $(3 - 6i)(2 + i)$

15   $(-7 + i)(-3 + i)$

16   $(5 + 2i)(5 - 2i)$

17   $(3 + 4i)(3 - 4i)$

18   $7i(13 + 8i)$

19   $-9i(4 - 8i)$

20   $(6i)(-2i)$

21   $4(8 - 11i)$

22   $-3(-6 + 12i)$

23   $(-3i)(5i)$

24   $(1 - i)(1 + i)$

25   $(\sqrt{7} + \sqrt{3}i)(\sqrt{7} - \sqrt{3}i)$

26   $(-7 + 3i)^2$

27   $(3 + 2i)^2$

28   $4i(2 + 5i)^2$

29   $i(3 - 2i)(5 + i)$

30   $(1 + i)^4$

31   $\left(-\dfrac{1}{2} - \dfrac{\sqrt{3}}{2}i\right)^3$

32   $\left(-\dfrac{1}{2} + \dfrac{\sqrt{3}}{2}i\right)^3$

33   i^{42}

34   i^{23}

35   i^{157}

36   $(-i)^{50}$

In Exercises 37–40 solve for x and y, if x and y are real.

37   $5x + 6i = -8 + 2yi$

38   $7 - 4yi = 9x + 3i$

39   $i(2x - 4y) = 4x + 2 + 3yi$

40   $(2x + y) + (3x - 4y)i = (x - 2) + (2y - 5)i$

2   CONJUGATES AND INVERSES

The complex number $a - bi$ is called the **conjugate** of the complex number $a + bi$. Since

(7.5)
$$(a + bi)(a - bi) = a^2 + b^2$$

we see that the product of a complex number and its conjugate is a real number. If $a^2 + b^2 \neq 0$, then multiplying both sides of (7.5) by $1/(a^2 + b^2)$ and rearranging terms on the left-hand side gives us

$$\left(\frac{1}{a^2 + b^2}\right)(a - bi)(a + bi) = 1.$$

Hence, if $a + bi \neq 0$, then the complex number $a + bi$ has a multiplicative inverse denoted by $(a + bi)^{-1}$, or $1/(a + bi)$, where

(7.6)
$$\frac{1}{a + bi} = \left(\frac{1}{a^2 + b^2}\right)(a - bi)$$

If $c + di \neq 0$, we define the **quotient**

$$\frac{a + bi}{c + di}$$

to be the product of $a + bi$ and $1/(c + di)$. We can write this quotient in the form $u + vi$, where u and v are real numbers, by multiplying numerator and denominator by the conjugate $c - di$ of the denominator as follows:

$$\frac{a + bi}{c + di} = \frac{a + bi}{c + di} \cdot \frac{c - di}{c - di}$$

$$= \frac{(ac + bd) + (bc - ad)i}{c^2 + d^2}$$

$$= \left(\frac{ac + bd}{c^2 + d^2}\right) + \left(\frac{bc - ad}{c^2 + d^2}\right)i.$$

The technique above may also be used to find the multiplicative inverse of $(a + bi)$. Specifically, we multiply numerator and denominator of $1/(a + bi)$ by $a - bi$ as follows:

$$\frac{1}{a + bi} = \frac{1}{a + bi} \cdot \frac{a - bi}{a - bi}$$

$$= \frac{a - bi}{a^2 + b^2} = \frac{1}{a^2 + b^2}(a - bi)$$

which is the same as (7.6).

Example 1 Express each of the following in the form $a + bi$.

(a) $\dfrac{1}{9 + 2i}$ (b) $\dfrac{7 - i}{3 - 5i}$

Solutions   (a)

$$\frac{1}{9 + 2i} = \frac{1}{9 + 2i} \cdot \frac{9 - 2i}{9 - 2i} = \frac{9 - 2i}{81 + 4} = \frac{9}{85} - \frac{2}{85}i$$

(b)

$$\frac{7 - i}{3 - 5i} = \frac{7 - i}{3 - 5i} \cdot \frac{3 + 5i}{3 + 5i}$$

$$= \frac{21 - 3i + 35i - 5i^2}{9 - 25i^2}$$

$$= \frac{26 + 32i}{34} = \frac{13}{17} + \frac{16}{17}i$$

Conjugates of complex numbers have several interesting and useful properties. To simplify the notation, if $z = a + bi$ is a complex number, then its conjugate will be denoted by $\bar{z}$, that is, $\bar{z} = a - bi$.

(7.7) **Theorem on Conjugates**

If z and w are complex numbers, then

(i) $\overline{z + w} = \overline{z} + \overline{w}$

(ii) $\overline{z \cdot w} = \overline{z} \cdot \overline{w}$

(iii) $\overline{z^n} = \overline{z}^n$, for every positive integer n

(iv) $\overline{z} = z$ if and only if z is real.

Proof. Let $z = a + bi$ and $w = c + di$, where $a, b, c,$ and d are real numbers. Since $z + w = (a + c) + (b + d)i$, we have, by the definition of conjugate and properties of addition of complex numbers,

$$\overline{z + w} = (a + c) - (b + d)i$$
$$= (a - bi) + (c - di)$$
$$= \overline{z} + \overline{w}.$$

That proves (i).

By (7.2), $z \cdot w = (ac - bd) + (ad + bc)i$ and hence the conjugate is

$$\overline{z \cdot w} = (ac - bd) - (ad + bc)i = (a - bi)(c - di) = \overline{z} \cdot \overline{w}$$

which proves (ii).

If we set $w = z$ in (ii), then $\overline{z \cdot z} = \overline{z} \cdot \overline{z}$, that is, $\overline{z^2} = \overline{z}^2$. We may then write $\overline{z^3} = \overline{z^2 \cdot z} = \overline{z^2} \cdot \overline{z} = \overline{z}^2 \cdot \overline{z} = \overline{z}^3$. Continuing in this manner, it appears that $\overline{z^n} = \overline{z}^n$ for all positive integers n. A complete proof of (iii) requires the method of mathematical induction discussed in Chapter Nine.

Finally, let us prove (iv). If $z = a + bi$ is real, then $b = 0$ and $\overline{z} = a - 0i = a + 0i = z$. Conversely, if $\overline{z} = z$, then $a - bi = a + bi$ and by (7.3), $-b = b$. It follows that $b = 0$; that is, z is real.

It is not difficult to extend (i) and (ii) of (7.7) to more than two complex numbers. For example, if $z, w,$ and u are complex numbers, then applying (i) twice we have

$$\overline{(z + w) + u} = \overline{z + w} + \overline{u} = \overline{z} + \overline{w} + \overline{u}.$$

The analogous result holds for more than three complex numbers. This fact may be stated: "The conjugate of a sum of complex numbers equals the sum of the conjugates." A similar result is true for products. An important application of (7.7) will be made in the next chapter.

Real numbers may be represented geometrically by means of points on a coordinate line. We can also obtain geometric representations for complex numbers by using points in a coordinate plane. Specifically, each complex number $a + bi$ determines a unique ordered pair (a, b). The corresponding point $P(a, b)$ in a coordinate plane is called the **geometric representation of** $a + bi$. To emphasize that we are assigning complex numbers to points in a plane, the point $P(a, b)$ will be labeled $a + bi$. A coordinate plane with a complex number assigned to each point is referred to as the **complex plane** instead of the xy-plane. Also, according to this scheme, the x-axis is called the **real axis** and the y-axis the **imaginary axis**. In Figure 7.1 we have indicated the geometric representations of several complex numbers.

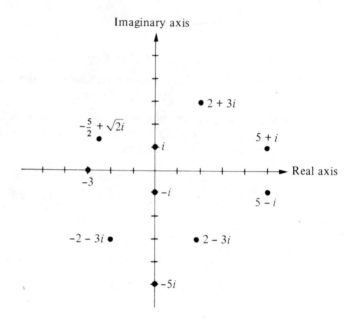

Figure 7.1

Note that to obtain the point corresponding to the conjugate $a - bi$ of any complex number $a + bi$ we simply reflect through the real axis.

In Chapter One we defined the concept of the absolute value $|a|$ of a real number a and noted that geometrically, a is the distance between the origin and the point on a coordinate line which corresponds to a. It is natural, therefore, to interpret the absolute value $a + bi$ of a complex number as the distance $\sqrt{a^2 + b^2}$ between the origin of a complex plane and the point (a, b) corresponding to $a + bi$. This motivates the following definition.

(7.8)

> The **absolute value** of a complex number $a + bi$ is denoted by $|a + bi|$ and is defined to be the nonnegative real number $\sqrt{a^2 + b^2}$.

Example 2   Find (a) $|2 - 6i|$;   (b) $|3i|$.

Solutions   Using (7.8) we obtain

(a) $|2 - 6i| = \sqrt{4 + 36} = \sqrt{40} = 2\sqrt{10}$

(b) $|3i| = \sqrt{0 + 9} = 3$

It is worth noting that the points which correspond to all of the complex numbers having a fixed absolute value lie on a circle with center at the origin in the complex plane. For example, the points corresponding to the complex numbers z with $|z| = 1$ lie on a unit circle. Also observe that if we take the square root of both sides of equation (7.5), then

$$\sqrt{(a + bi)(a - bi)} = \sqrt{a^2 + b^2} = |a + bi|.$$

If $b = 0$, the equation reduces to $\sqrt{a^2} = |a|$, which agrees with (1.27).

We have made considerable use of properties (1.1)–(1.5) in Chapter One. Notice, however, that inequalities have not been mentioned, because it is impossible to define positive elements in $\mathbb{C}$ which have the same properties as the positive elements in $\mathbb{R}$. In particular, we never consider a complex number with nonzero imaginary part as less than or greater than another complex number.

EXERCISES

In each of Exercises 1–28 express the given number in the form $a + bi$.

1 $\dfrac{1}{3 + 2i}$

2 $\dfrac{1}{5 + 8i}$

3 $\dfrac{7}{5 - 6i}$

4 $\dfrac{-3}{2 - 5i}$

5 $\dfrac{4 - 3i}{2 + 4i}$

6 $\dfrac{4 + 3i}{-1 + 2i}$

7 $\dfrac{6 + 4i}{1 - 5i}$

8 $\dfrac{7 - 6i}{-5 - i}$

9 $\dfrac{21 - 7i}{i}$

10 $\dfrac{10 + 9i}{-3i}$

11 $\dfrac{2 - 3i}{1 + i} + \dfrac{7 + 4i}{3 + 5i}$

12 $\dfrac{6 - 2i}{3 + i} - \dfrac{3 - 7i}{i}$

13 $8 + \dfrac{4 - i}{1 + 2i}$

14 $\dfrac{1}{10 - i} + 5i$

15 $\dfrac{1}{(1 + i)^3}$

16 $\dfrac{(1 - i)^3}{1 + i}$

17 $\dfrac{4 - i^2}{i - 2}$

18 $\dfrac{4i^2 - 25}{5 + 2i}$

19 $\left(\dfrac{1}{5i}\right)^3$

20 $\dfrac{1}{(3 + 2i)^2}$

21 $|3 - 4i|$

22 $|5 + 8i|$

23 $|-6 - 7i|$

24 $|1 - i|$

25 $|8i|$

26 $|i^7|$

27 $|i^{500}|$

28 $|-15i|$

Represent the complex numbers in Exercises 29–38 geometrically.

29   $4 + 2i$

30   $-5 + 3i$

31   $3 - 5i$

32   $-2 - 6i$

33   $-(3 - 6i)$

34   $(1 + 2i)^2$

35   $2i(2 + 3i)$

36   $(-3i)(2 - i)$

37   $(1 + i)^2$

38   $4(-1 + 2i)$

In each of Exercises 39–42 find all complex numbers z that satisfy the given equation, and express z in the form $a + bi$.

39   $5z + 3i = 2iz + 4$

40   $2iz - 6 = 9i + z$

41   $(z - 2i)^2 = (z + 3i)^2$

42   $z(z + 4i) = (z + i)(z - 3i)$

If $z = a + bi$ and $w = c + di$, verify the identities in Exercises 43–50.

43   $\bar{\bar{z}} = z$

44   $\overline{z - w} = \bar{z} - \bar{w}$

45   $|z| = \sqrt{z\bar{z}}$

46   $|-z| = |z|$

47   $|\bar{z}| = |z|$

48   $|z| = 0$ if and only if $z = 0$

49   $|zw| = |z||w|$

50   $|z/w| = |z|/|w|,\quad w \neq 0$

3   COMPLEX ROOTS OF EQUATIONS

It is easy to see that if p is any positive real number, then the equation $x^2 = -p$ has solutions in $\mathbb{C}$. As a matter of fact, one solution is $\sqrt{p}\,i$, since

$$(\sqrt{p}\,i)^2 = (\sqrt{p})^2 i^2 = p(-1) = -p.$$

Similarly, $-\sqrt{p}\,i$ is also a solution. Moreover, they are the only solutions, for if a complex number z is a solution, then $z^2 + p = 0$ and hence

$$(z + \sqrt{p}\,i)(z - \sqrt{p}\,i) = 0.$$

This implies that either $z = -\sqrt{p}\,i$ or $z = \sqrt{p}\,i$.

The next definition is motivated by the fact that $(\sqrt{p}\,i)^2 = -p$.

(7.9)

> If p is a positive real number, then the **principal square root** of $-p$ is denoted by $\sqrt{-p}$ and is defined to be the complex number $\sqrt{p}\,i$.

As illustrations of (7.9) we have

$$\sqrt{-9} = \sqrt{9}\,i = 3i, \quad \sqrt{-5} = \sqrt{5}\,i, \quad \sqrt{-1} = \sqrt{1}\,i = i.$$

Care must be taken in using the radical sign if the radicand is negative. For example, the formula $\sqrt{a}\sqrt{b} = \sqrt{ab}$ which holds for positive real numbers is not true when both a and b are negative. To illustrate,

$$\sqrt{-3}\sqrt{-3} = (\sqrt{3}i)(\sqrt{3}i) = (\sqrt{3})^2 i^2 = 3(-1) = -3$$

whereas

$$\sqrt{(-3)(-3)} = \sqrt{9} = 3.$$

Hence,

$$\sqrt{-3}\sqrt{-3} \neq \sqrt{(-3)(-3)}.$$

However, if only *one* of a or b is negative, then it can be shown that $\sqrt{a}\sqrt{b} = \sqrt{ab}$. In general, we shall not apply laws of radicals if radicands are negative. Instead, we shall change the form of radicals, using (7.9) before performing any operations.

Example 1   Express $(5 - \sqrt{-3})(-1 + \sqrt{-4})$ in the form $a + bi$.

Solution $(5 - \sqrt{-3})(-1 + \sqrt{-4}) = (5 - \sqrt{3}i)(-1 + 2i)$
$$= -5 - 2\sqrt{3}i^2 + 10i + \sqrt{3}i$$
$$= (-5 + 2\sqrt{3}) + (10 + \sqrt{3})i$$

In Chapter Two (see (2.12)) we proved that if a, b, and c are real numbers such that $b^2 - 4ac \geq 0$ and $a \neq 0$, then the solutions of the quadratic equation $ax^2 + bx + c = 0$ are

(7.10)
$$\frac{-b + \sqrt{b^2 - 4ac}}{2a} \quad \text{and} \quad \frac{-b - \sqrt{b^2 - 4ac}}{2a}.$$

We may now extend this fact to the case where $b^2 - 4ac < 0$. Indeed, the same manipulations used to obtain the quadratic formula, together with the developments in this chapter, show that if $b^2 - 4ac < 0$, then the solutions of $ax^2 + bx + c = 0$ are the two *complex* numbers given by (7.10). Notice that the solutions are conjugates of one another. This completes the theory of solutions of quadratic equations in one variable with real coefficients. The solutions are *always* given by the quadratic formula (2.2).

Example 2   Find the solutions of the equation $5x^2 + 2x + 1 = 0$.

Solution   By the quadratic formula we have

$$x = \frac{-2 \pm \sqrt{4 - 20}}{10} = \frac{-2 \pm \sqrt{-16}}{10} = \frac{-2 \pm 4i}{10}.$$

Dividing numerator and denominator by 2, we see that the solutions of the equation are $-\frac{1}{5} + (\frac{2}{5})i$ and $-\frac{1}{5} - (\frac{2}{5})i$.

Example 3   Find the roots of the equation $x^3 - 1 = 0$.

Solution   By (1.33) the given equation may be written as

$$(x - 1)(x^2 + x + 1) = 0.$$

Setting each factor equal to zero and solving the resulting equations, we obtain the solutions

$$1, \quad \frac{-1 \pm \sqrt{1 - 4}}{2}$$

which may be written as

$$1, \quad -\frac{1}{2} + \frac{\sqrt{3}}{2}i, \quad -\frac{1}{2} - \frac{\sqrt{3}}{2}i.$$

The three roots of $x^3 - 1 = 0$ are called the **cube roots of unity**. It can be shown that if n is any positive integer, then the equation $x^n - 1 = 0$ has n distinct complex roots. They are called the ***n*th roots of unity**.

It is easy to form a quadratic equation having complex roots z and w. We merely write

$$(x - z)(x - w) = 0.$$

The most important situation occurs if $w = \bar{z}$, the conjugate of z. In this case the equation has the form

$$(x - z)(x - \bar{z}) = 0$$

or equivalently

$$x^2 - (z + \bar{z})x + z\bar{z} = 0.$$

If $z = a + bi$, then $\bar{z} = a - bi$ and hence

$$z + \bar{z} = 2a \quad \text{and} \quad z\bar{z} = a^2 + b^2;$$

that is, the quadratic equation has real coefficients.

Example 4   Find a quadratic equation which has roots $3 + 2i$ and $3 - 2i$.

Solution   By the preceding remarks the equation is given by $(x - 3 - 2i)(x - 3 + 2i) = 0$. This simplifies to $x^2 - 6x + 13 = 0$.

Quadratic equations with complex coefficients may also be considered. It can be shown that the solutions are again given by the quadratic formula. Since in this case $b^2 - 4ac$ may be complex, the solution of such an equation may involve finding the square root of a complex number. We shall not discuss the general theory of roots of complex numbers here.

EXERCISES

In each of Exercises 1–12 express the given number in the form $a + bi$.

1 $\sqrt{-5}\sqrt{-5}$

2 $(-\sqrt{5})(\sqrt{-5})$

3 $(3 + \sqrt{-16}) - (6 - \sqrt{-9})$

4 $(4 - \sqrt{-4}) + (3 - \sqrt{-25})$

5 $(7 + \sqrt{-5})(7 - \sqrt{-5})$

6 $(\sqrt{-9} + 3)(\sqrt{-16} - 5)$

7 $\sqrt{-2}(\sqrt{2} + \sqrt{-2})$

8 $\sqrt{-3}(3 - \sqrt{-12})$

9 $(\sqrt{-10})^3$

10 $\sqrt{(-10)^3}$

11 $\dfrac{5 + \sqrt{-4}}{1 + \sqrt{-9}}$

12 $\dfrac{1}{6 + \sqrt{-3}}$

Find the solutions of the equations in Exercises 13–26.

13 $x^2 - 3x + 10 = 0$

14 $x^2 - 5x + 20 = 0$

15 $x^2 + 2x + 5 = 0$

16 $x^2 + 3x + 6 = 0$

17 $4x^2 + x + 3 = 0$

18 $-3x^2 + x - 5 = 0$

19 $x^3 - 125 = 0$

20 $x^3 + 27 = 0$

21 $x^6 - 64 = 0$

22 $x^4 = 81$

23 $4x^4 + 25x^2 + 36 = 0$

24 $27x^4 + 21x^2 + 4 = 0$

25 $x^3 + 3x^2 + 4x = 0$

26 $8x^3 - 12x^2 + 2x - 3 = 0$

In each of Exercises 27–38 find a quadratic equation with the given roots.

27 $4 + i,\ \ 4 - i$

28 $1 + 3i,\ \ 1 - 3i$

29 $-3 + 2i,\ \ -3 - 2i$

30 $\frac{1}{2} + \frac{1}{3}i,\ \ \frac{1}{2} - \frac{1}{3}i$

31 $-5i,\ \ 5i$

32 $\sqrt{2}i,\ \ -\sqrt{2}i$

33 $10,\ \ i$

34 $6,\ \ -6i$

35 $i,\ \ 1/i$

36 $i,\ \ i^3$

37 $2i,\ \ 3i$

38 $1 + i,\ \ 2 + 2i$

39 Prove that if z is any complex number, then $z + \bar{z}$ is a real number. Is $z - \bar{z}$ necessarily a real number?

40 Prove that the sum and product of the roots of the equation $ax^2 + bx + c = 0$ are $-b/a$ and c/a, respectively.

4   REVIEW

Concepts

Define or discuss the following.

1   The system of complex numbers

2   The conjugate of a complex number

3   The absolute value of a complex number

4   The nth roots of unity

5   Geometric representation of a complex number

6   The complex plane

Exercises

Express each number in Exercises 1–20 in the form $a + bi$ where a and b are real numbers.

1   $(7 + 5i) + (-8 + 3i)$ 2 $(-4 + 7i) - (3 - 3i)$

3   $(4 + 2i)(-5 + 4i)$ 4 $(-8 + 2i)(-7 + 5i)$

5   $8i(7 - 5i)$ 6 $(-5 + 6i)(-5 - 6i)$

7   $(9 + 4i)(9 - 4i)$ 8 $(3 + 8i)^2$

9   $i(6 + i)(2 - 3i)$ 10 $(6 - 3i)/(2 + 7i)$

11   $(4 + 6i)/(5 - i)$ 12 $(20 - 8i)/4i$

13   $1/(10 - 8i)$ 14 $3/(-2i)$

15   $|6 - 10i|$ 16 $|-11 + 10i|$

17   $-5(3 + i) - (4 + 2i)$ 18 $(2 + i)^4$

19   $(1 + i)^{-1} - (1 - i)^{-1}$ 20 i^{99}

Find the solutions of the equations in Exercises 21–24.

21   $5x^2 - 2x + 3 = 0$ 22 $3x^2 + x + 6 = 0$

23   $6x^4 + 29x^2 + 28 = 0$ 24 $x^4 + x^2 + 1 = 0$

In each of Exercises 25–28 find a quadratic equation with the given roots.

25   $7 + i,\ \ 7 - i$ 26 $12i,\ \ -12i$

27   $-1 - 4i,\ \ -1 + 4i$ 28 $1 + i,\ \ 3i$

29   Find the three cube roots of -27.

30   Find the solutions of the equation $x^5 - 32 = 0$.

CHAPTER 8

Zeros of Polynomials

The concept of polynomial was introduced in Chapter One. At that time we were interested primarily in basic manipulations such as adding, multiplying, and factoring. In this chapter we shall explore the theory of polynomials more deeply and discuss equations which involve polynomials of any degree.

1 PROPERTIES OF DIVISION

As in (1.30) every polynomial in a variable x, with real coefficients, can be written in the form

(8.1)
$$a_n x^n + a_{n-1} x^{n-1} + \cdots + a_1 x + a_0,$$

where n is a nonnegative integer and the coefficients $a_0, a_1, \ldots, a_n$ are real numbers. Recall that if $a_n \neq 0$ the polynomial has **degree n**. If all the coefficients of a polynomial are zero it is called the **zero polynomial** and is denoted by 0. It is customary not to assign a degree to the zero polynomial. If we allow the coefficients a_i to be complex numbers, then (8.1) is called a polynomial in x with *complex* coefficients.

Throughout this chapter, unless we specify otherwise, whenever we consider arbitrary polynomials such as (8.1), we will assume that the coefficients a_i are either real or complex numbers. Similarly, when the term "number" is used, we mean either a real or complex number. Symbols such as $f(x)$ and $g(x)$ will be used to denote polynomials in x. This is a natural notation, since each polynomial determines a function f, where the value $f(c)$ of the function at the number c is the result obtained by substituting c for x in the given polynomial. If $f(c) = 0$, then c is called a **zero**, or **root**, of the polynomial $f(x)$.

If a polynomial $g(x)$ is a factor of a polynomial $f(x)$, we often say that $f(x)$ is **divisible** by $g(x)$. For example, $x^2 - 25$ is divisible by $x - 5$ and by $x + 5$. A division process may also be introduced if $g(x)$ is *not* a factor of $f(x)$. The process is similar to that used for integers. For example, the integer 24 has positive factors 1, 2, 3, 4, 6, 8, 12, and 24; that is, 24 is *divisible* by those numbers. A different situation exists if we divide 24 by nonfactors such as 5, 7, 15, and 32. In such cases a process called *long division* is used which yields a *quotient* and *remainder*. The reader undoubtedly remembers the process from elementary arithmetic where, for example, the work involved in dividing 4126 by 23 can be arranged as follows:

$$
\begin{array}{r}
179 \\
23\,\overline{)\,4126} \\
23 \\
\overline{182} \\
161 \\
\overline{216} \\
207 \\
\overline{9}
\end{array}
$$

Here the number 179 is called the *quotient* and 9 the *remainder*. To complete the terminology, 23 is called the *divisor* and 4126 the *dividend*. The remainder should always be less than the divisor, for otherwise the quotient can be increased. The above result is sometimes written as follows:

$$\frac{4126}{23} = 179 + \frac{9}{23}.$$

Multiplying by 23 gives us

$$4126 = (23)(179) + 9$$

The above form is very useful for theoretical purposes and can be generalized to arbitrary integers. Specifically, it can be shown that if a and b are integers with $b > 0$, then there exist unique integers q and r such that

$$a = bq + r$$

where $0 \le r < b$. The integer q is called the **quotient** and r the **remainder** in the division of a by b.

A similar discussion can be given for polynomials. For example, the polynomial $x^4 - 16$ is divisible by $x^2 - 4$, $x^2 + 4$, $x + 2$, and $x - 2$; but $x^2 + 3x + 1$ is not a factor of $x^4 - 16$. However, by another process called *long division*, we write

$$
\begin{array}{r}
x^2 - 3x + 8 \\
x^2 + 3x + 1\,\overline{)\,x^4 - 16} \\
x^4 + 3x^3 + x^2 \\
\overline{-3x^3 - x^2} \\
-3x^3 - 9x^2 - 3x \\
\overline{8x^2 + 3x - 16} \\
8x^2 + 24x + 8 \\
\overline{-21x - 24}
\end{array}
$$

which yields the quotient $x^2 - 3x + 8$ and the remainder $-21x - 24$. In this division we proceed as indicated until we arrive at a polynomial (the remainder) which is either 0 or has smaller degree than the divisor. We shall assume familiarity with this process and not attempt to justify it. As with integers, the result of this division process is often written as follows:

$$\frac{x^4 - 16}{x^2 + 3x + 1} = (x^2 - 3x + 8) + \left(\frac{-21x - 24}{x^2 + 3x + 1}\right).$$

Multiplying by $x^2 + 3x + 1$,

$$x^4 - 16 = (x^2 + 3x + 1)(x^2 - 3x + 8) + (-21x - 24).$$

This has the same general form $a = bq + r$ that was given for integers.

The preceding example illustrates the following theorem, which we state without proof.

(8.2)   Division Algorithm for Polynomials

> If $f(x)$ and $g(x)$ are polynomials and if $g(x) \neq 0$, then there exist unique polynomials $q(x)$ and $r(x)$ such that
>
> $$f(x) = g(x)q(x) + r(x)$$
>
> where either $r(x) = 0$ or the degree of $r(x)$ is less than the degree of $g(x)$. The polynomial $q(x)$ is called the **quotient** and $r(x)$ the **remainder** in the division of $f(x)$ by $g(x)$.

An interesting special case of (8.2) occurs if $f(x)$ is divided by a linear polynomial of the form $x - c$, where c is a number. If $x - c$ is a factor of $f(x)$, then

$$f(x) = (x - c)q(x)$$

for some polynomial $q(x)$; that is, the remainder $r(x)$ is 0. If the remainder is not 0, then by (8.2) its degree is less than the degree of the divisor $x - c$, and hence $r(x)$ must have degree 0. This in turn means that the remainder is a nonzero number. Consequently, in all cases we have

(8.3) $$f(x) = (x - c)q(x) + d$$

where d is some number. If c is substituted for x in (8.3), we obtain

$$f(c) = (c - c)q(c) + d$$

which reduces to $f(c) = d$. Substituting for d in (8.3) gives us

(8.4) $$f(x) = (x - c)q(x) + f(c).$$

We have proved the following theorem.

(8.5)   Remainder Theorem

> If a polynomial $f(x)$ is divided by $x - c$, then the remainder is $f(c)$.

Example 1   If $f(x) = x^3 - 3x^2 + x + 5$, use the Remainder Theorem to find $f(2)$.

Solution   According to (8.5), $f(2)$ is the remainder when $f(x)$ is divided by $x - 2$. By long division,

$$
\begin{array}{r}
x^2 - x - 1 \\
x - 2 \overline{\smash{)}\begin{aligned}x^3 - 3x^2 + x + 5\end{aligned}} \\
\underline{x^3 - 2x^2} \\
-x^2 + x \\
\underline{-x^2 + 2x} \\
-x + 5 \\
\underline{-x + 2} \\
3
\end{array}
$$

and hence $f(2) = 3$. To check our work we have, by direct substitution,

$$f(2) = 2^3 - 3(2)^2 + 2 + 5 = 3.$$

The following important result is a consequence of (8.5).

(8.6)   Factor Theorem

> A polynomial $f(x)$ has a factor $x - c$ if and only if $f(c) = 0$.

Proof. From (8.4) we have $f(x) = (x - c)q(x) + f(c)$. If $f(c) = 0$, then $f(x) = (x - c)q(x)$; that is, $x - c$ is a factor of $f(x)$. Conversely, if $x - c$ is a factor, then the remainder upon division of $f(x)$ must be 0 and hence by the Remainder Theorem, $f(c) = 0$.

The Factor Theorem is useful for finding factors of polynomials, as illustrated in the next example.

Example 2   Show that $x - 2$ is a factor of the polynomial

$$f(x) = x^3 - 4x^2 + 3x + 2.$$

Solution   Since $f(2) = 8 - 16 + 6 + 2 = 0$, it follows from the Factor Theorem that $x - 2$ is a factor of $f(x)$. Of course, another method of solution would be to divide $f(x)$ by $x - 2$ and show that the remainder is 0. The quotient in the division would be another factor of $f(x)$.

Example 3   Find a polynomial $f(x)$ of degree 3 that has zeros 2, -1, and 3.

Solution   By the Factor Theorem, $f(x)$ has factors $x - 2$, $x + 1$, and $x - 3$. We may then write

$$f(x) = a(x - 2)(x + 1)(x - 3)$$

where any nonzero value may be assigned to a. If we let $a = 1$ and multiply, we obtain

$$f(x) = x^3 - 4x^2 + x + 6.$$

EXERCISES

In Exercises 1–6 find the quotient $q(x)$ and the remainder $r(x)$ if $f(x)$ is divided by $g(x)$.

1 $f(x) = x^4 + 3x^3 - 2x + 5$, $g(x) = x^2 + 2x - 4$

2 $f(x) = 4x^3 - x^2 + x - 3$, $g(x) = x^2 - 5x$

3 $f(x) = 5x^3 - 2x$, $g(x) = 2x^2 + 1$

4 $f(x) = 3x^4 - x^3 - x^2 + 3x + 4$, $g(x) = 2x^3 - x + 4$

5 $f(x) = 7x^3 - 5x + 2$, $g(x) = 2x^4 - 3x^2 + 9$

6 $f(x) = 10x - 4$, $g(x) = 8x^2 - 5x + 17$

In Exercises 7–12 use the Remainder Theorem to find $f(c)$.

7 $f(x) = 2x^3 - x^2 - 5x + 3$, $c = 4$

8 $f(x) = 4x^3 - 3x^2 + 7x + 10$, $c = 3$

9 $f(x) = x^4 + 5x^3 - x^2 + 5$, $c = -2$

10 $f(x) = x^4 - 7x^2 + 2x - 8$, $c = -3$

11 $f(x) = x^6 - 3x^4 + 4$, $c = \sqrt{2}$

12 $f(x) = x^5 - x^4 + x^3 - x^2 + x - 1$, $c = i$

13 Determine k so that $f(x) = x^3 + kx^2 - kx + 10$ is divisible by $x + 3$.

14 Determine all values of k such that $f(x) = k^2x^3 - 4kx - 3$ is divisible by $x - 1$.

15 Use the Factor Theorem to show that $x - 2$ is a factor of $f(x) = x^4 - 3x^3 - 2x^2 + 5x + 6$.

16 Show that $x + 2$ is a factor of $f(x) = x^{12} - 4096$.

17 Prove that $f(x) = 3x^4 + x^2 + 5$ has no factor of the form $x - c$ where c is a real number.

18 Find the remainder if the polynomial $3x^{100} + 5x^{85} - 4x^{38} + 2x^{17} - 6$ is divided by $x + 1$.

19 Use the Factor Theorem to prove that $x - y$ is a factor of $x^n - y^n$, for all positive integers n. If n is even, show that $x + y$ is also a factor of $x^n - y^n$.

20 If n is an odd positive integer, prove that $x + y$ is a factor of $x^n + y^n$.

21 Prove that a polynomial $f(x)$ has a zero if and only if $f(x)$ has a first-degree polynomial as a factor.

22   Comment on the following "proof" that if $f(x)$ is divided by $x - c$, then the remainder is $f(c)$.

$$
\begin{array}{r}
f \\
x - c \,\overline{\big)\, f(x)} \\
\underline{f(x) - f(c)} \\
f(c)
\end{array}
$$

2   SYNTHETIC DIVISION

When applying the Remainder Theorem, it is necessary to divide by polynomials of the form $x - c$. The process referred to as *synthetic division* simplifies the work if divisors are of that form. We shall illustrate the process by means of examples.

If the polynomial $3x^4 - 8x^3 + 9x + 5$ is divided by $x - 2$, we obtain

$$
\begin{array}{r}
3x^3 - 2x^2 - 4x + 1 \\
x - 2 \,\overline{\big)\, 3x^4 - 8x^3 + 0x^2 + 9x + 5} \\
\underline{3x^4 - 6x^3} \\
-2x^3 + 0x^2 \\
\underline{-2x^3 + 4x^2} \\
-4x^2 + 9x \\
\underline{-4x^2 + 8x} \\
x + 5 \\
\underline{x - 2} \\
7
\end{array}
$$

where the term $0x^2$ has been inserted in the dividend so that *all* powers of x are accounted for. Since this technique of long division seems to involve a great deal of labor for so simple a problem, we look for a means of simplifying the notation. After arranging the terms which involve like powers of x in vertical columns as above, it is seen that the repeated expressions $3x^4$, $-2x^3$, $-4x^2$, and x may be deleted without too much chance of confusion. Also, it appears unnecessary to "bring down" the terms $0x^2$, $9x$, and 5 from the dividend as indicated. With the elimination of those repetitions, our work takes on this form:

$$
\begin{array}{r}
3x^3 - 2x^2 - 4x + 1 \\
x - 2 \,\overline{\big)\, 3x^4 - 8x^3 + 0x^2 + 9x + 5} \\
\underline{-6x^3} \\
-2x^3 \\
\underline{} \\
4x^2 \\
\underline{-4x^2} \\
8x \\
\underline{x} \\
-2 \\
\underline{} \\
7
\end{array}
$$

If we take care to keep like powers of x under one another and if we account for missing terms by means of zero coefficients as above, then some labor can be saved by omitting the symbol x. Doing this in the preceding display, we obtain the following:

$$
\begin{array}{r}
\;\;3 \quad -2 \quad -4 \quad\;\; 1 \\
1-2\,\overline{\big|\;3 \quad -8 \quad\;\; 0 \quad\;\; 9 \quad\;\; 5} \\
\underline{-6} \\
-2 \\
\underline{4} \\
-4 \\
\underline{8} \\
1 \\
\underline{-2} \\
7
\end{array}
$$

Since the divisor is a polynomial of the form $x - c$, the two coefficients in the far left position are always $1 - c$. With this in mind we shall discard the coefficient 1. Moreover, to make our notation more compact let us move the numbers up in the following way:

$$
\begin{array}{r}
\;\;3 \quad -2 \quad -4 \quad\;\; 1 \\
-2\,\overline{\big|\;3 \quad -8 \quad\;\; 0 \quad\;\; 9 \quad\;\; 5} \\
-6 \quad\;\; 4 \quad\;\; 8 \quad -2 \\
\hline
-2 \quad -4 \quad\;\; 1 \quad\;\; 7
\end{array}
$$

If we now insert the leading coefficient 3 in the first position of the last row, the first four numbers of that row are the coefficients $3, -2, -4,$ and 1 of the quotient, and the final number 7 is the remainder. Since there is no need to write the coefficients of the quotient two times, we discard the first row in our scheme, obtaining

(8.7)
$$
\begin{array}{r}
-2\,\overline{\big|\;3 \quad -8 \quad\;\; 0 \quad\;\; 9 \quad\;\; 5} \\
-6 \quad\;\; 4 \quad\;\; 8 \quad -2 \\
\hline
\;\;3 \quad -2 \quad -4 \quad\;\; 1 \quad\;\; 7
\end{array}
$$

where the top line has also been deleted since there is no longer any need for it.

There is a simple way of interpreting (8.7). Note that every number in the second row can be obtained by multiplying the number in the third row of the *preceding* column by -2. Moreover, each number in the third row can be found by subtracting the number above it in the second row from the corresponding number in the first row. This suggests a procedure for carrying out (8.7) without actually thinking of the division process. After arranging the terms of the polynomial in decreasing powers of x, we write the coefficients in a row, supplying 0 for any missing term. Next we write $-c$ (in the above case, -2) to the left of this row, as indicated in (8.7). Next we bring down the leading coefficient 3 to the third row. Then we multiply that number by -2 to obtain the first number, -6, in the second row. We subtract -6 from -8 to obtain the second number, -2, in the third row, and then we multiply by $-c$ (in our case, -2) to obtain the second number, 4, in the second row. Again we subtract to get the third number, −4, in the third row. This process is continued until the final number in the third row (the remainder) is obtained.

It is possible to avoid the subtractions performed above if the number c is used in place of $-c$ in the far left position of the first row. In this event, when the process above is used the signs of the elements in the second row are changed, and hence to find elements of the third row, we *add* the number above it in the second row to the corresponding number in the first row. With this change (8.7) becomes

$$
\begin{array}{r|rrrrr}
2 & 3 & -8 & 0 & 9 & 5 \\
  &   & 6 & -4 & -8 & 2 \\
\hline
  & 3 & -2 & -4 & 1 & 7
\end{array}
$$

The latter scheme is called the process of **synthetic division**.

Example 1 Use synthetic division to find the quotient and remainder if $2x^4 + 5x^3 - 2x - 8$ is divided by $x + 3$.

Solution Since we are to divide by $x + 3$, the c in the expression $x - c$ is -3. Hence the synthetic division takes this form:

$$
\begin{array}{r|rrrrr}
-3 & 2 & 5 & 0 & -2 & -8 \\
   &   & -6 & 3 & -9 & 33 \\
\hline
   & 2 & -1 & 3 & -11 & 25
\end{array}
$$

The first four numbers in the third row are the coefficients of the quotient and the last number is the remainder. Hence the quotient is $2x^3 - x^2 + 3x - 11$ and the remainder is 25.

Synthetic division can be used to find values of polynomial functions, as illustrated in the next example.

Example 2 If $f(x) = 3x^5 - 38x^3 + 5x^2 - 1$, use synthetic division to find $f(4)$.

Solution By the Remainder Theorem, $f(4)$ is the remainder when $f(x)$ is divided by $x - 4$. Dividing synthetically we obtain

$$
\begin{array}{r|rrrrrr}
4 & 3 & 0 & -38 & 5 & 0 & -1 \\
  &   & 12 & 48 & 40 & 180 & 720 \\
\hline
  & 3 & 12 & 10 & 45 & 180 & 719
\end{array}
$$

Consequently, $f(4) = 719$

Synthetic division may be employed to help find zeros of polynomials. By the method illustrated in the preceding example, $f(c) = 0$ if and only if the remainder in the synthetic division by $x - c$ is 0.

Example 3   Show that -11 is a zero of the polynomial

$$f(x) = x^3 + 8x^2 - 29x + 44.$$

Solution   Dividing synthetically by $x + 11$ gives us

$$
\begin{array}{r|rrrr}
-11 & 1 & 8 & -29 & 44 \\
    &   & -11 & 33 & -44 \\
\hline
    & 1 & -3 & 4 & 0
\end{array}
$$

Thus $f(-11) = 0$.

The preceding example shows that -11 is a solution of the equation $x^3 + 8x^2 - 29x + 44 = 0$. In the next section we shall use synthetic division in this way to find solutions of equations.

EXERCISES

In each of Exercises 1–10 use synthetic division to find the quotient and remainder if the first polynomial is divided by the second.

1   $2x^3 - 3x^2 + 4x - 5$, $x - 2$

2   $3x^3 - 4x^2 - x + 8$, $x + 4$

3   $x^3 - 8x - 5$, $x + 3$

4   $5x^3 - 6x^2 + 15$, $x - 4$

5   $3x^5 + 6x^2 + 7$, $x + 2$

6   $-2x^4 + 10x - 3$, $x - 3$

7   $4x^4 - 5x^2 + 1$, $x - \frac{1}{2}$

8   $9x^3 - 6x^2 + 3x - 4$, $x - \frac{1}{3}$

9   $x^n - 1$, $x - 1$, where n is any positive integer

10   $x^n + 1$, $x + 1$, where n is any positive integer

Use synthetic division to solve Exercises 11–16.

11   If $f(x) = x^4 - 4x^3 + x^2 - 3x - 5$, find $f(2)$ and $f(-2)$.

12   If $f(x) = 0.3x^3 + 0.04x - 0.034$, find $f(0.2)$ and $f(-0.2)$.

13   If $f(x) = x^6 - x^5 + x^4 - x^3 + x^2 - x + 1$, find $f(4)$.

14   If $f(x) = 8x^5 - 3x^2 + 7$, find $f(1/2)$.

15   If $f(x) = x^2 + 3x - 5$, find $f(2 + \sqrt{3})$.

16   If $f(x) = x^3 - 3x^2 - 8$, find $f(1 + \sqrt{2})$.

In Exercises 17–20 use synthetic division to show that c is a zero of $f(x)$.

17   $f(x) = 3x^4 + 8x^3 - 2x^2 - 10x + 4$, $c = -2$

18   $f(x) = 4x^3 - 9x^2 - 8x - 3$, $c = 3$

19   $f(x) = 4x^3 - 6x^2 + 8x - 3$, $c = \frac{1}{2}$

20   $f(x) = 27x^4 - 9x^3 + 3x^2 + 6x + 1$, $c = -\frac{1}{3}$

3 FACTORIZATION THEORY

The Factor Theorem (8.6) indicates that there is a close relationship between the zeros of a polynomial $f(x)$ and the factors of $f(x)$. Indeed, if a number c can be found such that $f(c) = 0$, then a factor $x - c$ is obtained immediately. Unfortunately, except in special cases, zeros of polynomials are very difficult to find. For example, given the polynomial $f(x) = x^5 - 3x^4 + 4x^3 + 4x - 10$, there are no obvious zeros. Moreover, there is no device such as the quadratic formula which can be used to produce the zeros. In spite of the practical difficulty of determining zeros of polynomials, it is possible to make some headway concerning the *theory* of such zeros. The next result is basic for the development of this theory.

(8.8) **Fundamental Theorem of Algebra**

> If $f(x)$ is a polynomial of positive degree, then $f(x)$ has at least one complex zero.

The usual proof of Theorem (8.8) requires results from the field of mathematics called *functions of a complex variable*. In turn, a prerequisite for studying the latter field is a strong background in calculus. The first proof of the Fundamental Theorem of Algebra was given by the German mathematician Carl Friedrich Gauss (1777–1855), who is considered by many to be the greatest mathematician of all time.

As a special case of (8.8), if all the coefficients of $f(x)$ are real, then $f(x)$ has at least one complex zero. We should also remark that if $a + bi$ is a complex zero of a polynomial, it may happen that $b = 0$, in which case we refer to the number as a **real zero**. If (8.8) is combined with the Factor Theorem, the following useful result is obtained.

(8.9) **Theorem**

> Every polynomial of positive degree has a factor of the form $x - c$, where c is a real or complex number.

Theorem (8.9) enables us, at least in theory, to express every polynomial $f(x)$ of positive degree as a product of polynomials of degree 1. If $f(x)$ has a degree $n > 0$, then applying (8.9),

$$f(x) = (x - c_1)f_1(x)$$

where c_1 is some number and $f_1(x)$ is a polynomial of degree $n - 1$. If $n - 1 > 0$, we may apply (8.9) again, obtaining

$$f_1(x) = (x - c_2)f_2(x)$$

where c_2 is some number and $f_2(x)$ is a polynomial of degree $n - 2$. Hence,

$$f(x) = (x - c_1)(x - c_2)f_2(x).$$

Continuing this process, after n steps we arrive at a polynomial $f_n(x)$ of degree 0. Thus, $f_n(x) = a$ for some nonzero number a and we may write

$$f(x) = a(x - c_1)(x - c_2) \cdots (x - c_n)$$

where each c_i is a zero of $f(x)$. Evidently, the leading coefficient of the polynomial on the right in the last equation is a. It follows that a is the leading coefficient of $f(x)$. We have proved the following theorem.

(8.10)   Theorem

> If $f(x)$ is a polynomial of degree $n > 0$, then there exist n numbers $c_1, c_2, \ldots, c_n$ such that
>
> $$f(x) = a(x - c_1)(x - c_2) \cdots (x - c_n)$$
>
> where a is the leading coefficient of $f(x)$.

(8.11)   Corollary

> A polynomial of degree $n > 0$ has at most n different zeros.

Proof.   We shall give an indirect proof. Suppose $f(x)$ has *more* than n different zeros. Let us choose $n + 1$ of these zeros and label them $c_1, c_2, \ldots, c_n$, and c. We may use the c_i as in the proof of (8.9) to obtain the factorization in (8.10). Substituting c for x and using the fact that $f(c) = 0$, we obtain

$$0 = a(c - c_1)(c - c_2) \cdots (c - c_n).$$

However, each factor on the right side is different from zero because $c \neq c_i$ for every i. Since the product of nonzero numbers cannot equal zero, we have a contradiction.

Example 1   Find a polynomial $f(x)$ of degree 3 with zeros 2, -1, and 3, which has the value 5 at $x = 1$.

Solution   By the Factor Theorem, $f(x)$ has factors $x - 2$, $x + 1$, and $x - 3$. No other factors of degree 1 exist, since by the Factor Theorem another linear factor $x - c$ would produce a fourth zero of $f(x)$ in violation of (8.11). Hence $f(x)$ has the form

$$f(x) = a(x - 2)(x + 1)(x - 3)$$

for some number a. If $f(x)$ has the value 5 at $x = 1$, then $f(1) = 5$, that is,

$$a(1 - 2)(1 + 1)(1 - 3) = 5, \quad \text{or} \quad 4a = 5.$$

Consequently $a = 5/4$, and

$$f(x) = \frac{5}{4}(x - 2)(x + 1)(x - 3).$$

Multiplying the four factors we obtain

$$f(x) = \frac{5}{4}x^3 - 5x^2 + \frac{5}{4}x + \frac{15}{2}.$$

The numbers $c_1, c_2, \ldots, c_n$ in (8.10) are not necessarily all different. To illustrate, the polynomial $f(x) = x^3 + x^2 - 5x + 3$ has the factorization

$$f(x) = (x + 3)(x - 1)(x - 1).$$

If a factor $x - c$ occurs m times in the factorization (8.10), then c is called a **zero of multiplicity** m of $f(x)$. In the preceding illustration, 1 is a zero of multiplicity 2, and -3 is a zero of multiplicity 1.

As another example, if

$$f(x) = (x - 2)(x - 4)^3(x + 1)^2$$

then f has degree 6 and possesses three distinct zeros 2, 4, and -1, where 2 has multiplicity 1, 4 has multiplicity 3, and -1 has multiplicity 2.

If a zero of multiplicity m is counted as m zeros, then (8.10) tells us that a polynomial $f(x)$ of degree $n > 0$ has *at least* n zeros (not necessarily all different). Combining this with the fact that $f(x)$ has *at most* n zeros, we obtain the next result.

(8.12)   Theorem

> If $f(x)$ is a polynomial of degree $n > 0$ and if a zero of multiplicity m is counted m times, then $f(x)$ has precisely n zeros.

Example 2   Express $f(x) = x^5 - 4x^4 + 13x^3$ as a product of linear factors and list the five zeros of $f(x)$.

Solution   We begin by writing

$$f(x) = x^3(x^2 - 4x + 13).$$

By the quadratic formula, the zeros of the polynomial $x^2 - 4x + 13$ are given by

$$\frac{4 \pm \sqrt{16 - 52}}{2} = \frac{4 \pm \sqrt{-36}}{2} = \frac{4 \pm 6i}{2} = 2 \pm 3i.$$

Hence by the Factor Theorem, $x^2 - 4x + 13$ has factors $x - (2 + 3i)$ and $x - (2 - 3i)$ and we obtain the desired factorization

$$f(x) = x \cdot x \cdot x \cdot (x - 2 - 3i)(x - 2 + 3i).$$

Since $x - 0$ occurs as a factor three times, the number 0 is a zero of multiplicity three, and the five zeros of $f(x)$ are 0, 0, 0, $2 + 3i$, and $2 - 3i$.

In order to solve certain problems in calculus it is necessary to express quotients of polynomials as sums of simpler quotients. To illustrate, it can be verified by addition that

$$\frac{2x^2 + 3x - 8}{x(x + 1)(x - 2)} = \frac{4}{x} + \frac{-3}{x + 1} + \frac{1}{x - 2}.$$

The expression on the right side of the above equation is called the **partial fraction decomposition** of the quotient which appears on the left side. Generally, if $f(x)$ and $g(x)$ are any polynomials with real coefficients *and the degree of $f(x)$ is less than the degree of $g(x)$*, then it can be shown that the quotient $f(x)/g(x)$ may be expressed as a sum of other quotients (called **partial fractions**) each having one of the forms

$$\frac{A}{(cx + d)^m} \quad \text{or} \quad \frac{Cx + D}{(ax^2 + bx + c)^n}$$

for some nonnegative integers m, n, and where $ax^2 + bx + c$ is irreducible over the reals, in the sense that $b^2 - 4ac < 0$. We shall not prove this result but will, instead, state some rules for obtaining decompositions and illustrate several special cases.

If the denominator $g(x)$ of $f(x)/g(x)$ factors into distinct linear factors as in the above illustration, then for each factor $cx + d$ there corresponds a partial fraction of the form $A/(cx + d)$, where A is a real number. A technique for finding the numerators of the partial fractions is illustrated in the next example.

Example 3 Find the partial fraction decomposition of

$$\frac{4x^2 + 13x - 9}{x^3 + 2x^2 - 3x}.$$

Solution The denominator of the quotient has the factored form $x(x + 3)(x - 1)$. Each of the linear factors is handled as mentioned above. Thus, for the factor x there corresponds a partial fraction of the form A/x. Similarly, for the factors $x + 3$ and $x - 1$ there correspond partial fractions $B/(x + 3)$ and $C/(x - 1)$, respectively. The partial fraction decomposition has the form

$$\frac{4x^2 + 13x - 9}{x(x + 3)(x - 1)} = \frac{A}{x} + \frac{B}{x + 3} + \frac{C}{x - 1}.$$

Multiplying by the lowest common denominator gives us

(8.13) $4x^2 + 13x - 9 = A(x + 3)(x - 1) + Bx(x - 1) + Cx(x + 3).$

In a case such as this, in which the factors are all linear and nonrepeated, the values for A, B, and C can be found by substituting values for x which make the various factors zero. If we let $x = 0$ in (8.13), then $-9 = -3A$, or $A = 3$. Letting $x = 1$ in (8.13) gives us $8 = 4C$, or $C = 2$. Finally, if $x = -3$, then $-12 = 12B$ or $B = -1$. The partial fraction decomposition is, therefore,

$$\frac{4x^2 + 13x - 9}{x(x + 3)(x - 1)} = \frac{3}{x} + \frac{-1}{x + 3} + \frac{2}{x - 1}.$$

Another technique for finding A, B, and C is to compare coefficients of x. If the right-hand side of (8.13) is expanded and like powers of x are collected, then

$$4x^2 + 13x - 9 = (A + B + C)x^2 + (2A - B + 3C)x - 3A.$$

If two polynomials are equal for all x, then coefficients of like powers are the same (see Exercise 23). Consequently

$$\begin{cases} A + B + C = 4 \\ 2A - B + 3C = 13 \\ -3A = -9. \end{cases}$$

It is left to the reader to show that the solution of this system of equations is $A = 3$, $B = -1$, and $C = 2$.

If the denominator $g(x)$ in the discussion preceding Example 3 has a squared factor of the form $(cx + d)^2$, then there corresponds a sum of two partial fractions

$$\frac{A}{cx + d} + \frac{B}{(cx + d)^2}.$$

If a cubed factor $(cx + d)^3$ occurs, then a third fraction $C/(cx + d)^3$ is added to the above sum. This rule may be extended to any power $(cx + d)^m$.

Example 4   Find the partial fraction decomposition of

$$\frac{x^2 + 10x - 36}{x(x - 3)^2}.$$

Solution   Corresponding to the factor x in the denominator there is a partial fraction A/x. As mentioned above, for the squared factor $(x - 3)^2$ there corresponds a sum of two partial fractions $B/(x - 3) + C/(x - 3)^2$. Thus the partial fraction decomposition has the form

$$\frac{x^2 + 10x - 36}{x(x - 3)^2} = \frac{A}{x} + \frac{B}{x - 3} + \frac{C}{(x - 3)^2}.$$

Multiplying both sides by $x(x - 3)^2$ gives us

(8.14) $$x^2 + 10x - 36 = A(x - 3)^2 + Bx(x - 3) + Cx.$$

Two of the unknown constants may be determined easily. Letting $x = 3$ in the last equation gives us

$$9 + 30 - 36 = A(0) + B(0) + 3C.$$

This reduces to $3 = 3C$ and hence $C = 1$. In like manner, letting $x = 0$ we get

$$0 + 0 - 36 = A(-3)^2 + B(0) + C(0).$$

Hence $-36 = 9A$, or $A = -4$. The remaining constant can be found by comparing coefficients. If the right side of (8.14) is expanded and like powers of x collected we see that the coefficient of x^2 is $A + B$. This must equal the coefficient of x^2 on the left, that is, $A + B = 1$. Since $A = -4$, it follows that $B = 1 - A = 1 - (-4) = 5$. The partial fraction decomposition is, therefore,

$$\frac{x^2 + 10x - 36}{x(x - 3)^2} = \frac{-4}{x} + \frac{5}{x - 3} + \frac{1}{(x - 3)^2}.$$

The form of the partial fractions for the case where $g(x)$ has irreducible quadratic factors is illustrated in Exercises 33–35.

EXERCISES

In each of Exercises 1–6, find a polynomial $f(x)$ of degree 3 with the indicated zeros and satisfying the given conditions.

1   $5, -2, -3;\ f(2) = 4$

2   $4, 1, -6;\ f(5) = 2$

3   $2, 3, 1;\ f(0) = 12$

4   $\sqrt{2}, \pi, 0;\ f(0) = 0$

5   $2 + i, 2 - i, -4;\ f(1) = 3$

6   $1 + 2i, 1 - 2i, 5;\ f(-2) = 1$

7   Find a polynomial of degree 4 such that both -2 and 3 are zeros of multiplicity 2.

8   Find a polynomial of degree 5 such that -2 is a zero of multiplicity 3 and 4 is a zero of multiplicity 2.

9   Find a polynomial $f(x)$ of degree 8 such that 2 is a zero of multiplicity 3, 0 is a zero of multiplicity 5, and $f(3) = 54$.

10   Find a polynomial $f(x)$ of degree 7 such that 1 is a zero of multiplicity 2, -1 is a zero of multiplicity 2, 0 is a zero of multiplicity 3, and $f(2) = 36$.

In Exercises 11–18 find the zeros of the polynomials and state the multiplicity of each zero.

11   $f(x) = (x + 4)^3(3x - 4)$

12   $f(x) = (x - 5)^2(4x + 7)^3$

13   $f(x) = 2x^5 - 8x^4 - 10x^3$

14   $f(x) = (4x^2 - 5)^2$

15   $f(x) = (9x^2 - 25)^4(x^2 + 16)$

16   $f(x) = (2x^2 + 13x - 7)^3$

17   $f(x) = (x^2 + x - 2)^2(x^2 - 4)$

18   $f(x) = 4x^6 + x^4$

19   Show that -3 is a zero of multiplicity 2 of the polynomial $f(x) = x^4 + 7x^3 + 13x^2 - 3x - 18$ and express $f(x)$ as a product of linear factors.

20 Show that 4 is a zero of multiplicity 2 of the polynomial $f(x) = x^4 - 9x^3 + 22x^2 - 32$ and express $f(x)$ as a product of linear factors.

21 Show that 1 is a zero of multiplicity 5 of $f(x) = x^6 - 4x^5 + 5x^4 - 5x^2 + 4x - 1$ and express $f(x)$ as a product of linear factors.

22 Show that -1 is a zero of multiplicity 4 of $f(x) = x^5 + x^4 - 6x^3 - 14x^2 - 11x - 3$ and express $f(x)$ as a product of linear factors.

23 Let $f(x)$ and $g(x)$ be polynomials of degree not greater than n, where n is a positive integer. Show that if $f(x)$ and $g(x)$ are equal in value for more than n distinct values of x, then $f(x)$ and $g(x)$ are identical, that is, coefficients of like powers are the same. (*Hint:* Write

$$f(x) = a_n x^n + a_{n-1} x^{n-1} + \cdots + a_1 x + a_0$$
$$g(x) = b_n x^n + b_{n-1} x^{n-1} + \cdots + b_1 x + b_0$$

and consider $h(x) = f(x) - g(x) = (a_n - b_n)x^n + \cdots + (a_0 - b_0)$. Then show that $h(x)$ has more than n distinct zeros and conclude from (8.11) that $a_i = b_i$ for all i.)

24 Determine real numbers A and B such that $A(3x + 1) + B(x - 2) = 6x + 5$ is an identity. (*Hint:* First write the equation in the form

$$(3A + B)x + (A - 2B) = 6x + 5.$$

Next, by Exercise 23, $3A + B = 6$ and $A - 2B = 5$. Now solve for A and B.)

Find the partial fraction decompositions in Exercises 25–32.

25 $\dfrac{8x - 1}{(x - 2)(x + 3)}$

26 $\dfrac{x - 29}{(x - 4)(x + 1)}$

27 $\dfrac{4x^2 - 15x - 1}{(x - 1)(x + 2)(x - 3)}$

28 $\dfrac{x^2 + 19x + 20}{x(x + 2)(x - 5)}$

29 $\dfrac{2x + 3}{(x - 1)^2}$

30 $\dfrac{5x^2 - 4}{x^2(x + 2)}$

31 $\dfrac{x^2 - 6}{(x + 2)^2(2x - 1)}$

32 $\dfrac{2x^2 + x}{(x - 1)^2(x + 1)^2}$

In Exercises 33–36 find A, B, C, D.

33 $\dfrac{x^2 + 4x - 5}{(x^2 + 1)(x - 1)} = \dfrac{Ax + B}{x^2 + 1} + \dfrac{C}{x - 1}$

34 $\dfrac{x^2 - x - 21}{(x^2 + 4)(2x - 1)} = \dfrac{Ax + B}{x^2 + 4} + \dfrac{C}{2x - 1}$

35 $\dfrac{5x^3 - 3x^2 + 7x - 3}{(x^2 + 1)^2} = \dfrac{Ax + B}{x^2 + 1} + \dfrac{Cx + D}{(x^2 + 1)^2}$

36 $\dfrac{3x^3 - 18x^2 + 29x - 4}{(x + 1)(x - 2)^3} = \dfrac{A}{x + 1} + \dfrac{B}{x - 2} + \dfrac{C}{(x - 2)^2} + \dfrac{D}{(x - 2)^3}$

4   ZEROS OF POLYNOMIALS WITH REAL COEFFICIENTS

In this section we shall concentrate on polynomials with real coefficients. An interesting fact about such polynomials is illustrated in Example 2 of the preceding section, where the two complex zeros of $x^5 - 4x^4 + 13x^3$ were conjugates of one another. That is no accident, since the following general result is true.

(8.15)   **Theorem**

> If $f(x)$ is a polynomial of degree $n > 0$, with real coefficients, and if z is a complex zero of $f(x)$, then the conjugate $\bar{z}$ is also a zero of $f(x)$.

Proof.   We may write

$$f(x) = a_n x^n + a_{n-1} x^{n-1} + \cdots + a_1 x + a_0$$

where the a_i are real numbers and $a_n \neq 0$. If $f(z) = 0$, then

$$a_n z^n + a_{n-1} z^{n-1} + \cdots + a_1 z + a_0 = 0.$$

If two complex numbers are equal, then so are their conjugates. Consequently, the conjugates of each side of the latter equation are equal, that is,

$$\overline{a_n z^n + a_{n-1} z^{n-1} + \cdots + a_1 z + a_0} = \bar{0} = 0.$$

(The fact that $\bar{0} = 0$ follows from (iv) of (7.7).) As pointed out in the preceding chapter, the conjugate of a sum of complex numbers equals the sum of the conjugates, and therefore

(8.16) $$\overline{a_n z^n} + \overline{a_{n-1} z^{n-1}} + \cdots + \overline{a_1 z} + \overline{a_0} = 0.$$

Using (7.7), we have for each i,

$$\overline{a_i z^i} = \overline{a_i} \cdot \overline{z^i} = \overline{a_i} \cdot \bar{z}^i = a_i \bar{z}^i.$$

Hence, (8.16) may be written as follows:

$$a_n \bar{z}^n + a_{n-1} \bar{z}^{n-1} + \cdots + a_1 \bar{z} + a_0 = 0.$$

The last equation states that $f(\bar{z}) = 0$, which completes the proof.

Example 1   Find a polynomial $f(x)$ of degree 4 that has real coefficients and zeros $2 + i$ and $-3i$.

Solution   By (8.15), $f(x)$ must also have zeros $2 - i$ and $3i$, and hence by the Factor Theorem, $f(x)$ has factors $x - (2 + i)$, $x - (2 - i)$, $x - (-3i)$, and $x - (3i)$.

Multiplying those factors gives us a polynomial of the required type. Thus,

$$f(x) = [x - (2 + i)][x - (2 - i)](x - 3i)(x + 3i)$$
$$= (x^2 - 4x + 5)(x^2 + 9)$$
$$= x^4 - 4x^3 + 14x^2 - 36x + 45.$$

If a polynomial with real coefficients is factored as in (8.10), some of the factors $x - c_i$ may have a complex coefficient c_i. However, it is possible to obtain a factorization into polynomials with real coefficients, as stated in the next theorem.

(8.17)   Theorem

> Every polynomial of degree $n > 0$ with real coefficients can be expressed as a product of linear and quadratic polynomials with real coefficients, where the quadratic factors have no real zeros.

Proof. By (8.12), $f(x)$ has precisely n zeros $c_1, c_2, \ldots, c_n$, and as in (8.10) we obtain the factorization

$$f(x) = a(x - c_1)(x - c_2) \cdots (x - c_n).$$

Of course, some of the c_i may be real; that is, their imaginary parts may be zero. In such cases we obtain the linear factors referred to in the statement of the theorem. If a zero c_i is not real, then by (8.15) the conjugate $\bar{c}_i$ is also a zero of $f(x)$ and hence must be one of the numbers in the set $c_1, c_2, \ldots, c_n$. This implies that both $x - c_i$ and $x - \bar{c}_i$ appear in the factorization of $f(x)$. If those factors are multiplied, we obtain

$$(x - c_i)(x - \bar{c}_i) = x^2 - (c_i + \bar{c}_i)x + c_i\bar{c}_i$$

which has *real* coefficients, since $c_i + \bar{c}_i$ and $c_i\bar{c}_i$ are real numbers. Thus, the complex zeros of $f(x)$ and their conjugates give rise to quadratic polynomials which are irreducible over $\mathbb{R}$. This completes the proof.

Example 2 Express $x^4 - 2x^2 - 3$ (a) as a product of linear polynomials, and (b) as a product of linear and quadratic polynomials with real coefficients which are irreducible over $\mathbb{R}$.

Solutions (a)   The zeros of the given polynomial can be found by solving the equation $x^4 - 2x^2 - 3 = 0$, which may be regarded as quadratic in x^2. Solving for x^2 by means of the quadratic formula, we obtain

$$x^2 = \frac{2 \pm \sqrt{4 + 12}}{2} = \frac{2 \pm 4}{2}$$

or $x^2 = 3$ and $x^2 = -1$. Hence the zeros are $\sqrt{3}, -\sqrt{3}, i$, and $-i$, and so we obtain the factorization

$$x^4 - 2x^3 - 3 = (x - \sqrt{3})(x + \sqrt{3})(x - i)(x + i).$$

(b)   Multiplying the last two factors in the preceding factorization gives us

$$x^4 - 2x^2 - 3 = (x - \sqrt{3})(x + \sqrt{3})(x^2 + 1)$$

which is of the form stated in (8.17).

The solution of this example could also have been obtained by factoring the original expression without first finding the zeros. Thus,

$$x^4 - 2x^2 - 3 = (x^2 - 3)(x^2 + 1)$$
$$= (x + \sqrt{3})(x - \sqrt{3})(x + i)(x - i).$$

We have already pointed out that it is generally very difficult to find zeros of polynomials of high degree. However, if all the coefficients are integers or rational numbers, there is a method for finding the *rational* zeros, if they exist. The method is a consequence of the following theorem.

(8.18)   **Theorem on Rational Zeros**

> Suppose that $f(x) = a_n x^n + a_{n-1} x^{n-1} + \cdots + a_1 x + a_0$ is a polynomial with integer coefficients. If c/d is a rational zero of $f(x)$, where c and d have no common prime factors and $c > 0$, then c is a factor of a_0 and d is a factor of a_n.

Proof.   Let us show that c is a factor of a_0. If $c = 1$, the theorem follows at once, since 1 is a factor of *any* number. Now suppose that $c \neq 1$. In this case $c/d \neq 1$, for if $c/d = 1$, we obtain $c = d$, and since c and d have no prime factor in common, this implies that $c = d = 1$, a contradiction. Hence in the following discussion we have $c \neq 1$ and $c \neq d$.

Since $f(c/d) = 0$,

$$a_n(c^n/d^n) + a_{n-1}(c^{n-1}/d^{n-1}) + \cdots + a_1(c/d) + a_0 = 0.$$

Multiplying by d^n and then adding $-a_0 d^n$ to both sides, we obtain

$$a_n c^n + a_{n-1} c^{n-1} d + \cdots + a_1 c d^{n-1} = -a_0 d^n,$$

or

$$c(a_n c^{n-1} + a_{n-1} c^{n-2} d + \cdots + a_1 d^{n-1}) = -a_0 d^n.$$

This shows that c is a factor of the integer $a_0 d^n$. Hence if c is factored into primes, say $c = p_1 p_2 \cdots p_k$, then each prime p_i is also a factor of $a_0 d^n$. However, by hypothesis, none of the p_i is a factor of d. This implies that each p_i is a factor of a_0, that is, c is a factor of a_0. A similar argument may be used to prove that d is a factor of a_n.

The technique of using (8.18) for finding rational solutions of equations with integer coefficients is illustrated in the following example.

Example 1   Find all rational solutions of the equation

$$3x^4 + 14x^3 + 14x^2 - 8x - 8 = 0.$$

Solution   The problem is equivalent to finding the rational zeros of the indicated polynomial. According to (8.18), if c/d is a rational zero and $c > 0$, then c is a divisor of -8 and d is a divisor of 3. Hence the possible choices for c are 1, 2, 4, and 8, and the choices for d are ± 1 and ± 3. Consequently, any rational roots are included among the numbers ± 1, ± 2, ± 4, ± 8, $\pm\frac{1}{3}$, $\pm\frac{2}{3}$, $\pm\frac{4}{3}$, and $\pm\frac{8}{3}$. Of these sixteen possibilities, not more than four can be zeros by (8.11). It is necessary to check to see which, if any, are zeros. The method of synthetic division is recommended for carrying out this task. We find that

$$
\begin{array}{r|rrrrr}
-2 & 3 & 14 & 14 & -8 & -8 \\
   &   & -6 & -16 & 4 & 8 \\
\hline
   & 3 & 8 & -2 & -4 & 0
\end{array}
$$

which shows that -2 is a zero. Moreover, the synthetic division provides the coefficients of the quotient in the division of the polynomial by $x + 2$. Hence we have the following factorization of the given polynomial:

$$(x + 2)(3x^3 + 8x^2 - 2x - 4).$$

The remaining solutions of the equation must be zeros of the second factor, and therefore we may use the latter polynomial to check for solutions. Dividing by $x + \frac{2}{3}$ synthetically gives us

$$
\begin{array}{r|rrrr}
-\frac{2}{3} & 3 & 8 & -2 & -4 \\
   &   & -2 & -4 & 4 \\
\hline
   & 3 & 6 & -6 & 0
\end{array}
$$

and therefore $-\frac{2}{3}$ is a zero.

The remaining zeros are solutions of the equation $3x^2 + 6x - 6 = 0$, or equivalently, $x^2 + 2x - 2 = 0$. By the quadratic formula, the equation has solutions

$$\frac{-2 \pm \sqrt{4 - 4(-2)}}{2} = \frac{-2 \pm \sqrt{12}}{2} = \frac{-2 \pm 2\sqrt{3}}{2}.$$

Hence the given polynomial has two rational roots, -2 and $-\frac{2}{3}$, and two irrational roots, $-1 + \sqrt{3}$ and $-1 - \sqrt{3}$.

Theorem (8.18) may also be applied to equations with rational coefficients. We merely multiply both sides of the equation by the least common denominator of all the coefficients to obtain an equation with integral coefficients and then proceed as above.

Example 2   Find all rational solutions of the equation

$$(\tfrac{2}{3})x^4 + (\tfrac{1}{2})x^3 - (\tfrac{5}{4})x^2 - x - (\tfrac{1}{6}) = 0.$$

Solution   Multiplying both sides of the equation by 12 produces the equivalent equation

$$8x^4 + 6x^3 - 15x^2 - 12x - 2 = 0.$$

According to (8.18), if c/d is a rational solution, then the choices for c are 1 and 2 and the choices for d are ± 1, ± 2, ± 4, and ± 8. Hence the only possible rational roots are ± 1, ± 2, $\pm \tfrac{1}{2}$, $\pm \tfrac{1}{4}$, and $\pm \tfrac{1}{8}$. By trial we have

$$
\begin{array}{r|rrrrr}
-\tfrac{1}{2} & 8 & 6 & -15 & -12 & -2 \\
 &   & -4 & -1 & 8 & 2 \\
\hline
 & 8 & 2 & -16 & -4 & 0
\end{array}
$$

and hence $-\tfrac{1}{2}$ is a solution. Using synthetic division on the coefficients of the quotient, we obtain

$$
\begin{array}{r|rrrr}
-\tfrac{1}{4} & 8 & 2 & -16 & -4 \\
 &   & -2 & 0 & 4 \\
\hline
 & 8 & 0 & -16 & 0
\end{array}
$$

and consequently $-\tfrac{1}{4}$ is a solution. The last synthetic division gave us the quotient $8x^2 - 16$. Setting this equal to zero and solving, we obtain $x^2 = 2$, or $x = \pm\sqrt{2}$. Thus the given equation has rational solutions $-\tfrac{1}{2}$, $-\tfrac{1}{4}$ and irrational solutions $\sqrt{2}$, $-\sqrt{2}$.

The discussion in this section gives no practical information about finding the irrational zeros of polynomials. The examples we have worked are not typical of problems encountered in applications. Indeed, it is not unusual for a polynomial with rational coefficients to have *no* rational zeros. Except in the simplest cases, the best that can be accomplished is to find decimal approximations to the irrational zeros. There exist methods which may be used to approximate irrational zeros to any degree of accuracy. A standard way is to use a technique studied in calculus called Newton's Method. In practice, computers have taken over the task of approximating irrational solutions of equations to a large extent.

If only rough approximations to the real solutions of an equation $f(x) = 0$ are required, then graphical methods are available. For example, we could sketch the graph of $y = f(x)$ and estimate where $y = 0$; that is, we could approximate the x-intercepts. Needless to say, the accuracy of the approximation depends on the care with which the graph is sketched.

The results of this chapter form the basis for work in the *Theory of Equations*, where additional techniques for finding roots of equations are developed.

EXERCISES

In each of Exercises 1–8, find a polynomial with real coefficients which has the given zeros and degree.

1 $4 + i$; degree 2   **2** $3 - 3i$; degree 2

3 $4, 3 - 2i$; degree 3

4 $-2, 1 + 5i$; degree 3

5 $1 + 4i, 2 - i$; degree 4

6 $3, 0, 8 - 7i$; degree 4

7 $5i, 1 + i, 0$; degree 5

8 $i, 2i, 3i$; degree 6

9 Does there exist a polynomial of degree 3 with real coefficients having zeros $1, -1$, and i? Justify your answer.

10 The complex number i is a zero of the polynomial $f(x) = x^3 - ix^2 + 2ix + 2$; however, the conjugate $-i$ of i is not a zero. Why doesn't this contradict (8.15)?

In Exercises 11–23 find all solutions of the given equations.

11 $2x^3 - 7x^2 - 10x + 24 = 0$

12 $2x^3 - 3x^2 - 8x - 3 = 0$

13 $6x^3 + 19x^2 + x - 6 = 0$

14 $6x^3 + 5x^2 - 17x - 6 = 0$

15 $8x^3 + 18x^2 + 45x + 27 = 0$

16 $3x^3 - x^2 + 11x - 20 = 0$

17 $x^4 - x^3 - 9x^2 - 3x - 36 = 0$

18 $3x^4 + 16x^3 + 28x^2 + 31x + 30 = 0$

19 $9x^4 + 15x^3 - 20x^2 - 20x + 16 = 0$

20 $15x^4 + 4x^3 + 11x^2 + 4x - 4 = 0$

21 $4x^5 + 12x^4 - 41x^3 - 99x^2 + 10x + 24 = 0$

22 $4x^5 + 24x^4 - 13x^3 - 174x^2 + 9x + 270 = 0$

Prove that the equations in Exercises 23–26 have no rational roots.

23 $3x^4 + 7x^3 + 3x^2 - 8x - 10 = 0$

24 $2x^4 - 3x^3 + 6x^2 - 24x + 5 = 0$

25 $8x^4 + 16x^3 - 26x^2 - 12x + 15 = 0$

26 $5x^5 + 2x^4 + x^3 - 10x^2 - 4x - 2 = 0$

27 If n is an odd positive integer, prove that a polynomial of degree n with real coefficients has at least one real zero.

28 Show that (8.15) is not necessarily true if $f(x)$ has complex coefficients.

29 Complete the proof of (8.18) by showing that d is a factor of a_n.

30 If a polynomial of the form $x^n + a_{n-1}x^{n-1} + \cdots + a_1 x + a_0$, where each a_i is an integer, has a rational root r, show that r is an integer and is a factor of a_0.

5 REVIEW

Concepts

Define or discuss each of the following.

1 The division algorithm for polynomials

2 The Remainder Theorem

3 The Factor Theorem

4 Synthetic division

5 The Fundamental Theorem of Algebra

6 The multiplicity of a zero of a polynomial

7 The number of zeros of a polynomial

8 The relation between rational zeros and coefficients of a polynomial with integer coefficients

9 Partial fraction decompositions.

Exercises

In each of Exercises 1–4 find the quotient and remainder if $f(x)$ is divided by $g(x)$.

1 $f(x) = 3x^5 - 4x^3 + x + 5, \quad g(x) = x^3 - 2x + 7$

2 $f(x) = 7x^2 + 3x - 10, \quad g(x) = x^3 - x^2 + 10$

3 $f(x) = 9x + 4, \quad g(x) = 2x - 5$

4 $f(x) = 4x^3 - x^2 + 2x - 1, \quad g(x) = x^2$

5 If $f(x) = -4x^4 + 3x^3 - 5x^2 + 7x - 10$, use the Remainder Theorem to find $f(-2)$.

6 Use the Remainder Theorem to prove that $x - 3$ is a factor of $2x^4 - 5x^3 - 4x^2 + 9$.

In Exercises 7 and 8 use synthetic division to find the quotient and remainder if $f(x)$ is divided by $g(x)$.

7 $f(x) = 6x^5 - 4x^2 + 8, \quad g(x) = x + 2$

8 $f(x) = 2x^3 + 5x^2 - 2x + 1, \quad g(x) = x - \sqrt{2}$

In Exercises 9 and 10 find polynomials with real coefficients which have the indicated zeros, degrees, and satisfy the given conditions.

9 $-3 + 5i, -1; \quad$ degree 3; $\quad f(1) = 4$

10 $1 - i, 3, 0;$ degree 4; $\quad f(2) = -1$

11 Find a polynomial of degree 5 such that -3 is a zero of multiplicity 2 and 0 is a zero of multiplicity 5.

12 Show that 2 is a zero of multiplicity 3 of the polynomial $x^5 - 4x^4 - 3x^3 + 34x^2 - 52x + 24$ and express this polynomial as a product of linear factors.

In Exercises 13 and 14 find the zeros of the polynomials and state the multiplicity of each zero.

13 $(x^2 - 2x + 1)^2(x^2 + 2x - 3)$ **14** $x^6 + 2x^4 + x^2$

In Exercises 15 and 16 find the partial fraction decomposition.

15 $\dfrac{-2x^2 + 9x - 2}{x(x - 2)(x + 1)}$ **16** $\dfrac{7x^2 + 8x - 23}{(x - 5)(x + 3)^2}$

17 Prove that $7x^6 + 2x^4 + 3x^2 + 10$ has no real roots.

Find all solutions of the equations in Exercises 18–20.

18 $x^4 + 9x^3 + 31x^2 + 49x + 30 = 0$

19 $16x^3 - 20x^2 - 8x + 3 = 0$

20 $x^4 - 7x^2 + 6 = 0$

CHAPTER 9

Topics
in
Algebra

The method of proof called mathematical induction *considered in the first section of this chapter is very important in all branches of mathematics. In particular, in Section 2 it is used to prove the famous* Binomial Theorem. *Section 3 contains a discussion of* sequences *and* summation notation. *Of special interest to us will be* arithmetic *and* geometric sequences. *The final two sections contain a discussion of counting processes that arise frequently in mathematics and in everyday life. Among these are the notions of* permutations *and* combinations.

1 MATHEMATICAL INDUCTION

In several earlier parts of our work we pointed out that the method of mathematical induction was required in order to complete a proof. The method may be used to show that certain statements or formulas are true for all positive integers. For example, if n is a positive integer, let P_n denote the statement

$$(xy)^n = x^n y^n$$

where x and y are real numbers. Thus, P_1 represents the statement $(xy)^1 = x^1 y^1$, P_2 denotes $(xy)^2 = x^2 y^2$, P_3 is $(xy)^3 = x^3 y^3$, and so on. It is easy to show that P_1, P_2, and P_3 are *true* statements. However, since the set of positive integers is infinite, it is impossible to check the validity of P_n for every positive integer n. In order to give a proof, the method provided by (9.2) is required. This method is based on the following fundamental axiom.

(9.1) Axiom of Mathematical Induction

Suppose a set S of positive integers has the following two properties:
(i) S contains the integer 1.
(ii) Whenever S contains a positive integer k, S also contains $k + 1$.
Then S contains every positive integer.

The reader should have little reluctance about accepting (9.1). If S is a set of positive integers satisfying property (ii), then whenever S contains an arbitrary positive integer k, it must also contain the next positive integer, $k + 1$. If S also satisfies property (i), then S contains 1 and hence by (ii), S contains $1 + 1$, or 2. Applying (ii) again, we see that S contains $2 + 1$, or 3. Once again, S must contain $3 + 1$, or 4. If we continue in this manner, it can be argued that if n is any *specific* positive integer, then n is in S, since we can proceed a step at a time as above, eventually reaching n. Although this argument does not *prove* (9.1), it certainly makes it plausible.

We shall use (9.1) to establish the following fundamental principle.

(9.2)   Principle of Mathematical Induction

If with each positive integer n there is associated a statement P_n, then all the statements P_n are true provided the following two conditions hold;

(i) P_1 is true.

(ii)   Whenever k is a positive integer such that P_k is true, then P_{k+1} is also true.

Proof.   Assume that (i) and (ii) of (9.2) hold and let S denote the set of all positive integers n such that P_n is true. By assumption, P_1 is true and consequently 1 is in S. Thus S satisfies property (i) of (9.1). Whenever S contains a positive integer k, then by the definition of S, P_k is true and hence from assumption (ii) of (9.2), P_{k+1} is also true. This means that S contains $k + 1$. We have shown that whenever S contains a positive integer k, then S also contains $k + 1$. Consequently, property (ii) of (9.1) is true. Hence by (9.1), S contains every positive integer; that is, P_n is true for every positive integer n.

There are other variations of the principle of mathematical induction. One variation appears in (9.10). In most of our work the statement P_n will usually be given in the form of an equation involving the arbitrary positive integer n, as in our illustration $(xy)^n = x^n y^n$.

When applying (9.2), the following two steps should always be followed:

(9.3)

Step (i)   Prove that P_1 is true.

Step (ii)   Assume that P_k is true and prove that P_{k+1} is true.

Step (ii) is usually the most confusing for the beginning student. We do not *prove* that P_k is true (except for $k = 1$). Instead, we show that *if* P_k is true, then the statement P_{k+1} is true. That is all that is necessary according to (9.2). The assumption that P_k is true is referred to as the **induction hypothesis**.

Many interesting formulas about positive integers can be established by using mathematical induction, two of which are illustrated in Examples 1 and 2. Others appear in the Exercises.

Example 1 Prove that for every positive integer n, the sum of the first n positive integers is $n(n + 1)/2$.

Solution For any positive integer n, let P_n denote the statement

(9.4)
$$1 + 2 + 3 + \cdots + n = \frac{n(n + 1)}{2}$$

where by convention, when $n \leq 4$, the left side is adjusted so that there are precisely n terms in the sum. The following are some special cases of P_n: If $n = 2$, then P_2 is

$$1 + 2 = \frac{2(2 + 1)}{2}, \quad \text{or} \quad 3 = 3.$$

If $n = 3$, then P_3 is

$$1 + 2 + 3 = \frac{3(3 + 1)}{2}, \quad \text{or} \quad 6 = 6.$$

If $n = 5$, then P_5 is

$$1 + 2 + 3 + 4 + 5 = \frac{5(5 + 1)}{2}, \quad \text{or} \quad 15 = 15.$$

We wish to show that P_n is true for every n. Although it is instructive to check (9.4) for several values of n as we did above, it is unnecessary to do so. We need only follow steps (i) and (ii) of (9.3).

(i) If we substitute $n = 1$ in (9.4), then by convention the left side collapses to 1 and the right side is $\dfrac{1(1 + 1)}{2}$, which also equals 1. This proves that P_1 is true.

(ii) *Assume* that P_k is true. Thus the induction hypothesis is

(9.5)
$$1 + 2 + 3 + \cdots + k = \frac{k(k + 1)}{2}.$$

Our goal is to prove that P_{k+1} is true, that is,

(9.6)
$$1 + 2 + 3 + \cdots + (k + 1) = \frac{(k + 1)[(k + 1) + 1]}{2}.$$

By the induction hypothesis we already have a formula for the sum of the first k positive integers. Hence a formula for the sum of the first $k + 1$ positive integers may be found simply by adding $(k + 1)$ to both sides of (9.5). Doing so and simplifying, we obtain

$$1 + 2 + 3 + \cdots + k + (k + 1) = \frac{k(k + 1)}{2} + (k + 1)$$
$$= \frac{k(k + 1) + 2(k + 1)}{2}$$

$$= \frac{k^2 + 3k + 2}{2}$$

$$= \frac{(k + 1)(k + 2)}{2}$$

$$= \frac{(k + 1)[(k + 1) + 1]}{2}.$$

We have shown that P_{k+1} is true, and therefore the proof by mathematical induction is complete.

Example 2   Prove that for each positive integer n,

$$1^2 + 3^2 + \cdots + (2n - 1)^2 = \frac{n(2n - 1)(2n + 1)}{3}.$$

Solution   For each positive integer n, let P_n denote the given statement. Note that this is a formula for the sum of the squares of the first n odd positive integers. We again follow the two-step procedure in (9.3).

   (i)   Substituting 1 for n in P_n, we obtain

$$1^2 = \frac{(1)(2 - 1)(2 + 1)}{3} = \frac{3}{3} = 1$$

which shows that P_1 is true.

   (ii)   *Assume* that P_k is true. Thus the induction hypothesis is

$$1^2 + 3^2 + \cdots + (2k - 1)^2 = \frac{k(2k - 1)(2k + 1)}{3}.$$

We wish to prove that P_{k+1} is true, that is,

$$1^2 + 3^2 + \cdots + [2(k + 1) - 1]^2$$
$$= \frac{(k + 1)[2(k + 1) - 1][2(k + 1) + 1]}{3}.$$

The latter equation simplifies to

(9.7) $$1^2 + 3^2 + \cdots + (2k + 1)^2 = \frac{(k + 1)(2k + 1)(2k + 3)}{3}.$$

Observe that the second from the last term on the left-hand side of this equation is $(2k - 1)^2$. (Why?) In a manner similar to the solution of Example 1, we may obtain (9.7) by adding $(2k + 1)^2$ to both sides of the equation stated in the induction hypothesis. This gives us

$$1^2 + 3^2 + \cdots + (2k - 1)^2 + (2k + 1)^2$$
$$= \frac{k(2k - 1)(2k + 1)}{3} + (2k + 1)^2.$$

We leave it as an exercise for the reader to show that the right side of the preceding equation may be written in the form of the right side of (9.7). This proves that P_{k+1} is true, and hence by (9.3), P_n is true for every n.

The Laws of Exponents can be proved by mathematical induction. In order to apply (9.3), we shall use the following definition of exponents.

(9.8)   Definition

> If x is any real number, then
> (i) $x^1 = x$
> (ii)   whenever k is a positive integer for which x^k is defined, let
> $x^{k+1} = x^k \cdot x$.

A definition such as (9.8) is called a **recursive definition**. In general, if a concept is defined for every positive integer n in such a way that the case corresponding to $n = 1$ is given, and if it is also stated how any case after the first is obtained from the preceding one, then the definition is a recursive definition. For example, by (i) of (9.8) we have $x^1 = x$. Next, applying (ii) of (9.8) we obtain

$$x^2 = x^{1+1} = x^1 \cdot x = x \cdot x.$$

Since x^2 is now defined, we may employ (ii) again (with $k = 2$), obtaining

$$x^3 = x^{2+1} = x^2 \cdot x = (x \cdot x) \cdot x.$$

This defines x^3, and hence (ii) of (9.8) may be used again to obtain x^4. Thus,

$$x^4 = x^{3+1} = x^3 \cdot x = [(x \cdot x) \cdot x] \cdot x.$$

Observe that this agrees with our previous formulation of x^n as a product of x by itself n times. It can be shown (by mathematical induction) that (9.8) defines x^n for every positive integer n.

Example 3   If x is a real number, prove that $x^m \cdot x^n = x^{m+n}$ for all positive integers m and n.

Solution   Let m be an arbitrary positive integer. For each positive integer n, let P_n denote the statement

(9.9)
$$x^m \cdot x^n = x^{m+n}.$$

We shall use (9.3) to prove that P_n is true for every positive integer n.
    (i)   To show that P_1 is true we may use (i) and (ii) of (9.8) as follows:

$$x^m \cdot x^1 = x^m \cdot x$$
$$= x^{m+1}$$

which is formula (9.9) with $n = 1$. Hence P_1 is true.

(ii) Assume that P_k is true. Thus the induction hypothesis is

$$x^m \cdot x^k = x^{m+k}.$$

We wish to prove that P_{k+1} is true, that is,

$$x^m \cdot x^{k+1} = x^{m+(k+1)}.$$

The proof may be arranged as follows, where reasons are stated to the right of each step.

$$
\begin{aligned}
x^m \cdot x^{k+1} &= x^m \cdot (x^k \cdot x) & \text{(ii) of (9.8)} \\
&= (x^m \cdot x^k) \cdot x & \text{(associative law in } \mathbb{R}) \\
&= x^{m+k} \cdot x & \text{(induction hypothesis)} \\
&= x^{(m+k)+1} & \text{(ii) of (9.8)} \\
&= x^{m+(k+1)} & \text{(associative law for integers).}
\end{aligned}
$$

By (9.3) the proof by induction is complete.

Consider a positive integer j and suppose that with each integer $n \geq j$ there is associated a statement P_n. For example, if $j = 6$, then the statements are numbered P_6, P_7, P_8, and so on. The principle of mathematical induction may be extended to cover this situation. Just as before, two steps are used. Specifically, to prove that the statements S_n are true for $n \geq j$, we use the following two steps.

(9.10)

> (i′) Prove that S_j is true,
>
> (ii′) *Assume* that S_k is true for $k \geq j$ and *prove* that S_{k+1} is true.

Example 4 Let a be a nonzero real number such that $a > -1$. Prove that $(1 + a)^n > 1 + na$ for every integer $n \geq 2$.

Solution For each positive integer n, let P_n denote the inequality $(1 + a)^n > 1 + na$. Note that P_1 is *false*, since $(1 + a)^1 = 1 + (1)(a)$. However, we can show that P_n is true for $n \geq 2$ by using (9.10) with $j = 2$.

(i′) We first note that $(1 + a)^2 = 1 + 2a + a^2$. Since $a \neq 0$, we have $a^2 > 0$ and therefore $1 + 2a + a^2 > 1 + 2a$. This gives us $(1 + a)^2 > 1 + 2a$, and hence P_2 is true.

(ii′) Assume that P_k is true. Thus the induction hypothesis is

$$(1 + a)^k > 1 + ka.$$

We wish to show that P_{k+1} is true, that is,

$$(1 + a)^{k+1} > 1 + (k + 1)a.$$

Since $a > -1$, we have $a + 1 > 0$, and hence multiplying both sides of the induction hypothesis by $1 + a$ will not change the inequality sign. Consequently,

$$(1 + a)^k (1 + a) > (1 + ka)(1 + a)$$

which may be rewritten as

$$(1 + a)^{k+1} > 1 + ka + a + ka^2$$

or as

$$(1 + a)^{k+1} > 1 + (k + 1)a + ka^2.$$

Since $ka^2 > 0$, we have

$$1 + (k + 1)a + ka^2 > 1 + (k + 1)a$$

and therefore

$$(1 + a)^{k+1} > 1 + (k + 1)a.$$

Thus, P_{k+1} is true and the proof is complete.

EXERCISES

In each of Exercises 1–18, prove that the given formula is true for every positive integer n.

1 $2 + 4 + 6 + \cdots + 2n = n(n + 1)$

2 $1 + 4 + 7 + \cdots + (3n - 2) = \dfrac{n(3n - 1)}{2}$

3 $1 + 3 + 5 + \cdots + (2n - 1) = n^2$

4 $3 + 9 + 15 + \cdots + (6n - 3) = 3n^2$

5 $2 + 7 + 12 + \cdots + (5n - 3) = \dfrac{n}{2}(5n - 1)$

6 $2 + 6 + 18 + \cdots + 2 \cdot 3^{n-1} = 3^n - 1$

7 $1 + 2 \cdot 2 + 3 \cdot 2^2 + 4 \cdot 2^3 + \cdots + n \cdot 2^{n-1} = 1 + (n - 1) \cdot 2^n$

8 $(-1)^1 + (-1)^2 + (-1)^3 + \cdots + (-1)^n = \dfrac{(-1)^n - 1}{2}$

9 $1^2 + 2^2 + 3^2 + \cdots + n^2 = \dfrac{n(n + 1)(2n + 1)}{6}$

10 $1^3 + 2^3 + 3^3 + \cdots + n^3 = \left[\dfrac{n(n + 1)}{2}\right]^2$

11 $\dfrac{1}{1 \cdot 2} + \dfrac{1}{2 \cdot 3} + \dfrac{1}{3 \cdot 4} + \cdots + \dfrac{1}{n(n + 1)} = \dfrac{n}{n + 1}$

12 $\dfrac{1}{1 \cdot 2 \cdot 3} + \dfrac{1}{2 \cdot 3 \cdot 4} + \dfrac{1}{3 \cdot 4 \cdot 5} + \cdots + \dfrac{1}{n(n + 1)(n + 2)} = \dfrac{n(n + 3)}{4(n + 1)(n + 2)}$

13 $3 + 3^2 + 3^3 + \cdots + 3^n = \frac{3}{2}(3^n - 1)$

14 $1^3 + 3^3 + 5^3 + \cdots + (2n - 1)^3 = n^2(2n^2 - 1)$

15 $n < 2^n$ **16** $1 + 2n \leq 3^n$

17 $1 + 2 + 3 + \cdots + n < \frac{1}{8}(2n + 1)^2$ **18** If $0 < a < b$, then $\left(\dfrac{a}{b}\right)^{n+1} < \left(\dfrac{a}{b}\right)^n$.

Prove that the statements in Exercises 19–22 are true for every positive integer n.

19 3 is a factor of $n^3 - n + 3$.   **20** 2 is a factor of $n^2 + n$.

21 4 is a factor of $5^n - 1$.   **22** 9 is a factor of $10^{n+1} + 3 \cdot 10^n + 5$.

23 Use mathematical induction to prove that if a is any real number greater than 1, then $a^n > 1$ for every positive integer n.

24 If $a \neq 1$, prove that

$$1 + a + a^2 + \cdots + a^{n-1} = \frac{a^n - 1}{a - 1}$$

for every positive integer n.

25 If a and b are real numbers, use mathematical induction to prove that $(ab)^n = a^n b^n$ for every positive integer n.

26 If a is a real number, prove that $(a^m)^n = a^{mn}$ for all positive integers m and n.

27 Use mathematical induction to prove that $a - b$ is a factor of $a^n - b^n$ for every positive integer n. (*Hint:* $a^{k+1} - b^{k+1} = a^k(a - b) + (a^k - b^k)b$.)

28 Prove that $a + b$ is a factor of $a^{2n-1} + b^{2n-1}$ for every positive integer n.

29 If z is a complex number and $\bar{z}$ is its conjugate, prove that $\overline{z^n} = \bar{z}^n$ for every positive integer n (see (7.7)).

30 Prove that for every positive integer n, if $z_1, z_2, \ldots, z_n$ are complex numbers, then
$$\overline{z_1 z_2 \cdots z_n} = \bar{z}_1 \bar{z}_2 \cdots \bar{z}_n$$

31 Prove that

$$\log(a_1 a_2 \cdots a_n) = \log a_1 + \log a_2 + \cdots + \log a_n$$

for all $n \geq 2$, where each a_i is a positive real number.

32 Prove the **Generalized Distributive Law**

$$a(b_1 + b_2 + \cdots + b_n) = ab_1 + ab_2 + \cdots + ab_n$$

for all $n \geq 2$, where a and each b_i are real numbers.

33 Prove that

$$a + ar + ar^2 + \cdots + ar^{n-1} = \frac{a(1 - r^n)}{1 - r}$$

where n is any positive integer and a and r are real numbers with $r \neq 1$.

34 Prove that

$$a + (a + d) + (a + 2d) + \cdots + [a + (n - 1)d] = (n/2)[2a + (n - 1)d]$$

where n is any positive integer and a and d are real numbers.

35 If a and b are real numbers and n is any positive integer, prove that

$$(a - b)(a^{n-1} + a^{n-2}b + \cdots + ab^{n-2} + b^{n-1}) = a^n - b^n.$$

36 Prove that for every positive integer $n \geq 3$, the sum of the interior angles of a polygon of n sides is $(n - 2) \cdot 180°$.

37 Prove, by mathematical induction, that if a finite set S consists of n elements, then the number of subsets of S is 2^n.

2   THE BINOMIAL THEOREM

It is often necessary to work with expressions of the form $(a + b)^n$, where a and b are mathematical expressions of some type and n is a large positive integer. There exists a general formula for *expanding* $(a + b)^n$, that is, for expressing it as a sum. The theorem which gives us the formula is called the **Binomial Theorem**. In order to obtain the formula let us first consider cases where n is a small positive integer. By examining the patterns which emerge, we shall make an educated guess about the nature of the general formula. Finally we shall prove by means of mathematical induction that our guess is correct.

If we actually perform the multiplications, the following expansions of $(a + b)^n$ are obtained for the cases $n = 2, 3, 4$, and 5:

$$(a + b)^2 = a^2 + 2ab + b^2$$
$$(a + b)^3 = a^3 + 3a^2b + 3ab^2 + b^3$$
$$(a + b)^4 = a^4 + 4a^3b + 6a^2b^2 + 4ab^3 + b^4$$
$$(a + b)^5 = a^5 + 5a^4b + 10a^3b^2 + 10a^2b^3 + 5ab^4 + b^5.$$

Let us now make some observations regarding the above expansions of $(a + b)^n$. We see that there are always $n + 1$ terms, the first being a^n and the last b^n. Each intermediate term contains a product of the form a^ib^j, where $i + j = n$. Moreover, as we move from one term to the next, the exponent associated with a decreases by 1 whereas the exponent associated with b increases by 1. The pattern for the coefficients is rather interesting. If we start at one end of the expansion and consider successive terms, the coefficients match those obtained by starting at the other end of the expansion and proceeding in the reverse direction.

It requires some ingenuity to find a formula for the general term in the expansion. The second term of the special cases above for $(a + b)^n$ is always $na^{n-1}b$. The third term is seen to be

$$\frac{n(n - 1)}{2}a^{n-2}b^2.$$

If there are more than three terms, the fourth term is

(9.11)
$$\frac{n(n-1)(n-2)}{3\cdot 2}a^{n-3}b^3.$$

It appears that for each term, *if we take the product of the coefficient and the exponent of a and then divide by the number of the term, we obtain the coefficient of the next term* in the expansion. For example, applying this rule to (9.11) gives us the following fifth term (when it exists):

(9.12)
$$\frac{n(n-1)(n-2)(n-3)}{4\cdot 3\cdot 2}a^{n-4}b^4.$$

The denominators in (9.11) and (9.12) may be abbreviated by employing the **factorial notation**. If n is any positive integer, then the symbol $n!$ (read "n factorial") is defined by

(9.13)
$$n! = n(n-1)(n-2)\cdots 1$$

where there are n factors on the right side. As special cases we have $1! = 1$, $2! = 2\cdot 1, 3! = 3\cdot 2\cdot 1 = 6, 4! = 4\cdot 3\cdot 2\cdot 1 = 24, 5! = 5\cdot 4\cdot 3\cdot 2\cdot 1 = 120$, etc. To ensure that certain formulas will be true for all *nonnegative* integers, we define $0! = 1$. The factorial notation may also be defined recursively by writing $1! = 1$, and for any positive integer k, $(k+1)! = (k+1)k!$.

If we compare the coefficients with the exponents in (9.11) and (9.12), we are led to believe that for a positive integer r, the $(r+1)$st term in the expansion of $(a+b)^n$ is given by

(9.14)
$$\frac{n(n-1)(n-2)\cdots(n-r+1)}{r!}a^{n-r}b^r.$$

Note that if $r = n$, then the coefficient reduces to $n!/n!$ and we obtain b^n, the last term in the expansion. If $r = n-1$, then we obtain nab^{n-1}, which is the second from the last term, and so on. Hence we *conjecture* that the following formula is true for every positive integer n and all real or complex numbers a and b.

(9.15)
$$(a+b)^n = a^n + na^{n-1}b + \frac{n(n-1)}{2!}a^{n-2}b^2 + \cdots$$
$$+ \frac{n(n-1)(n-2)\cdots(n-r+1)}{r!}a^{n-r}b^r$$
$$+ \cdots + nab^{n-1} + b^n$$

In order to prove (9.15) we shall use mathematical induction as follows. For each positive integer n, let P_n denote the statement (9.15).

(i) If $n = 1$, then (9.15) reduces to $(a+b)^1 = a^1 + b^1$. Consequently, P_1 is true.

(ii)   Assume that P_k is true. Thus the induction hypothesis is

(9.16)
$$(a + b)^k = a^k + ka^{k-1}b + \frac{k(k-1)}{2!}a^{k-2}b^2 + \cdots$$

$$+ \frac{k(k-1)(k-2)\cdots(k-r+2)}{(r-1)!}a^{k-r+1}b^{r-1}$$

$$+ \frac{k(k-1)(k-2)\cdots(k-r+1)}{r!}a^{k-r}b^r$$

$$+ \cdots + kab^{k-1} + b^k$$

where we have shown both the rth and the $(r+1)$st terms in the expansion. If we multiply both sides of equation (9.16) by $(a + b)$, we obtain

$$(a+b)^{k+1} = \left[a^{k+1} + ka^kb + \frac{k(k-1)}{2!}a^{k-1}b^2 + \cdots \right.$$

$$+ \frac{k(k-1)\cdots(k-r+1)}{r!}a^{k-r+1}b^r + \cdots + ab^k \left. \right]$$

$$+ \left[a^kb + ka^{k-1}b^2 + \cdots + \frac{k(k-1)\cdots(k-r+2)}{(r-1)!}a^{k-r+1}b^r \right.$$

$$+ \cdots + kab^k + b^{k+1} \left. \right]$$

where the terms in the first pair of brackets result from multiplying the right side of (9.16) by a and the terms in the second pair of brackets result from multiplying by b. Rearranging and combining terms, we have

$$(a+b)^{k+1} = a^{k+1} + (k+1)a^kb + \left[\frac{k(k-1)}{2!} + k \right]a^{k-1}b^2 + \cdots$$

$$+ \left[\frac{k(k-1)\cdots(k-r+1)}{r!} + \frac{k(k-1)\cdots(k-r+2)}{(r-1)!} \right]a^{k-r+1}b^r$$

$$+ \cdots + (1+k)ab^k + b^{k+1}.$$

It is left to the reader to show that if the coefficients are simplified, then we obtain (9.15) with $k + 1$ substituted for n. Thus, P_{k+1} is true and therefore (9.15) holds for every positive integer n.

Example 1   Find the binomial expansion of $(2x + 3y^2)^4$.

Solution   Using (11.15) with $a = 2x$, $b = 3y^2$, and $n = 4$, we obtain

$$(2x + 3y^2)^4 = (2x)^4 + 4(2x)^3(3y^2) + \frac{4 \cdot 3}{2!}(2x)^2(3y^2)^2$$

$$+ \frac{4 \cdot 3 \cdot 2}{3!}(2x)(3y^2)^3 + \frac{4 \cdot 3 \cdot 2 \cdot 1}{4!}(3y^2)^4.$$

This simplifies to

$$(2x + 3y^2)^4 = 16x^4 + 96x^3y^2 + 216x^2y^4 + 216xy^6 + 81y^8.$$

The symbol $\binom{n}{r}$ is often used to denote the coefficient of $a^{n-r}b^r$ in (9.14); that is,

(9.17)

$$\binom{n}{r} = \frac{n(n-1)\cdots(n-r+1)}{r!}$$

An equivalent formula is

$$\binom{n}{r} = \frac{n!}{r!(n-r)!}$$

where r may be assigned any integral value between 0 and n. If we substitute n for r in (9.17), we obtain

$$\binom{n}{n} = \frac{n!}{n!} = 1$$

It is also convenient to *define*

$$\binom{n}{0} = 1$$

Formula (9.17) may then be used if r is any integer such that $0 \le r \le n$. Using this notation, (9.15) takes on the form

(9.18)

$$(a + b)^n = \binom{n}{0}a^nb^0 + \binom{n}{1}a^{n-1}b + \binom{n}{2}a^{n-2}b^2 + \cdots + \binom{n}{r}a^{n-r}b^r$$

$$+ \cdots + \binom{n}{n-1}a^{n-(n-1)}b^{n-1} + \binom{n}{n}a^0b^n$$

The numbers $\binom{n}{r}$ are called **binomial coefficients**. As a special case of (9.18), when $n = 4$, we obtain

$$(a + b)^4 = \binom{4}{0}a^4b^0 + \binom{4}{1}a^3b + \binom{4}{2}a^2b^2 + \binom{4}{3}ab^3 + \binom{4}{4}a^0b^4.$$

By (9.17) the last formula reduces to the expansion of $(a + b)^4$ given at the beginning of this section.

The next example illustrates the fact that if one of a or b is negative, then the terms of the expansion are alternately positive and negative.

Example 2   Expand $\left(\dfrac{1}{x} - 2\sqrt{x}\right)^5$.

Solution   Letting $a = 1/x, b = -2\sqrt{x}$, and $n = 5$ in (9.15), we obtain

$$\left(\frac{1}{x} - 2\sqrt{x}\right)^5 = \left(\frac{1}{x}\right)^5 + 5\left(\frac{1}{x}\right)^4(-2\sqrt{x}) + \frac{5\cdot4}{2!}\left(\frac{1}{x}\right)^3(-2\sqrt{x})^2$$

$$+ \frac{5\cdot4\cdot3}{3!}\left(\frac{1}{x}\right)^2(-2\sqrt{x})^3 + \frac{5\cdot4\cdot3\cdot2}{4!}\left(\frac{1}{x}\right)(-2\sqrt{x})^4$$

$$+ \frac{5\cdot4\cdot3\cdot2\cdot1}{5!}(-2\sqrt{x})^5.$$

This simplifies to

$$\left(\frac{1}{x} - 2\sqrt{x}\right)^5 = \frac{1}{x^5} - \frac{10}{x^{7/2}} + \frac{40}{x^2} - \frac{80}{x^{1/2}} + 80x - 32x^{5/2}.$$

For certain problems it is only required to find a specific term in the expansion of $(a + b)^n$. To work such problems we first find the exponent r that is to be assigned to b. Notice that by (9.15) or (9.18) *the exponent of b is always one less than the number of the term.* Once r is found, the exponent of a is $n - r$. Referring to (9.14), we see that the coefficient of the term involving $a^{n-r}b^r$ is of the form $p/r!$, *where p is the product of r factors and where the first factor is n and each factor is one less than the preceding factor.*

Example 3   Find the fifth term in the expansion of $(4x^2 + 3/y)^{13}$.

Solution   We let $a = 4x^2$ and $b = 3/y$. The exponent of b in the fifth term is 4 and hence the exponent of a is 9. From the discussion of the preceding paragraph we obtain

$$\frac{13\cdot12\cdot11\cdot10}{4!}(4x^2)^9(3/y)^4.$$

Finally, there is an interesting triangular array of numbers, called **Pascal's Triangle**, which can be used to obtain the binomial coefficients. The numbers are arranged as follows.

$$
\begin{array}{ccccccccccc}
 & & & & & 1 & & & & & \\
 & & & & 1 & & 1 & & & & \\
 & & & 1 & & 2 & & 1 & & & \\
 & & 1 & & 3 & & 3 & & 1 & & \\
 & 1 & & 4 & & 6 & & 4 & & 1 & \\
1 & & 5 & & 10 & & 10 & & 5 & & 1 \\
\end{array}
$$

$$1 \quad 6 \quad 15 \quad 20 \quad 15 \quad 6 \quad 1$$

$$\cdot \quad \cdot \quad \cdot \quad \cdot \quad \cdot \quad \cdot \quad \cdot \quad \cdot$$

$$\cdot \quad \cdot \quad \cdot \quad \cdot \quad \cdot \quad \cdot \quad \cdot \quad \cdot \quad \cdot$$

The numbers in the second row are the coefficients in the expansion of $(a + b)^1$; those in the third row are the coefficients determined by $(a + b)^2$; those in the fourth row are obtained from $(a + b)^3$, and so on. Each number in the array which is different from 1 can be found by adding the two numbers in the previous row which appear above and immediately to the left and right of the number.

EXERCISES

In each of Exercises 1–12, expand and simplify the given expression.

1 $(a + b)^6$ 2 $(a + b)^7$

3 $(a - b)^8$ 4 $(a - b)^9$

5 $(3x - 5y)^4$ 6 $(2t - s)^5$

7 $(u^2 + 4v)^5$ 8 $(\frac{1}{2}c + d^3)^4$

9 $(r^{-2} - 2r)^6$ 10 $(x^{1/2} - y^{-1/2})^6$

11 $(1 + x)^{10}$ 12 $(1 - x)^{10}$

13 Find the first four terms in the binomial expansion of $(3c^{2/5} + c^{4/5})^{25}$.

14 Find the first three terms and the last three terms in the binomial expansion of $(x^3 + 5x^{-2})^{20}$.

15 Find the last two terms in the expansion of $(4b^{-1} - 3b)^{15}$.

16 Find the last three terms in the expansion of $(s - 2t^3)^{12}$.

Solve Exercises 17–28 without expanding completely.

17 Find the fifth term in the expansion of $(3a^2 + \sqrt{b})^9$.

18 Find the sixth term in the expansion of $\left(\dfrac{2}{c} + \dfrac{c^2}{3}\right)^7$.

19 Find the seventh term in the expansion of $(\frac{1}{2}u - 2v)^{10}$.

20 Find the fourth term in the expansion of $(2x^3 - y^2)^6$.

21 Find the middle term in the expansion of $(x^{1/3} + y^{1/3})^{12}$.

22 Find the two middle terms in the expansion of $(rs + t)^7$.

23 Find the term which does not contain x in the expansion of $\left(6x - \dfrac{1}{2x}\right)^{10}$.

24 Find the term involving x^8 in the expansion of $(y + 3x^2)^6$.

25 Find the term containing y^6 in the expansion of $(x - 2y^3)^4$.

26 Find the term containing b^9 in the expansion of $(5a + 2b^3)^4$.

27 Find the term containing c^3 in the expansion of $(\sqrt{c} + \sqrt{d})^{10}$.

28 In the expansion of $(xy - 2y^{-3})^8$ find the term which does not contain y.

29 Use the first four terms in the binomial expansion of $(1 + 0.02)^{10}$ to approximate $(1.02)^{10}$. Also approximate $(1.02)^{10}$ by using logarithms and then compare answers.

30   Use the first four terms in the binomial expansion of $(1 - 0.01)^4$ to approximate $(0.99)^4$. Compare with the answer obtained by approximating $(0.99)^4$ through the use of logarithms.

3   INFINITE SEQUENCES AND SUMMATION NOTATION

An **infinite sequence** is a collection of real or complex numbers

$$a_1, a_2, a_3, \ldots, a_n, \ldots$$

which is in one-to-one correspondence with the positive integers. For convenience we often refer to an infinite sequence merely as a **sequence**. The number a_1 is called the **first term** of the sequence, a_2 is the **second term** and, in general, a_n is the **nth term**. The dots at the end in the previous display indicate that there is no last number in the sequence. Throughout our work, all terms of sequences will be real numbers.

To specify an infinite sequence it is not sufficient merely to state the numbers in the collection. The *order* in which the numbers appear is also important. For example, the sequence whose first five terms are

$$1, \tfrac{1}{2}, \tfrac{1}{3}, \tfrac{1}{4}, \tfrac{1}{5}, \ldots$$

is different from the sequence which begins

$$1, \tfrac{1}{3}, \tfrac{1}{2}, \tfrac{1}{4}, \tfrac{1}{5}, \ldots$$

even if all the terms after the third are the same in both sequences. In general, the sequence

$$a_1, a_2, a_3, \ldots, a_n, \ldots$$

is said to be **equal** to the sequence

$$b_1, b_2, b_3, \ldots, b_n, \ldots$$

if and only if $a_i = b_i$ for every positive integer i. To avoid any misunderstanding about ordering, it is customary to specify the nth term for an arbitrary positive integer n. This is often done by means of a formula, as in the following example.

Example 1   List the first four terms and the tenth term of the sequence whose nth term is as follows.

(a) $a_n = \dfrac{n}{n + 1}$

(b) $a_n = 2 + (0.1)^n$

(c) $a_n = (-1)^{n+1} \dfrac{n^2}{3n - 1}$

(d) $a_n = 4$

Solutions To find the first four terms we substitute, successively, $n = 1, 2, 3$, and 4 in the formula for a_n. The tenth term is found by substituting 10 for n. Doing this and simplifying gives us the following:

First four terms	*Tenth term*
(a) $\frac{1}{2}, \frac{2}{3}, \frac{3}{4}, \frac{4}{5}$	$\frac{10}{11}$
(b) $2.1, 2.01, 2.001, 2.0001$	2.0000000001
(c) $\frac{1}{2}, -\frac{4}{5}, \frac{9}{8}, -\frac{16}{11}$	$-\frac{100}{29}$
(d) $4, 4, 4, 4$	4

It is not essential that a formula for a_n be given. Indeed, sometimes that is impossible, as illustrated in the next example.

Example 2 List the first seven terms of the sequence whose nth term a_n is the nth positive prime number.

Solution The first seven primes are 2, 3, 5, 7, 11, 13, and 17. Hence $a_1 = 2$, $a_2 = 3$, $a_3 = 5$, $a_4 = 7$, $a_5 = 11$, $a_6 = 13$, and $a_7 = 17$. No one has yet found a formula which yields the nth prime number. Our sequence is perfectly legitimate, however, since a_n is uniquely determined for every positive integer n.

An infinite sequence is sometimes defined by stating a recursive definition for the terms, as in the following example.

Example 3 Find the first four terms and the nth term of the infinite sequence defined by

$$a_1 = 3 \quad \text{and} \quad a_{k+1} = 2a_k, \quad \text{for} \quad k \geq 1.$$

Solution The sequence is defined recursively since the first term is given and, moreover, whenever a term a_k of the sequence is known, then the next term a_{k+1} can be found. Thus,

$$a_1 = 3$$
$$a_2 = 2a_1 = 2 \cdot 3 = 6$$
$$a_3 = 2a_2 = 2 \cdot 2 \cdot 3 = 2^2 \cdot 3 = 12$$
$$a_4 = 2a_3 = 2 \cdot 2 \cdot 2 \cdot 3 = 2^3 \cdot 3 = 24.$$

We have written the terms as products so as to gain some insight into the nature of the nth term. Continuing, we obtain $a_5 = 2^4 \cdot 3$ and $a_6 = 2^5 \cdot 3$; and it appears that

$$a_n = 2^{n-1} \cdot 3$$

for every positive integer n. We shall prove that this guess is correct by means of mathematical induction. If we let P_n denote the statement $a_n = 2^{n-1} \cdot 3$, then P_1 is true since $a_1 = 2^0 \cdot 3 = 3$. Next, *assume* that P_k is true, that is, $a_k = 2^{k-1} \cdot 3$. We then have

$$
\begin{aligned}
a_{k+1} &= 2a_k &&\text{(definition of } a_{k+1}) \\
&= 2 \cdot 2^{k-1} \cdot 3 &&\text{(induction hypothesis)} \\
&= 2^k \cdot 3 &&\text{(a law of exponents)} \\
&= 2^{(k+1)-1} \cdot 3 &&\text{(Why?)}
\end{aligned}
$$

which shows that P_{k+1} is true. Hence, $a_n = 2^{n-1} \cdot 3$ for every positive integer n.

As a final remark, it is important to observe that if only the first few terms of an infinite sequence are known, then it is impossible to predict additional terms. For example, if we were given 3, 6, 9, ... and asked to find the fourth term, we could not proceed without further information. The infinite sequence with nth term

$$ a_n = 3n + (1 - n)^3 (2 - n)^2 (3 - n) $$

has for its first four terms 3, 6, 9, and 120. It is possible to describe sequences where the first three terms are 3, 6, and 9 and the fourth term is *any* given number. This shows that when we work with infinite sequences it is essential to have specific information about the nth term or to know a general scheme for obtaining each term from the preceding one.

It is often desirable to find the sum of many terms of an infinite sequence. For ease in expressing such sums we use the **summation notation** described below.

Given an infinite sequence

$$ a_1, a_2, a_3, \ldots, a_n, \ldots $$

the symbol $\sum_{i=1}^{m} a_i$ represents the sum of the first m terms, that is,

$$ \sum_{i=1}^{m} a_i = a_1 + a_2 + a_3 + \cdots + a_m $$

The Greek capital letter Σ (sigma) indicates a sum and the symbol a_i represents the ith term. The letter i is called the **index of summation** or the **summation variable**, and the numbers 1 and m indicate the extreme values of the summation variable.

Example 4 Find $\sum_{i=1}^{4} i^2(i - 3)$.

Solution In this case, $a_i = i^2(i - 3)$. To find the indicated sum we merely substitute, in succession, the integers 1, 2, 3 and 4 for i and add the resulting terms. Thus,

$$ \sum_{i=1}^{4} i^2(i - 3) = 1^2(1 - 3) + 2^2(2 - 3) + 3^2(3 - 3) + 4^2(4 - 3) $$

$$ = (-2) + (-4) + 0 + 16 = 10. $$

The letter used for the summation variable is arbitrary. To illustrate, if we use j for the summation variable, then

$$\sum_{j=1}^{m} a_j = a_1 + a_2 + a_3 + \cdots + a_m$$

which is the same as $\sum_{i=1}^{m} a_i$. Other symbols can be used similarly. As a numerical example, the sum in Example 4 can be written as

$$\sum_{k=1}^{4} k^2(k-3).$$

If n is a positive integer, then the sum of the first n terms of an infinite sequence will be denoted by S_n. For example, given $a_1, a_2, a_3, \ldots, a_n, \ldots,$

$$S_n = \sum_{i=1}^{n} a_i = a_1 + a_2 + \cdots + a_n$$

The number S_n is called the **nth partial sum** of the sequence $a_1, a_2, a_3, \ldots, a_n, \ldots,$ and the infinite sequence

$$S_1, S_2, S_3, \ldots, S_n, \ldots$$

is called a **sequence of partial sums**. Sequences of partial sums are very important in calculus.

Example 5   Find the first four terms and the nth term of the sequence of partial sums associated with the sequence $1, 2, 3, \ldots, n, \ldots$ of positive integers.

Solution   The first four terms of the partial sum sequence are
$$S_1 = 1$$
$$S_2 = 1 + 2 = 3$$
$$S_3 = 1 + 2 + 3 = 6$$
$$S_4 = 1 + 2 + 3 + 4 = 10.$$

From Example 1 of Section 1 we see that

$$S_n = 1 + 2 + 3 + \cdots + n = \frac{n(n+1)}{2}.$$

If a_i is the same for all positive integers i, say $a_i = c$, where c is a real number, then

$$\sum_{i=1}^{n} a_i = a_1 + a_2 + a_3 + \cdots + a_n$$
$$= c + c + c + \cdots + c$$
$$= nc.$$

It is with this in mind that we state the following rule:

(9.19)

$$\sum_{i=1}^{n} c = nc$$

The formula above could also be established by mathematical induction (see Exercise 60).

The domain of the summation variable does not have to begin at 1. For example, the following is self-explanatory:

$$\sum_{i=4}^{8} a_i = a_4 + a_5 + a_6 + a_7 + a_8.$$

As another variation, if the first term of an infinite sequence is a_0, as in

$$a_0, a_1, a_2, \ldots, a_n, \ldots$$

then sums of the form

$$\sum_{i=0}^{n} a_i = a_0 + a_1 + a_2 + \cdots + a_n$$

may be considered. Note that this is the sum of the first $n + 1$ terms of the given sequence.

Example 6 Find $\displaystyle\sum_{i=0}^{3} \frac{2^i}{(i + 1)}$.

Solution
$$\sum_{i=0}^{3} \frac{2^i}{(i + 1)} = \frac{2^0}{(0 + 1)} + \frac{2^1}{(1 + 1)} + \frac{2^2}{(2 + 1)} + \frac{2^3}{(3 + 1)}$$

$$= 1 + 1 + \frac{4}{3} + 2$$

$$= \frac{16}{3}.$$

The summation notation can be used to denote polynomials compactly. For example, in place of

$$f(x) = a_0 + a_1 x + a_2 x^2 + \cdots + a_n x^n$$

we may write

$$f(x) = \sum_{i=0}^{n} a_i x^i.$$

As another illustration, the rather cumbersome formula (9.18) for the binomial theorem can be written very simply as

$$(a + b)^n = \sum_{r=0}^{n} \binom{n}{r} a^{n-r} b^r.$$

The following theorem concerning sums is used in many advanced courses in mathematics.

(9.20)   **Theorem**

If $a_1, a_2, \ldots, a_n, \ldots$ and $b_1, b_2, \ldots, b_n, \ldots$ are infinite sequences, then for every positive integer n,

(i) $\displaystyle\sum_{i=1}^{n} (a_i + b_i) = \sum_{i=1}^{n} a_i + \sum_{i=1}^{n} b_i$

(ii) $\displaystyle\sum_{i=1}^{n} (a_i - b_i) = \sum_{i=1}^{n} a_i - \sum_{i=1}^{n} b_i$

(iii) $\displaystyle\sum_{i=1}^{n} ca_i = c\left(\sum_{i=1}^{n} a_i \right),$ for every number c.

Proof.   Although the theorem can be proved by mathematical induction, we shall use an argument that makes the truth of the formulas transparent. We begin as follows:

$$\sum_{i=1}^{n} (a_i + b_i) = (a_1 + b_1) + (a_2 + b_2) + (a_3 + b_3) + \cdots + (a_n + b_n).$$

Using the Commutative and Associative Properties many times, we may rearrange the terms on the right to produce

$$\sum_{i=1}^{n} (a_i + b_i) = (a_1 + a_2 + a_3 + \cdots + a_n) + (b_1 + b_2 + b_3 + \cdots + b_n).$$

Expressing the right side in summation notation gives us formula (i).

For formula (iii) we have

$$\sum_{i=1}^{n} (ca_i) = ca_1 + ca_2 + ca_3 + \cdots + ca_n$$

$$= c(a_1 + a_2 + a_3 + \cdots + a_n)$$

$$= c\left(\sum_{i=1}^{n} a_i \right).$$

The proof of (ii) is left as an exercise.

EXERCISES

In Exercises 1–20, find the first five terms and the eighth term of the sequence which has the given nth term.

1 $a_n = 12 - 3n$

2 $a_n = \dfrac{3}{5n - 2}$

3 $a_n = \dfrac{3n - 2}{n^2 + 1}$

4 $a_n = 10 + \dfrac{1}{n}$

5 $a_n = 9$

6 $a_n = (n - 1)(n - 2)(n - 3)$

7 $a_n = 2 + (-0.1)^n$

8 $a_n = 4 + (0.1)^n$

9 $a_n = (-1)^{n-1}\dfrac{n + 7}{2n}$

10 $a_n = (-1)^n\dfrac{6 - 2n}{\sqrt{n + 1}}$

11 $a_n = 1 + (-1)^{n+1}$

12 $a_n = (-1)^{n+1} + (0.1)^{n-1}$

13 $a_n = \dfrac{2^n}{n^2 + 2}$

14 $a_n = \sqrt{2}$

15 a_n is the square of the nth prime.

16 a_n is the nth prime greater than 50.

17 a_n is the largest prime less than $10n$.

18 a_n is the smallest prime greater than $15n$.

19 a_n is the number of decimal places in $(0.1)^n$.

20 a_n is the number of positive integers less than n^3.

Find the first five terms of the infinite sequences defined recursively in Exercises 21–28.

21 $a_1 = 2, a_{k+1} = 3a_k - 5$

22 $a_1 = 5, a_{k+1} = 7 - 2a_k$

23 $a_1 = -3, a_{k+1} = a_k^2$

24 $a_1 = 128, a_{k+1} = a_k/4$

25 $a_1 = 5, a_{k+1} = ka_k$

26 $a_1 = 3, a_{k+1} = 1/a_k$

27 $a_1 = 2, a_{k+1} = (a_k)^k$

28 $a_1 = 2, a_{k+1} = (a_k)^{1/k}$

29 A test question lists the first four terms of a sequence as 2, 4, 6, and 8 and asks for the fifth term. Show that the fifth term can be any real number a by finding the nth term of a sequence which has for its first five terms, 2, 4, 6, 8, and a.

30 The number of bacteria in a certain culture doubles every day. If the initial number of bacteria is 500, how many are present after one day? Two days? Three days? Find a formula for the number of bacteria present after n days.

Find the number given in each of Exercises 31–46.

31 $\displaystyle\sum_{k=1}^{5} (2k - 7)$

32 $\displaystyle\sum_{k=1}^{6} (10 - 3k)$

33 $\displaystyle\sum_{k=1}^{4} (k^2 - 5)$

34 $\displaystyle\sum_{k=1}^{10} [1 + (-1)^k]$

35 $\displaystyle\sum_{k=0}^{5} k(k - 2)$

36 $\displaystyle\sum_{k=0}^{4} (k - 1)(k - 3)$

37 $\displaystyle\sum_{k=3}^{6} \frac{k-5}{k-1}$　　　　　　**38** $\displaystyle\sum_{k=1}^{6} \frac{3}{k+1}$

39 $\displaystyle\sum_{k=1}^{5} (-3)^{k-1}$　　　　　　**40** $\displaystyle\sum_{k=0}^{4} 3(2^k)$

41 $\displaystyle\sum_{k=1}^{100} 100$　　　　　　**42** $\displaystyle\sum_{k=1}^{1000} 5$

43 $\displaystyle\sum_{k=1}^{n} (k^2 + 3k + 5)$ (*Hint:* Use (9.20) to write the sum as $\sum_{k=1}^{n} + 3 \sum_{k=1}^{n} k + \sum_{k=1}^{n} 5$. Next employ Exercise 9 and Example 1 of Section 1, together with (9.19).)

44 $\displaystyle\sum_{k=1}^{n} (3k^2 - 2k + 1)$　　　　　　**45** $\displaystyle\sum_{k=1}^{n} (2k - 3)^2$

46 $\displaystyle\sum_{k=1}^{n} (k^3 + 2k^2 - k + 4)$ (*Hint:* See Exercise 10 of Section 1.)

Use summation notation to express the sums in Exercises 47–56.

47 $1 + 5 + 9 + 13 + 17$　　　　　**48** $2 + 5 + 8 + 11 + 14$

49 $\dfrac{1}{2} + \dfrac{2}{5} + \dfrac{3}{8} + \dfrac{4}{11}$　　　　　**50** $\dfrac{1}{4} + \dfrac{2}{9} + \dfrac{3}{14} + \dfrac{4}{19}$

51 $1 - \dfrac{x^2}{2} + \dfrac{x^4}{4} - \dfrac{x^6}{6} + \cdots + (-1)^{n+1} \dfrac{x^{2n}}{2n}$

52 $2 - 4 + 8 - 16 + 32 - 64$　　　　　**53** $1 - \dfrac{1}{2} + \dfrac{1}{3} - \dfrac{1}{4} + \dfrac{1}{5} - \dfrac{1}{6} + \dfrac{1}{7}$

54 $1 + x + \dfrac{x^2}{2} + \dfrac{x^3}{3} + \cdots + \dfrac{x^n}{n}$　　　　　**55** $\dfrac{1}{1 \cdot 2} + \dfrac{1}{2 \cdot 3} + \dfrac{1}{3 \cdot 4} + \cdots + \dfrac{1}{99 \cdot 100}$

56 $\dfrac{1}{1 \cdot 2 \cdot 3} + \dfrac{1}{2 \cdot 3 \cdot 4} + \dfrac{1}{3 \cdot 4 \cdot 5} + \cdots + \dfrac{1}{98 \cdot 99 \cdot 100}$

57 Prove (ii) of (9.20).

58 Extend (i) of (9.20) to $\sum_{i=1}^{n} (a_i + b_i + c_i)$.

59 Prove (9.20) by mathematical induction.

60 Prove (9.19) by mathematical induction.

4 ARITHMETIC SEQUENCES

In this section and the next we shall concentrate on two special types of sequences. The first type may be defined as follows.

(9.21)　**Definition**

> An **arithmetic sequence** is a sequence such that successive terms differ by the same real number.

Arithmetic sequences are also called **arithmetic progressions**. By definition, a sequence

$$a_1, a_2, a_3, \ldots, a_n, \ldots$$

is arithmetic if and only if there is a real number d such that

(9.22)

$$a_{k+1} - a_k = d$$

for every positive integer k. The number d is called the **common difference** associated with the arithmetic sequence.

Example 1   Show that the sequence

$$1, 4, 7, 10, \ldots, 3n - 2, \ldots$$

is arithmetic and find the common difference.

Solution   Since $a_n = 3n - 2$ it follows that for every positive integer k,

$$\begin{aligned}
a_{k+1} - a_k &= [3(k + 1) - 2] - (3k - 2) \\
&= 3k + 3 - 2 - 3k + 2 = 3.
\end{aligned}$$

Hence by Definition (9.21), the given sequence is arithmetic with common difference 3.

Given an arithmetic sequence we have, by (9.22),

$$a_{k+1} = a_k + d$$

for every positive integer k. This provides a recursive formula for obtaining successive terms. Beginning with any real number a_1, we can generate an arithmetic sequence with common difference d simply by adding d to a_1, then to $a_1 + d$, and so on, obtaining

(9.23)

$$a_1, \quad a_1 + d, \quad a_1 + 2d, \quad a_1 + 3d, \quad a_1 + 4d, \quad \ldots.$$

It is evident that the nth term a_n of (9.23) is given by

(9.24)

$$a_n = a_1 + (n - 1)d$$

This formula can be proved by mathematical induction.

Example 2   Find the fifteenth term of the arithmetic sequence whose first three terms are 20, 16.5, and 13.

Solution   The common difference is -3.5. (Why?) Substituting $a_1 = 20$, $d = -3.5$, and $n = 15$ in (9.24), we obtain

$$a_{15} = 20 + (15 - 1)(-3.5) = 20 - 49 = -29.$$

Example 3   If the fourth term of an arithmetic sequence is 5 and the ninth term is 20, find the sixth term.

Solution   Substituting $n = 4$ and $n = 9$ in (9.24) and using the fact that $a_4 = 5$ and $a_9 = 20$, we get the following system of linear equations in the variables a_1 and d:

$$\begin{cases} 5 = a_1 + (4 - 1)d \\ 20 = a_1 + (9 - 1)d. \end{cases}$$

This system has the unique solution $d = 3$ and $a_1 = -4$. (Verify!) Substituting in (9.24), we obtain

$$a_6 = (-4) + (6 - 1)(3) = 11.$$

(9.25)   **Theorem**

> If $a_1, a_2, \ldots, a_n, \ldots$ is an arithmetic sequence with common difference d, then the nth partial sum S_n is given by both
>
> $$S_n = \frac{n}{2}[2a_1 + (n - 1)d] \quad \text{and} \quad S_n = \frac{n}{2}(a_1 + a_n).$$

Proof.   Using (9.23) and (9.24) we have

$$S_n = a_1 + (a_1 + d) + (a_1 + 2d) + \cdots + [a_1 + (n - 1)d].$$

Employing the Commutative and Associative Properties many times we may write

$$S_n = (a_1 + a_1 + a_1 + \cdots + a_1) + [d + 2d + \cdots + (n - 1)d]$$

where a_1 appears n times within the first parentheses. It follows that

$$S_n = na_1 + d[1 + 2 + \cdots + (n - 1)].$$

The expression within brackets is the sum of the first $n - 1$ positive integers. From Example 1 of Section 1 (with $n - 1$ in place of n), that sum equals

$$\frac{(n - 1)n}{2}.$$

Substituting in the last equation for S_n and factoring gives us

$$S_n = na_1 + d\frac{n(n - 1)}{2}$$

$$= \frac{n}{2}[2a_1 + (n - 1)d].$$

Since $a_n = a_1 + (n - 1)d$, we get

$$S_n = \frac{n}{2}(a_1 + a_n).$$

Example 4   Find the sum of all the even integers from 2 through 100.

Solution   The given problem is equivalent to finding the sum of the first fifty terms of the arithmetic sequence $2, 4, 6, \ldots, 2n, \ldots$. Substituting $n = 50$, $a_1 = 2$, and $a_{50} = 100$ in the second formula of (9.25), we obtain

$$S_{50} = \frac{50}{2}(2 + 100) = 2550.$$

In order to check our work we use the first formula in (9.25):

$$S_{50} = \frac{50}{2}[2 \cdot 2 + (50 - 1)2] = 25[4 + 98] = 2550.$$

The **arithmetic mean** of two numbers a and b is defined as $(a + b)/2$, which is also called the **average** of a and b. Note that the numbers

$$a, \quad \frac{a + b}{2}, \quad b$$

are terms of an arithmetic sequence. This concept may be generalized as follows. If $c_1, c_2, \ldots, c_k$ are real numbers such that

$$a, c_1, c_2, \ldots, c_k, b$$

is an arithmetic sequence, then $c_1, c_2, \ldots, c_k$ are called the k **arithmetic means** of the numbers a and b. The process of determining such numbers is referred to as *inserting* k arithmetic means between a and b.

Example 5   Insert three arithmetic means between 2 and 9.

Solution   We wish to find three numbers c_1, c_2, and c_3 such that $2, c_1, c_2, c_3, 9$ is an arithmetic sequence. The common difference d may be found by using (9.24) with $n = 5$, $a_5 = 9$, and $a_1 = 2$. This gives us

$$9 = 2 + (5 - 1)d, \quad \text{or} \quad d = \tfrac{7}{4}.$$

The three arithmetic means are

$$c_1 = a_1 + d = 2 + \tfrac{7}{4} = \tfrac{15}{4}$$
$$c_2 = c_1 + d = \tfrac{15}{4} + \tfrac{7}{4} = \tfrac{22}{4} = \tfrac{11}{2}$$
$$c_3 = c_2 + d = \tfrac{11}{2} + \tfrac{7}{4} = \tfrac{29}{4}.$$

EXERCISES

In each of Exercises 1–8 find the fifth term, the tenth term, and the nth term of the given arithmetic sequence.

1 $2, 6, 10, 14, \ldots$

2 $16, 13, 10, 7, \ldots$

3 $3, 2.7, 2.4, 2.1, \ldots$

4 $-6, -4.5, -3, -1.5, \ldots$

5 $-7, -3.9, -0.8, 2.3, \ldots$

6 $x - 8, x - 3, x + 2, x + 7, \ldots$

7 $\log 3, \log 9, \log 27, \log 81, \ldots$

8 $\log 1000, \log 100, \log 10, 0, \ldots$

9 Find the twelfth term of the arithmetic sequence whose first two terms are 9.1 and 7.5.

10 Find the eleventh term of the arithmetic sequence whose first two terms are $2 + \sqrt{2}$ and 3.

11 The sixth and seventh terms of an arithmetic sequence are 2.7 and 5.2. Find the first term.

12 Given an arithmetic sequence with $a_3 = 7$ and $a_{20} = 43$, find a_{15}.

In each of Exercises 13–16 find the sum of the arithmetic sequence which satisfies the given conditions.

13 $a_1 = 40, d = -3, n = 30$

14 $a_1 = 5, d = 0.1, n = 40$

15 $a_1 = -9, a_{10} = 15, n = 10$

16 $a_7 = 7/3, d = -2/3, n = 15$

Find the sums in Exercises 17–20.

17 $\displaystyle\sum_{k=1}^{20} (3k - 5)$

18 $\displaystyle\sum_{k=1}^{12} (7 - 4k)$

19 $\displaystyle\sum_{k=1}^{18} \left(\frac{1}{2}k + 7\right)$

20 $\displaystyle\sum_{k=1}^{10} \left(\frac{1}{4}k + 3\right)$

21 How many integers between 32 and 395 are divisible by 6? Find their sum.

22 How many negative integers greater than -500 are divisible by 33? Find their sum.

23 How many terms are in an arithmetic sequence with first term -2, common difference $\frac{1}{4}$, and sum 21?

24 How many terms are in an arithmetic sequence with sixth term -3, common difference 0.2, and sum -33?

25 Insert five arithmetic means between 2 and 10.

26 Insert three arithmetic means between 3 and -5.

27 A pile of logs has 24 logs in the first layer, 23 in the second, 22 in the third, and so on. The last layer contains 10 logs. Find the total number of logs in the pile.

28 A seating section in a certain athletic stadium has 30 seats in the first row, 32 seats in the second, 34 in the third, and so on, until the tenth row is reached, after which there are 10 more rows, each containing 50 seats. Find the total number of seats in the section.

29 A man wishes to construct a ladder with nine rungs which diminish uniformly from 24 inches at the base to 18 inches at the top. Determine the lengths of the seven intermediate rungs.

30 A boy on a bicycle coasts down a hill, covering 4 feet the first second and in each succeeding second 5 feet more than in the preceding second. If he reaches the bottom of the hill in 11 seconds, find the total distance traveled.

5 GEOMETRIC SEQUENCES

Another important type of infinite sequence is defined as follows.

(9.26)   **Definition**

> A **geometric sequence** is a sequence such that the quotient of any term after the first by the preceding term is the same nonzero real number.

Geometric sequences are also called **geometric progressions**. By definition, a sequence

$$a_1, a_2, \ldots, a_n, \ldots$$

is geometric if and only if there is a real number $r \neq 0$ such that

(9.27)

$$\frac{a_{k+1}}{a_k} = r$$

for every positive integer k. The number r is called the **common ratio** associated with the geometric sequence. We see from (9.27) that the indicated sequence is geometric (with common ratio r) if and only if

$$a_{k+1} = a_k r$$

for every positive integer k. As with arithmetic sequences, this provides a recursive method for obtaining terms. Beginning with any nonzero real number a_1, we *multiply* by the number r successively, obtaining

(9.28) $a_1, a_1 r, a_1 r^2, a_1 r^3, \ldots$.

It appears that the nth term a_n of (9.28) *is given by*

(9.29)

$$a_n = a_1 r^{n-1}$$

which can be proved by mathematical induction.

Example 1 Find the first five terms and the tenth term of the geometric sequence having first term 3 and common ratio $-1/2$.

Solution If we let $a_1 = 3$ and $r = -1/2$, then by (11.28) the first five terms are,

$$3, -\frac{3}{2}, \frac{3}{4}, -\frac{3}{8}, \frac{3}{16}.$$

Using (11.29) with $n = 10$, we obtain

$$a_{10} = 3\left(-\frac{1}{2}\right)^9 = -\frac{3}{512}.$$

Example 2   If the third term of a geometric sequence is 5 and the sixth term is -40, find the eighth term.

Solution   We are given $a_3 = 5$ and $a_6 = -40$. Substituting $n = 3$ and $n = 6$ in formula (9.29) leads to the following system of equations:

$$\begin{cases} 5 = a_1 r^2 \\ -40 = a_1 r^5. \end{cases}$$

Since $r \neq 0$, the first equation is equivalent to $a_1 = 5/r^2$. Substituting for a_1 in the second equation we obtain

$$-40 = \left(\frac{5}{r^2}\right) \cdot r^5 = 5r^3.$$

Hence $r^3 = -8$ and $r = -2$. If we now substitute -2 for r in the equation $5 = a_1 r^2$, we obtain $a_1 = 5/4$. Finally, applying (9.29) with $n = 8$ gives us

$$a_8 = \left(\frac{5}{4}\right)(-2)^7 = -160.$$

Let us find a formula for S_n, the nth partial sum of a geometric sequence. Using (9.28) and (9.29), we have

(9.30) $$S_n = a_1 + a_1 r + a_1 r^2 + \cdots + a_1 r^{n-2} + a_1 r^{n-1}.$$

If $r = 1$, this gives us $S_n = na_1$. Next suppose that $r \neq 1$. Multiplying both sides of (9.30) by r, we obtain

$$rS_n = a_1 r + a_1 r^2 + a_1 r^3 + \cdots + a_1 r^{n-1} + a_1 r^n.$$

If we subtract the preceding equation from (9.30), then many terms on the right side drop out, leaving

$$S_n - rS_n = a_1 - a_1 r^n \quad \text{or} \quad (1 - r)S_n = a_1(1 - r^n).$$

Since $r \neq 1$ we have $1 - r \neq 0$, and dividing both sides by $1 - r$ gives us

$$S_n = \frac{a_1(1 - r^n)}{1 - r}.$$

We have proved the following theorem.

(9.31)   **Theorem**

The nth partial sum of a geometric sequence with first term a_1 and common ratio $r \neq 1$ is

$$S_n = a_1 \frac{(1 - r^n)}{1 - r}.$$

Example 3   Find the sum of the first five terms of the geometric sequence which begins as follows: 1, 0.3, 0.09, 0.027, ...

Solution   We let $a_1 = 1$, $r = 0.3$, and $n = 5$ in (9.31), obtaining

$$S_5 = 1\left(\frac{1 - (0.3)^5}{1 - 0.3}\right)$$

which reduces to $S_5 = 1.4251$.

Example 4   A man wishes to save money by setting aside 1 cent the first day, 2 cents the second day, 4 cents the third day and so on, doubling the amount each day. If this is continued, how much must be set aside on the fifteenth day? Assuming he does not run out of money, what is the total amount saved at the end of 30 days?

Solution   The amount (in cents) set aside on successive days forms a geometric sequence

$$1, 2, 4, 8, \ldots$$

with first term 1 and common ratio 2. The amount needed for the fifteenth day is found by using (9.29) with $a_1 = 1$ and $n = 15$, which gives us $1 \cdot 2^{14}$, or \$163.84. To find the total amount set aside after 30 days, we use (9.31) with $n = 30$. Thus

$$S_{30} = 1\frac{(1 - 2^{30})}{1 - 2},$$

which simplifies to \$10,737,418.23.

Given the geometric series (9.28) with $r \neq 1$, we may use (9.31) to get

(9.32)
$$S_n = \frac{a_1}{1 - r} - \frac{a_1}{1 - r}r^n.$$

Although we shall not do it here, it can be shown that if $|r| < 1$, then r^n *approaches the number* 0 *as n increases* without bound; that is, we can make r^n as close as we wish to 0 by taking n sufficiently large. It follows from (9.32) that S_n approaches $a_1/(1 - r)$ as n increases without bound. The number $a_1/(1 - r)$ is called *the sum of the infinite geometric series*

$$a_1 + a_1 r + a_1 r^2 + \cdots + a_1 r^{n-1} + \cdots.$$

This gives us the following result.

(9.33)   **Theorem**

If $|r| < 1$, then the infinite geometric series

$$a_1 + a_1 r + a_1 r^2 + \cdots + a_1 r^{n-1} + \cdots$$

has the sum

$$\frac{a_1}{1 - r}.$$

Theorem (9.33) can be used to prove the fact mentioned in Chapter One that every real number represented by a repeating decimal is rational. This is illustrated in the next example.

Example 5 Find the rational number which corresponds to the infinite repeating decimal 5.4$\overline{27}$, where the bar means that the block of digits underneath is to be repeated indefinitely.

Solution From the decimal expression 5.4272727... we obtain the infinite series

$$5.4 + 0.027 + 0.00027 + 0.0000027 + \cdots.$$

The part of the expression after the first term is

$$0.027 + 0.00027 + 0.0000027 + \cdots$$

which has the form given in (9.33) with $a_1 = 0.027$ and $r = 0.01$. Hence the sum of this infinite geometric series is

$$\frac{0.027}{1 - 0.01} = \frac{0.027}{0.99} = \frac{27}{990} = \frac{3}{110}.$$

Thus it appears that the desired number is $5.4 + 3/110$, or $597/110$. A check by long division shows that $597/110$ does equal the given repeating decimal.

If the terms of an infinite sequence are alternately positive and negative and if we consider the expression

$$a_1 + (-a_2) + a_3 + (-a_4) + \cdots + [(-1)^{n+1}a_n] + \cdots$$

where all the a_i are positive real numbers, then this expression is referred to as an *alternating infinite series* and we write it in the form

$$a_1 - a_2 + a_3 - a_4 + \cdots + (-1)^{n+1}a_n + \cdots.$$

Illustrations of alternating infinite series can be obtained by using infinite geometric series with negative common ratio.

EXERCISES

In each of Exercises 1–12 find the fifth term, the eighth term, and the nth term of the given geometric sequence.

1 8, 4, 2, 1, ...

2 4, 1.2, 0.36, 0.108, ...

3 300, −30, 3, −0.3, ...

4 1, −$\sqrt{3}$, 3, −$\sqrt{27}$, ...

5 5, 25, 125, 625, ...

6 2, 6, 18, 54, ...

7 4, −6, 9, −13.5, ...

8 162, −54, 18, −6, ...

9 $1, -x^2, x^4, -x^6, \ldots$

10 $1, -\dfrac{x}{3}, \dfrac{x^2}{9}, -\dfrac{x^3}{27}, \ldots$

11 $2, 2^{x+1}, 2^{2x+1}, 2^{3x+1}, \ldots$

12 $10, 10^{2x-1}, 10^{4x-3}, 10^{6x-5}, \ldots$

13 Find the sixth term of the geometric sequence whose first two terms are 4 and 6.

14 Find the seventh term of the geometric sequence that has 2 and $-\sqrt{2}$ for its second and third terms respectively.

15 In a certain geometric sequence $a_5 = 1/16$ and $r = 3/2$. Find a_1 and S_5.

16 Given a geometric sequence in which $a_4 = 4$ and $a_7 = 12$, find r and a_{10}.

Find the sum in each of Exercises 17–20.

17 $\displaystyle\sum_{k=1}^{10} 3^k$

18 $\displaystyle\sum_{k=1}^{9} (-\sqrt{5})^k$

19 $\displaystyle\sum_{k=0}^{9} (-\tfrac{1}{2})^{k+1}$

20 $\displaystyle\sum_{k=1}^{7} (3^{-k})$

21 A vacuum pump removes one-half of the air in a container at each stroke. After 10 strokes, what percentage of the original amount of air remains in the container?

22 The yearly depreciation of a certain machine is 25 % of its value at the beginning of the year. If the original cost of the machine is \$20,000, find its value after 6 years.

23 A culture of bacteria increases 20 % every hour. If the original culture contains 10,000 bacteria, find a formula for the number of bacteria present after t hours. How many bacteria are in the culture at the end of 10 hours?

24 If an amount P of money is deposited in a savings account which pays interest at a rate of r percent per year compounded quarterly and if the principal and accumulated interest are left in the account, find a formula for the total amount in the account after n years.

Find the sums of the infinite geometric series in Exercises 25–30, whenever they exist.

25 $1 - \dfrac{1}{2} + \dfrac{1}{4} - \dfrac{1}{8} + \cdots$

26 $2 + \dfrac{2}{3} + \dfrac{2}{9} + \dfrac{2}{27} + \cdots$

27 $1.5 + 0.015 + 0.00015 + \cdots$

28 $1 - 0.1 + 0.01 - 0.001 + \cdots$

29 $\sqrt{2} - 2 + \sqrt{8} - 4 + \cdots$

30 $250 - 100 + 40 - 16 + \cdots$

In each of Exercises 31–38 find the rational number represented by the given repeating decimal.

31 $0.\overline{23}$

32 $0.0\overline{71}$

33 $2.4\overline{17}$

34 $10.5\overline{5}$

35 $5.1\overline{46}$

36 $3.2\overline{394}$

37 $1.6\overline{124}$

38 $123.61\overline{83}$

39 A rubber ball is dropped from a height of 10 meters. If it rebounds approximately one-half the distance after each fall, use an infinite geometric series to approximate the total distance the ball travels before coming to rest.

40 The bob of a pendulum swings through an arc 24 cm long on its first swing. If each successive swing is approximately five-sixths the length of the preceding swing, use an infinite geometric series to approximate the total distance it travels before coming to rest.

6 PERMUTATIONS

Suppose that four teams are involved in a tournament in which first, second, third, and fourth places will be determined. For identification purposes, we label the teams *a*, *b*, *c*, and *d*. Let us find the number of different ways that first and second place can be decided. It is convenient to use a **tree diagram** as in Figure 9.1. Beginning at the word "start," the four possibilities for first place are listed. From each of these an arrow points to a possible second-place finisher. In this

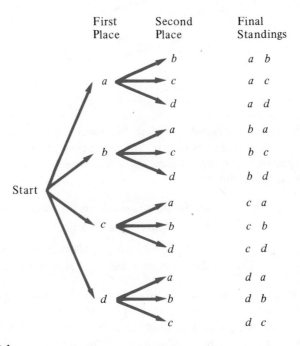

Figure 9.1

problem the second-place position is occupied by a team other than the one in the first-place position and hence *three* arrows are drawn from each first position. The final standings list the possible outcomes, from left to right. These are found by following all the different paths (*branches* of the tree) which lead from the word "start" to the second-place team. The total number of outcomes is 12, which we note is the product of the number of choices for first place and the number of choices for second place (after the first has been determined).

Let us now find the total number of ways that first, second, third, and fourth positions can be filled. To sketch a tree diagram we may begin by drawing arrows from the word "start" to each possible first-place finisher a, b, c, or d. Then we draw arrows from those to possible second-place finishers, as was done in Figure 9.1. Next, from each second-place position we draw arrows indicating the possible third-place positions. Finally, we draw arrows to the fourth-place team. If we consider only the case where team a finishes in first place, we have the diagram shown in Figure 9.2. Note that there are six possible final standings in

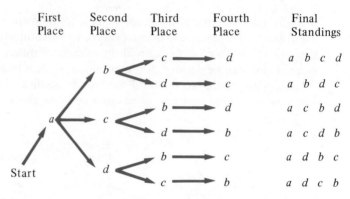

Figure 9.2

which team a occupies first place. In a complete tree diagram there would also be three other branches of this type corresponding to first place for b, c, and d, respectively. A complete diagram would display the following 24 possibilities for the final standings:

$$abcd, \quad abdc, \quad acbd, \quad acdb, \quad adbc, \quad adcb,$$
$$bacd, \quad badc, \quad bcad, \quad bcda, \quad bdac, \quad bdca,$$
$$cabd, \quad cadb, \quad cbad, \quad cbda, \quad cdab, \quad cdba,$$
$$dabc, \quad dacb, \quad dbac, \quad dbca, \quad dcab, \quad dcba.$$

Note that the number 24 is the product of the number of ways (4) that first place may occur, the number of ways (3) that second place may occur (after first place has been determined), the number of possible outcomes (2) for third place (after second place has been decided), and the number of ways (1) that fourth place can occur (after the first three places have been taken).

The discussion above illustrates the following general rule, which we accept as a basic axiom of counting.

(9.34)   Fundamental Counting Principle

> Let $E_1, E_2, \ldots, E_k$ be a sequence of k events. If for each i, E_i can occur in m_i ways, then the total number of ways the events may take place is the product $m_1 m_2 \cdots m_k$.

Returning to the first illustration above, we let E_1 represent the determination of the first-place team, so that $m_1 = 4$. If E_2 denotes the determination of the second-place team, then $m_2 = 3$. Hence by (9.34) the number of outcomes for the sequence E_1, E_2 is $4 \cdot 3 = 12$, which is the same as that found by means of the tree diagram. If we proceed to E_3, the determination of the third-place team, then $m_3 = 2$, and hence $m_1 m_2 m_3 = 24$. Finally, if E_1, E_2, and E_3 have occurred, there is only one possible outcome for E_4. Thus $m_4 = 1$ and $m_1 m_2 m_3 m_4 = 24$.

Instead of teams, let us now regard a, b, c, and d merely as symbols and consider the various *orderings* or *arrangements* that may be assigned to these symbols, taking them either two at a time, three at a time, or four at a time. By abstracting in this way we may apply our methods to other similar situations. For example, the problem of determining the number of two-digit numbers that can be formed from the digits 1, 2, 3, and 4 so that no digit occurs twice in any number is essentially the same as our illustration of listing first and second place in the tournament. Incidentally, the arrangements discussed above are called **arrangements without repetitions**, since a symbol may not be used twice in the same arrangement. In Example 1 below we consider arrangements in which repetitions *are* allowed.

Previously we defined ordered pairs and ordered triples. Similarly, an **ordered 4-tuple** is a set containing four elements x_1, x_2, x_3, x_4 in which an ordering has been specified, so that one of the elements may be referred to as the *first element*, another as the *second element*, and so on. The symbol (x_1, x_2, x_3, x_4) is used for the ordered 4-tuple having first element x_1, second element x_2, third element x_3, and fourth element x_4. In general, for any positive integer r, we speak of the **ordered r-tuple**

$$(9.35) \qquad (x_1, x_2, \ldots, x_r)$$

as a set of elements in which x_1 is designated as the first element, x_2 as the second element and so on.

Example 1   How many ordered triples can be obtained by using the letters $a, b, c,$ and d? How many ordered 4-tuples can be obtained? How many r-tuples?

Solution   To answer the first question we must determine the number of symbols of the form (x_1, x_2, x_3) that can be obtained by using only the letters $a, b, c,$ and d. This is not the same as listing first, second, and third place as in our previous illustration, since we have not ruled out the possibility of repetitions. For example, (a, b, a), (a, a, b), and (b, a, a) are different triples of the desired type. If for $i = 1, 2, 3$, we let E_i represent the determination of x_i in the triple (x_1, x_2, x_3), then since repetitions are allowed, there are four possibilities for each of $E_1, E_2,$ and E_3. Hence by (9.34) the total number of ordered triples is $4 \cdot 4 \cdot 4$, or 64. Similarly, the number of possible 4-tuples (x_1, x_2, x_3, x_4) is $4 \cdot 4 \cdot 4 \cdot 4 = 256$. Evidently, the number of r-tuples is the product $4 \cdot 4 \cdot \cdots \cdot 4$, where 4 appears as a factor r times. That product equals 4^r.

Example 2   A class consists of 60 girls and 40 boys. In how many ways can a president, vice-president, treasurer, and secretary be chosen if the treasurer must be a girl, the secretary must be a boy, and a student may not hold more than one office?

Solution   Let E_1 represent the choice of treasurer, E_2 the choice of secretary, and E_3 and E_4 the choice for president and vice-president respectively. As in (9.34), we let m_i denote the number of different ways E_i can occur, for $i = 1, 2, 3$, and 4. It follows that $m_1 = 60$, $m_2 = 40$, $m_3 = 98$, and $m_4 = 97$. (Why?) By (9.34) the total number of possibilities is $60 \cdot 40 \cdot 98 \cdot 97 = 22,814,400$.

The preceding example illustrates the fact that if some of the events E_i are specialized in any way, then in applying (9.34) those special E_i should be performed first in the sequence.

Let n be a positive integer and let S be a set consisting of n distinct elements. When we speak about sets we are generally not concerned about the order or arrangement of the elements. In the remainder of this section, however, the arrangement of the elements will be our main concern. In the following definition, S denotes a set containing n elements, and r is a positive integer such that $r \le n$.

(9.36)   Definition

> A **permutation** of r elements of a set S is an arrangement, without repetitions, of r elements of S.

We shall use the symbol $_nP_r$ to denote the number of different permutations of r elements that can be obtained from a set containing n elements. It is also common to refer to a permutation of r elements of a set containing n elements as a **permutation of n elements taken r at a time**. As a special case, $_nP_n$ denotes the number of arrangements of n elements of S, that is, $_nP_n$ is the number of ways of arranging *all* the elements of S.

In our first illustration involving the four teams a, b, c, d, we had $_4P_2 = 12$, since there were 12 different ways of arranging the four teams in groups of two. It was also shown that the number of ways to arrange all the elements a, b, c, d is 24. In the present notation this would be written $_4P_4 = 24$.

It is not difficult to find a general formula for $_nP_r$. If S denotes a set containing n elements, then the problem of determining $_nP_r$ is equivalent to determining the number of different r-tuples (11.35), where each x_i is an element of S and no element of S appears twice in the same r-tuple. We shall find this number by means of (11.34). For each $i = 1, 2, \ldots, r$, let E_i represent the determination of the element x_i in (11.35), and let m_i be the number of different ways of performing E_i. We wish to apply the sequence $E_1, E_2, \ldots, E_r$. There are n possible choices for x_1 and consequently we have $m_1 = n$. Since repetitions are not allowed, there are $n - 1$ choices for x_2, so that $m_2 = n - 1$. Continuing in this manner, we successively obtain $m_3 = n - 2$, $m_4 = m - 3$, and ultimately $m_r = n - (r - 1)$, or $m_r = n - r + 1$. Hence by (11.34) we have

(9.37)
$$_nP_r = n(n - 1)(n - 2) \cdots (n - r + 1)$$

where there are r factors on the right side of the equation. As illustrations, if we consider the special cases $r = 1, 2, 3$, and 4, then $n - r + 1$ equals $n, n - 1, n - 2$, and $n - 3$, respectively, and using (9.37) gives us

$$_nP_1 = n$$
$$_nP_2 = n(n - 1)$$
$$_nP_3 = n(n - 1)(n - 2)$$
$$_nP_4 = n(n - 1)(n - 2)(n - 3).$$

Example 3 Find $_nP_r$ if (a) $n = 5, r = 2$; (b) $n = 6, r = 4$; (c) $n = 5, r = 5$.

Solutions Using (9.37) we have

(a) $\qquad\qquad _5P_2 = 5 \cdot 4 = 20$

(b) $\qquad\qquad _6P_4 = 6 \cdot 5 \cdot 4 \cdot 3 = 360$

(c) $\qquad\qquad _5P_5 = 5 \cdot 4 \cdot 3 \cdot 2 \cdot 1 = 120.$

Example 4 A baseball team consists of nine players. Find the number of ways of arranging the first four positions in the batting order, if the pitcher is excluded.

Solution Using (9.37) with $n = 8$ and $r = 4$, we obtain

$$_8P_4 = 8 \cdot 7 \cdot 6 \cdot 5 = 1680.$$

As a special case of (9.37), if $r = n$ we obtain the number of different arrangements of *all* the elements of S. In this case, $n - r + 1 = n - n + 1 = 1$, and (9.37) becomes

(9.38) $\qquad\qquad _nP_n = n(n - 1)(n - 2) \cdots 3 \cdot 2 \cdot 1 = n!$

Consequently $_nP_n$ is the product of the first n positive integers. We may also obtain a form for $_nP_r$ which involves the factorial notation. If r and n are positive integers with $r \leq n$, then

$$\frac{n!}{(n - r)!} = \frac{n(n - 1)(n - 2) \cdots (n - r + 1) \cdot [(n - r)!]}{(n - r)!}$$

$$= n(n - 1)(n - 2) \cdots (n - r + 1).$$

Comparison with (9.37) gives us

(9.39) $\qquad\qquad \boxed{_nP_r = \frac{n!}{(n - r)!}}$

Substituting $r = n$ in (9.39) we obtain

$$_nP_n = \frac{n!}{(n - n)!} = \frac{n!}{0!} = \frac{n!}{1} = n!,$$

which is in agreement with (9.38).

EXERCISES

Solve Exercises 1–14 by using tree diagrams or (9.34).

1   How many three-digit numbers can be formed from the digits 1, 2, 3, 4, and 5 if (a) repetitions are not allowed?   (b) repetitions are allowed?

2   Repeat Exercise 1 for four-digit numbers.

3   How many numbers can be formed from the digits 1, 2, 3, and 4 if repetitions are not allowed? (*Note:* 42 and 231 are examples of such numbers.)

4   Work Exercise 3 if repetitions are allowed.

5   If eight basketball teams take part in a tournament, find the number of different ways that first, second, and third place can be decided, if ties are not allowed.

6   Repeat Exercise 5 for twelve teams.

7   A girl has four skirts and six blouses. How many different skirt–blouse combinations can she wear?

8   If the girl in Exercise 7 also has three sweaters, in how many different ways can she combine the three articles of clothing?

9   In a certain state, automobile license plates start with one letter of the alphabet followed by five numerals, using the digits 0, 1, 2, ..., 9. Find how many different license plates are possible if:

(a)   the first digit following the letter cannot be 0,
(b)   the first letter cannot be O or I and the first digit cannot be 0.

10   Two dice are tossed, one after the other. In how many different ways can they fall? List the number of different ways the sum of the dots can equal: (a) three,   (b) five,   (c) seven,   (d) nine,   (e) eleven.

11   A row of six seats in a classroom is to be filled by selecting individuals from a group of ten students. In how many different ways can the seats be occupied? If there are six boys and four girls in the group and if boys and girls are to be alternated, find the number of different seating arrangements.

12   A student in a certain school may take mathematics at 8, 10, 11, or 2 o'clock; English at 9, 10, 1, or 2; and History at 8, 11, 2, or 3. Find the number of different ways in which the student can schedule the three courses.

13   In how many different ways can a test consisting of ten true-or-false questions be answered?

14   A test consists of six multiple-choice questions, and the number of choices for each question is five. In how many different ways can the test be answered?

In each of Exercises 15–22, find the given number.

15 $_7P_3$ **16** $_8P_5$ **17** $_9P_6$ **18** $_5P_3$

19 $_5P_5$ **20** $_4P_4$ **21** $_6P_1$ **22** $_5P_1$

Use permutations to solve Exercises 23–30.

23   In how many different ways can eight people be seated in a row?

24   In how many different ways can ten books be arranged on a shelf?

25   A signal man has six different flags. How many different signals can be sent by placing three flags, one above the other, on a flag pole?

26   In how many different ways can five books be selected from a set of twelve different books?

27   How many three-digit numbers can be formed from the digits 2, 4, 6, 8, and 9 if (a) repetitions are not allowed?   (b) repetitions are allowed?

28   There are 24 letters in the Greek alphabet. How many fraternities may be specified by choosing three Greek letters if: (a) repetitions are not allowed?   (b) repetitions are allowed?

29   How many seven-digit phone numbers can be formed from the digits 0, 1, 2, 3, ..., 9 if the first digit may not be 0?

30   After selecting nine players for a baseball game, the manager of the team arranges the batting order so that the pitcher bats last and the best hitter bats fourth. In how many different ways can the remainder of the batting order be arranged?

7   DISTINGUISHABLE PERMUTATIONS AND COMBINATIONS

Certain problems involve finding different arrangements of objects, some of which are indistinguishable. For example, suppose we are given five discs of the same size, where three are black, one is white, and one is red. Let us find the number of ways they can be arranged in a row so that different color arrangements are obtained. If the discs were all different, then by (9.38) the number of arrangements would be 5!, or 120. However, since some of the discs have the same appearance, we cannot obtain 120 different arrangements. To clarify this point, let us write

$$B \ B \ B \ W \ R$$

for the arrangement having black discs in the first three positions in the row, the white disc in the fourth position, and the red disc in the fifth position. Now the first three discs can be arranged in 3!, or 6, different ways, but these arrangements cannot be distinguished from one another because the first three discs look alike. We say that those 3! permutations are **nondistinguishable**. Similarly, given any other arrangement, say

$$B \ R \ B \ W \ B$$

there are 3! different ways of arranging the three black discs, but again each such arrangement is nondistinguishable from the others. Let us call two arrangements of the five objects **distinguishable permutations** if one arrangement cannot be obtained from the other by rearranging the like objects. Thus B B B W R and B R B W B are distinguishable permutations. Let n denote the number of distinguishable permutations. Since with each such arrangement there corresponds 3! nondistinguishable permutations, we must have $3!n = 5!$, the number of permutations of five *different* objects. Hence $n = 5!/3! = 5 \cdot 4 = 20$. By the same type of reasoning we can obtain the following extension of the previous problem:

> If r objects in a collection of n objects are alike, and if the remaining objects are different from each other and also from the r objects, then the number of distinguishable permutations of the n objects is $n!/r!$.

That result can be generalized to the case in which there are several subcollections of indistinguishable objects. For example, consider eight discs as above, four of which are black, three white, and one red. In this case, with each arrangement, such as

(9.40)   B   W   B   W   B   W   B   R

there are 4! arrangements of the black discs and 3! arrangements of the white discs which have no effect on the color arrangement. Hence there are 4!3! arrangements of discs that can be made in (9.40) which will not produce distinguishable permutations. If we let n denote the number of *distinguishable* permutations of the objects, then it follows that $4!3!\,n = 8!$, since 8! is the number of permutations we would obtain if the objects were all different. This gives us

$$n = \frac{8!}{4!3!} = \frac{8 \cdot 7 \cdot 6 \cdot 5}{3!} \cdot \frac{4!}{4!} = 280.$$

The following general result can be proved.

(9.41)   **Theorem**

> Consider a collection of n objects in which n_1 are alike, n_2 are alike of another kind, ..., n_k are alike of a further kind, and let
>
> $$n = n_1 + n_2 + \cdots + n_k.$$
>
> Then the number of distinguishable permutations of the n objects is
>
> $$\frac{n!}{n_1!\,n_2!\cdots n_k!}.$$

Example 1   Find the number of distinguishable permutations of the letters in the word "MISSISSIPPI."

Solution   In this example we are given a collection of eleven objects in which four are of one kind (the letter S), four are of another kind (I), two are of a third kind (P), and one is a fourth kind (M). Hence by (9.41) the number of distinguishable permutations is

$$\frac{11!}{4!4!2!1!}$$

which simplifies to 34,650.

When we work with permutations our concern is with the orderings or arrangements of elements. Let us now ignore the order or arrangement of elements and consider the following question: Given a set containing n distinct elements, in how many ways can a subset of r elements be chosen, where $r \leq n$? Before answering, let us state a definition.

(9.42)   **Definition**

> A **combination** of r elements of a set S is a subset of S which contains r distinct elements.

We shall use the symbol $_nC_r$ to denote the number of combinations of r elements that can be obtained from a set containing n elements. The phrase "a combination of n elements taken r at a time" is often used instead of the phrase in (9.42).

If S contains n elements, then to find $_nC_r$ we must find the total number of subsets of the form

(9.43) $$\{x_1, x_2, \ldots, x_r\}$$

where the x_i are different elements of S. Since the elements of the subset (9.43) can be arranged in $r!$ different ways, each such subset produces $r!$ different r-tuples. Hence the total number of different r-tuples is $r!\,_nC_r$. However, in the previous section we found that the number of r-tuples is $_nP_r$, which by (9.39) equals $\dfrac{n!}{(n-r)!}$. This implies that

$$r!\,_nC_r = \frac{n!}{(n-r)!}$$

and hence

(9.44) $$_nC_r = \frac{n!}{(n-r)!\,r!}.$$

Example 2   A baseball squad has six outfielders, seven infielders, five pitchers, and two catchers. In how many different ways can a team of nine players be chosen?

Solution   The number of ways of choosing three outfielders from the six candidates is

$$_6C_3 = \frac{6!}{3!(6-3)!} = \frac{6!}{3!3!} = 20.$$

The number of ways of choosing the four infielders is

$$_7C_4 = \frac{7!}{4!(7-4)!} = \frac{7 \cdot 6 \cdot 5}{3!} = 35.$$

There are five ways of choosing a pitcher and two choices for the catcher. It follows from (9.34) that the total number of ways to choose a team is $20 \cdot 35 \cdot 5 \cdot 2 = 7000$.

It is worth noting that if $r = n$, then the right side of (9.44) becomes

$$\frac{n!}{(n-n)!n!} = \frac{n!}{0!n!} = 1.$$

Moreover, $_nC_n$ is the number of subsets consisting of n elements that can be obtained from a set of n elements. This is also equal to 1. Hence (9.44) is true for $r = n$.

It is convenient to assign a meaning to $_nC_r$ when $r = 0$. If (9.44) is to be true in this case, then

$$_nC_0 = \frac{n!}{n!0!} = 1.$$

Hence we *define* $_nC_0 = 1$, which is the same as $_nC_n$. Finally, for consistency we also *define* $_0C_0 = 1$. Thus $_nC_r$ has meaning for all nonnegative integers n and r with $r \leq n$, and formula (9.44) is valid in all those cases.

As a final observation, note that

$$_nC_r = \frac{n!}{(n-r)!r!} = \frac{n(n-1)\cdots(n-r+1)}{r!}.$$

Comparison with (9.17) shows that $_nC_r$ *is identical with the binomial coefficient* $\binom{n}{r}$. An interesting application of that fact is given in the following example.

Example 3   If a set S contains n elements, find the number of distinct subsets of S.

Solution   Let r be any nonnegative integer such that $r \leq n$. From our previous work the number of subsets of S which contain r elements is $_nC_r$, *or* $\binom{n}{r}$. Hence, to find the total number of subsets we form the sum

$$\binom{n}{0} + \binom{n}{1} + \binom{n}{2} + \binom{n}{3} + \cdots + \binom{n}{n}.$$

By (9.18) this is precisely the binomial expansion of $(1 + 1)^n$. Thus there are 2^n subsets of a set of n elements. In particular, a set of 3 elements has 2^3, or 8 different subsets. A set of 4 elements has 2^4, or 16 subsets. A set of 10 elements has 2^{10}, or 1024 subsets.

EXERCISES

In Exercises 1–8 find the given number.

1 $_7C_3$	**2** $_8C_4$	**3** $_9C_8$	**4** $_6C_2$
5 $_nC_{n-1}$	**6** $_nC_1$	**7** $_7C_0$	**8** $_5C_5$

9 If five black, three red, two white, and two green discs are to be arranged in a row, find the number of possible color arrangements for the discs.

10 Work Exercise 9 if there are three black, three red, three white, and three green discs.

11 Find the number of distinguishable permutations of the letters in the word "BOOKKEEPER."

12 Find the number of distinguishable permutations of the letters in the word "MOON." List all of the permutations.

13 Ten boys wish to play a basketball game. In how many different ways can two teams consisting of five players each be chosen?

14 A student may answer any six of ten questions in an examination. In how many ways can a choice be made? How many choices are possible if the first two questions must be answered?

15 How many straight lines are determined by eight points if no three of the points are collinear? How many triangles are determined?

16 A committee of five persons is to be chosen from a group of twelve men and eight women. If the committee is to consist of three men and two women, determine the number of ways of selecting the committee.

17 A student has five mathematics books, four history books, and eight fiction books. In how many ways can they be arranged on a shelf if books in the same category are kept next to one another?

18 A basketball squad consists of twelve individuals. Disregarding positions, in how many ways can a team of five be selected? If the center of a team must be selected from two specific individuals on the squad and the other four members of the team from the remaining ten players, find the number of different teams possible.

19 A football squad consists of three centers, ten linemen who can play either guard or tackle, three quarterbacks, six halfbacks, four ends, and four fullbacks. In how many ways can a team consisting of one center, two guards, two tackles, two ends, two halfbacks, a quarterback, and a fullback be selected?

20 In how many different ways can seven keys be arranged on a key ring if the keys can slide completely around the ring?

8 REVIEW

Concepts

Define or discuss each of the following.

1 The axiom of mathematical induction

2 The principle of mathematical induction

3 The Binomial Theorem

4 Binomial coefficients

5 Infinite sequence

6 Summation notation

7 The nth partial sum of an infinite sequence

8 Arithmetic sequence

9 Geometric sequence

10 Arithmetic mean of two numbers

11 Permutation

12 Combination

Exercises

Prove that the statements in Exercises 1–5 are true for every positive integer n.

1 $2 + 5 + 8 + \cdots + (3n - 1) = \dfrac{n(3n + 1)}{2}$

2 $2^2 + 4^2 + 6^2 + \cdots + (2n)^2 = \dfrac{2n(2n + 1)(n + 1)}{3}$

3 $\dfrac{1}{1 \cdot 3} + \dfrac{1}{3 \cdot 5} + \dfrac{1}{5 \cdot 7} + \cdots + \dfrac{1}{(2n - 1)(2n + 1)} = \dfrac{n}{2n + 1}$

4 $1 \cdot 2 + 2 \cdot 3 + 3 \cdot 4 + \cdots + n(n + 1) = \dfrac{n(n + 1)(n + 2)}{3}$

5 3 is a factor of $n^3 + 2n$.

6 Prove that $2^n > n^2$ for every positive integer $n \geq 5$.

7 Expand and simplify $(x^2 - 3y)^6$.

8 Find the first four terms in the binomial expansion of $(a^{2/5} + 2a^{-3/5})^{20}$.

9 Find the sixth term in the expansion of $(b^3 - \frac{1}{2}c^2)^9$.

10 In the expansion of $(2c^3 + 5c^{-2})^{10}$ find the term which does not contain c.

In each of Exercises 11–14 find the first four terms and the seventh term of the sequence that has the given nth term.

11 $a_n = \dfrac{5n}{3 - 2n^2}$ **12** $a_n = (-1)^{n+1} - (0.1)^n$

13 $a_n = 1 + (-\frac{1}{2})^{n-1}$ **14** $a_n = \dfrac{2^n}{(n + 1)(n + 2)(n + 3)}$

Find the first five terms of the infinite sequence defined recursively in each of Exercises 15–18.

15 $a_1 = 10, \; a_{k+1} = 1 + (1/a_k)$ **16** $a_1 = 2, \; a_{k+1} = a_k!$

17 $a_1 = 9, \; a_{k+1} = \sqrt{a_k}$ **18** $a_1 = 1, \; a_{k+1} = (1 + a_k)^{-1}$

Find the number represented by the sum in each of Exercises 19–22.

19 $\displaystyle\sum_{k=1}^{5} (k^2 + 4)$

20 $\displaystyle\sum_{k=2}^{6} \frac{2k - 8}{k - 1}$

21 $\displaystyle\sum_{k=1}^{100} 10$

22 $\displaystyle\sum_{i=1}^{4} (2^i - 10)$

In Exercises 23–26 use summation notation to express the sums.

23 $3 + 6 + 9 + 12 + 15$

24 $2 + 4 + 8 + 16 + 32 + 64 + 128$

25 $100 - 95 + 90 - 85 + 80$

26 $a_0 + a_4 x^4 + a_8 x^8 + \cdots + a_{100} x^{100}$

27 Find the tenth term and the sum of the first ten terms of the arithmetic sequence whose first two terms are $4 + \sqrt{3}$ and 3.

28 Find the sum of the first eight terms of an arithmetic sequence in which the fourth term is 9 and the common difference is -5.

29 The fifth and thirteenth terms of an arithmetic sequence are 5 and 77 respectively. Find the first term and the tenth term.

30 Insert four arithmetic means between 20 and -10.

31 Find the tenth term of the geometric sequence whose first two terms are 1/8 and 1/4.

32 If a geometric sequence has 3 and -0.3 as its third and fourth terms, find the eighth term.

33 Find a positive number c such that $4, c, 8$ are successive terms of a geometric sequence.

34 In a certain geometric sequence the eighth term is 100 and the common ratio is $-3/2$. Find the first term.

Find the sums in Exercises 29–32.

35 $\displaystyle\sum_{k=1}^{20} (3k - 5)$

36 $\displaystyle\sum_{k=1}^{10} (6 - \tfrac{1}{2}k)$

37 $\displaystyle\sum_{k=1}^{10} (2^k - \tfrac{1}{2})$

38 $\displaystyle\sum_{k=1}^{8} (\tfrac{1}{2} - 2^k)$

39 Find the sum of the infinite geometric series

$$1 - \frac{2}{5} + \frac{4}{25} - \frac{8}{125} + \cdots.$$

40 Find the rational number whose decimal representation is $6.\overline{274}$.

41 In how many ways can thirteen cards be selected from a deck of 52 cards? In how many ways can thirteen cards be selected if one wishes to obtain five spades, three hearts, three clubs, and two diamonds?

42 How many four-digit numbers can be formed from the digits 1, 2, 3, 4, 5, and 6 if (a) repetitions are not allowed? (b) repetitions are allowed?

43 If a student must answer eight of twelve questions on an examination, in how many ways can a choice be made? How many choices are possible if the first three questions must be answered?

44 If six black, five red, four white, and two green discs are to be arranged in a row, find the number of possible color arrangements.

Appendix: Tables

Table 1. Powers and Roots

n	n^2	$\sqrt{n}$	n^3	$\sqrt[3]{n}$	n	n^2	$\sqrt{n}$	n^3	$\sqrt[3]{n}$
1	1	1.000	1	1.000	51	2,601	7.141	132,651	3.708
2	4	1.414	8	1.260	52	2,704	7.211	140,608	3.733
3	9	1.732	27	1.442	53	2,809	7.280	148,877	3.756
4	16	2.000	64	1.587	54	2,916	7.348	157,464	3.780
5	25	2.236	125	1.710	55	3,025	7.416	166,375	3.803
6	36	2.449	216	1.817	56	3,136	7.483	175,616	3.826
7	49	2.646	343	1.913	57	3,249	7.550	185,193	3.849
8	64	2.828	512	2.000	58	3,364	7.616	195,112	3.871
9	81	3.000	729	2.080	59	3,481	7.681	205,379	3.893
10	100	3.162	1,000	2.154	60	3,600	7.746	216,000	3.915
11	121	3.317	1,331	2.224	61	3,721	7.810	226,981	3.936
12	144	3.464	1,728	2.289	62	3,844	7.874	238,328	3.958
13	169	3.606	2,197	2.351	63	3,969	7.937	250,047	3.979
14	196	3.742	2,744	2.410	64	4,096	8.000	262,144	4.000
15	225	3.873	3,375	2.466	65	4,225	8.062	274,625	4.021
16	256	4.000	4,096	2.520	66	4,356	8.124	287,496	4.041
17	289	4.123	4,913	2.571	67	4,489	8.185	300,763	4.062
18	324	4.243	5,832	2.621	68	4,624	8.246	314,432	4.082
19	361	4.359	6,859	2.668	69	4,761	8.307	328,509	4.102
20	400	4.472	8,000	2.714	70	4,900	8.367	343,000	4.121
21	441	4.583	9,261	2.759	71	5,041	8.426	357,911	4.141
22	484	4.690	10,648	2.802	72	5,184	8.485	373,248	4.160
23	529	4.796	12,167	2.844	73	5,329	8.544	389,017	4.179
24	576	4.899	13,824	2.884	74	5,476	8.602	405,224	4.198
25	625	5.000	15,625	2.924	75	5,625	8.660	421,875	4.217
26	676	5.099	17,576	2.962	76	5,776	8.718	438,976	4.236
27	729	5.196	19,683	3.000	77	5,929	8.775	456,533	4.254
28	784	5.292	21,952	3.037	78	6,084	8.832	474,552	4.273
29	841	5.385	24,389	3.072	79	6,241	8.888	493,039	4.291
30	900	5.477	27,000	3.107	80	6,400	8.944	512,000	4.309
31	961	5.568	29,791	3.141	81	6,561	9.000	531,441	4.327
32	1,024	5.657	32,768	3.175	82	6,724	9.055	551,368	4.344
33	1,089	5.745	35,937	3.208	83	6,889	9.110	571,787	4.362
34	1,156	5.831	39,304	3.240	84	7,056	9.165	592,704	4.380
35	1,225	5.916	42,875	3.271	85	7,225	9.220	614,125	4.397
36	1,296	6.000	46,656	3.302	86	7,396	9.274	636,056	4.414
37	1,369	6.083	50,653	3.332	87	7,569	9.327	658,503	4.431
38	1,444	6.164	54,872	3.362	88	7,744	9.381	681,472	4.448
39	1,521	6.245	59,319	3.391	89	7,921	9.434	704,969	4.465
40	1,600	6.325	64,000	3.420	90	8,100	9.487	729,000	4.481
41	1,681	6.403	68,921	3.448	91	8,281	9.539	753,571	4.498
42	1,764	6.481	74,088	3.476	92	8,464	9.592	778,688	4.514
43	1,849	6.557	79,507	3.503	93	8,649	9.644	804,357	4.531
44	1,936	6.633	85,184	3.530	94	8,836	9.695	830,584	4.547
45	2,025	6.708	91,125	3.557	95	9,025	9.747	857,375	4.563
46	2,116	6.782	97,336	3.583	96	9,216	9.798	884,736	4.579
47	2,209	6.856	103,823	3.609	97	9,409	9.849	912,673	4.595
48	2,304	6.928	110,592	3.634	98	9,604	9.899	941,192	4.610
49	2,401	7.000	117,649	3.659	99	9,801	9.950	970,299	4.626
50	2,500	7.071	125,000	3.684	100	10,000	10.000	1,000,000	4.642

Table 2. Common Logarithms

N	0	1	2	3	4	5	6	7	8	9
1.0	.0000	.0043	.0086	.0128	.0170	.0212	.0253	.0294	.0334	.0374
1.1	.0414	.0453	.0492	.0531	.0569	.0607	.0645	.0682	.0719	.0755
1.2	.0792	.0828	.0864	.0899	.0934	.0969	.1004	.1038	.1072	.1106
1.3	.1139	.1173	.1206	.1239	.1271	.1303	.1335	.1367	.1399	.1430
1.4	.1461	.1492	.1523	.1553	.1584	.1614	.1644	.1673	.1703	.1732
1.5	.1761	.1790	.1818	.1847	.1875	.1903	.1931	.1959	.1987	.2014
1.6	.2041	.2068	.2095	.2122	.2148	.2175	.2201	.2227	.2253	.2279
1.7	.2304	.2330	.2355	.2380	.2405	.2430	.2455	.2480	.2504	.2529
1.8	.2553	.2577	.2601	.2625	.2648	.2672	.2695	.2718	.2742	.2765
1.9	.2788	.2810	.2833	.2856	.2878	.2900	.2923	.2945	.2967	.2989
2.0	.3010	.3032	.3054	.3075	.3096	.3118	.3139	.3160	.3181	.3201
2.1	.3222	.3243	.3263	.3284	.3304	.3324	.3345	.3365	.3385	.3404
2.2	.3424	.3444	.3464	.3483	.3502	.3522	.3541	.3560	.3579	.3598
2.3	.3617	.3636	.3655	.3674	.3692	.3711	.3729	.3747	.3766	.3784
2.4	.3802	.3820	.3838	.3856	.3874	.3892	.3909	.3927	.3945	.3962
2.5	.3979	.3997	.4014	.4031	.4048	.4065	.4082	.4099	.4116	.4133
2.6	.4150	.4166	.4183	.4200	.4216	.4232	.4249	.4265	.4281	.4298
2.7	.4314	.4330	.4346	.4362	.4378	.4393	.4409	.4425	.4440	.4456
2.8	.4472	.4487	.4502	.4518	.4533	.4548	.4564	.4579	.4594	.4609
2.9	.4624	.4639	.4654	.4669	.4683	.4698	.4713	.4728	.4742	.4757
3.0	.4771	.4786	.4800	.4814	.4829	.4843	.4857	.4871	.4886	.4900
3.1	.4914	.4928	.4942	.4955	.4969	.4983	.4997	.5011	.5024	.5038
3.2	.5051	.5065	.5079	.5092	.5105	.5119	.5132	.5145	.5159	.5172
3.3	.5185	.5198	.5211	.5224	.5237	.5250	.5263	.5276	.5289	.5302
3.4	.5315	.5328	.5340	.5353	.5366	.5378	.5391	.5403	.5416	.5428
3.5	.5441	.5453	.5465	.5478	.5490	.5502	.5514	.5527	.5539	.5551
3.6	.5563	.5575	.5587	.5599	.5611	.5623	.5635	.5647	.5658	.5670
3.7	.5682	.5694	.5705	.5717	.5729	.5740	.5752	.5763	.5775	.5786
3.8	.5798	.5809	.5821	.5832	.5843	.5855	.5866	.5877	.5888	.5899
3.9	.5911	.5922	.5933	.5944	.5955	.5966	.5977	.5988	.5999	.6010
4.0	.6021	.6031	.6042	.6053	.6064	.6075	.6085	.6096	.6107	.6117
4.1	.6128	.6138	.6149	.6160	.6170	.6180	.6191	.6201	.6212	.6222
4.2	.6232	.6243	.6253	.6263	.6274	.6284	.6294	.6304	.6314	.6325
4.3	.6335	.6345	.6355	.6365	.6375	.6385	.6395	.6405	.6415	.6425
4.4	.6435	.6444	.6454	.6464	.6474	.6484	.6493	.6503	.6513	.6522
4.5	.6532	.6542	.6551	.6561	.6571	.6580	.6590	.6599	.6609	.6618
4.6	.6628	.6637	.6646	.6656	.6665	.6675	.6684	.6693	.6702	.6712
4.7	.6721	.6730	.6739	.6749	.6758	.6767	.6776	.6785	.6794	.6803
4.8	.6812	.6821	.6830	.6839	.6848	.6857	.6866	.6875	.6884	.6893
4.9	.6902	.6911	.6920	.6928	.6937	.6946	.6955	.6964	.6972	.6981
5.0	.6990	.6998	.7007	.7016	.7024	.7033	.7042	.7050	.7059	.7067
5.1	.7076	.7084	.7093	.7101	.7110	.7118	.7126	.7135	.7143	.7152
5.2	.7160	.7168	.7177	.7185	.7193	.7202	.7210	.7218	.7226	.7235
5.3	.7243	.7251	.7259	.7267	.7275	.7284	.7292	.7300	.7308	.7316
5.4	.7324	.7332	.7340	.7348	.7356	.7364	.7372	.7380	.7388	.7396

TABLE 2. COMMON LOGARITHMS A3

N	0	1	2	3	4	5	6	7	8	9
5.5	.7404	.7412	.7419	.7427	.7435	.7443	.7451	.7459	.7466	.7474
5.6	.7482	.7490	.7497	.7505	.7513	.7520	.7528	.7536	.7543	.7551
5.7	.7559	.7566	.7574	.7582	.7589	.7597	.7604	.7612	.7619	.7627
5.8	.7634	.7642	.7649	.7657	.7664	.7672	.7679	.7686	.7694	.7701
5.9	.7709	.7716	.7723	.7731	.7738	.7745	.7752	.7760	.7767	.7774
6.0	.7782	.7789	.7796	.7803	.7810	.7818	.7825	.7832	.7839	.7846
6.1	.7853	.7860	.7868	.7875	.7882	.7889	.7896	.7903	.7910	.7917
6.2	.7924	.7931	.7938	.7945	.7952	.7959	.7966	.7973	.7980	.7987
6.3	.7993	.8000	.8007	.8014	.8021	.8028	.8035	.8041	.8048	.8055
6.4	.8062	.8069	.8075	.8082	.8089	.8096	.8102	.8109	.8116	.8122
6.5	.8129	.8136	.8142	.8149	.8156	.8162	.8169	.8176	.8182	.8189
6.6	.8195	.8202	.8209	.8215	.8222	.8228	.8235	.8241	.8248	.8254
6.7	.8261	.8267	.8274	.8280	.8287	.8293	.8299	.8306	.8312	.8319
6.8	.8325	.8331	.8338	.8344	.8351	.8357	.8363	.8370	.8376	.8382
6.9	.8388	.8395	.8401	.8407	.8414	.8420	.8426	.8432	.8439	.8445
7.0	.8451	.8457	.8463	.8470	.8476	.8482	.8488	.8494	.8500	.8506
7.1	.8513	.8519	.8525	.8531	.8537	.8543	.8549	.8555	.8561	.8567
7.2	.8573	.8579	.8585	.8591	.8597	.8603	.8609	.8615	.8621	.8627
7.3	.8633	.8639	.8645	.8651	.8657	.8663	.8669	.8675	.8681	.8686
7.4	.8692	.8698	.8704	.8710	.8716	.8722	.8727	.8733	.8739	.8745
7.5	.8751	.8756	.8762	.8768	.8774	.8779	.8785	.8791	.8797	.8802
7.6	.8808	.8814	.8820	.8825	.8831	.8837	.8842	.8848	.8854	.8859
7.7	.8865	.8871	.8876	.8882	.8887	.8893	.8899	.8904	.8910	.8915
7.8	.8921	.8927	.8932	.8938	.8943	.8949	.8954	.8960	.8965	.8971
7.9	.8976	.8982	.8987	.8993	.8998	.9004	.9009	.9015	.9020	.9025
8.0	.9031	.9036	.9042	.9047	.9053	.9058	.9063	.9069	.9074	.9079
8.1	.9085	.9090	.9096	.9101	.9106	.9112	.9117	.9122	.9128	.9133
8.2	.9138	.9143	.9149	.9154	.9159	.9165	.9170	.9175	.9180	.9186
8.3	.9191	.9196	.9201	.9206	.9212	.9217	.9222	.9227	.9232	.9238
8.4	.9243	.9248	.9253	.9258	.9263	.9269	.9274	.9279	.9284	.9289
8.5	.9294	.9299	.9304	.9309	.9315	.9320	.9325	.9330	.9335	.9340
8.6	.9345	.9350	.9355	.9360	.9365	.9370	.9375	.9380	.9385	.9390
8.7	.9395	.9400	.9405	.9410	.9415	.9420	.9425	.9430	.9435	.9440
8.8	.9445	.9450	.9455	.9460	.9465	.9469	.9474	.9479	.9484	.9489
8.9	.9494	.9499	.9504	.9509	.9513	.9518	.9523	.9528	.9533	.9538
9.0	.9542	.9547	.9552	.9557	.9562	.9566	.9571	.9576	.9581	.9586
9.1	.9590	.9595	.9600	.9605	.9609	.9614	.9619	.9624	.9628	.9633
9.2	.9638	.9643	.9647	.9652	.9657	.9661	.9666	.9671	.9675	.9680
9.3	.9685	.9689	.9694	.9699	.9703	.9708	.9713	.9717	.9722	.9727
9.4	.9731	.9736	.9741	.9745	.9750	.9754	.9759	.9763	.9768	.9773
9.5	.9777	.9782	.9786	.9791	.9795	.9800	.9805	.9809	.9814	.9818
9.6	.9823	.9827	.9832	.9836	.9841	.9845	.9850	.9854	.9859	.9863
9.7	.9868	.9872	.9877	.9881	.9886	.9890	.9894	.9899	.9903	.9908
9.8	.9912	.9917	.9921	.9926	.9930	.9934	.9939	.9943	.9948	.9952
9.9	.9956	.9961	.9965	.9969	.9974	.9978	.9983	.9987	.9991	.9996

Answers
to
Odd-Numbered Exercises

CHAPTER 1

Section 1.2, page 8

1 (1.1)   **3** (1.2)   **5** (1.3)   **7** (1.3)   **9** (1.4)   **11** $ab + 3a + 2b + 6$
13 $ab + 3a + 2b + 6$ **15** $2xy + 4x - 3y - 6$ **17** $2xy + 4x - 3y - 6$
19 $12rs + 15s + 24r + 30$ **21** $29/12$ **23** $5/9$ **25** $11/12$
27 $13/10$ **29** $10/21$
31 Two examples are $2 - 3 \neq 3 - 2$ and $2 - (3 - 4) \neq (2 - 3) - 4$.
33 Not true; for example, $(1/2) + (1/3) \neq 1/(2 + 3)$
35 If $a = -a$, then $a + a = 0$, or $2a = 0$. Multiplying both sides by $1/2$ produces $a = 0$.
     Conversely, if $a = 0$, then also $(-1)a = (-1)0 = 0$ and hence $-a = 0$. Thus $a = 0 = -a$.
37 If $ab = 0$ and $a \neq 0$, then $(1/a)ab = (1/a)0 = 0$, and hence $b = 0$.
39 Since $(a/b)(b/a) = ab/ba = 1$, b/a is the multiplicative inverse of a/b; that is, $b/a = (a/b)^{-1}$.
41 To prove (1.8) we shall use the fact that if $x + y = 0$, then $y = -x$.
     (i)   Since $-a + a = 0$ we may take $x = -a$ and $y = a$ above, obtaining $a = -(-a)$.
     (ii)   Since $ab + (-a)b = [a + (-a)]b = (0)b = 0$, we take $x = ab$ and $y = (-a)b$, obtaining
           $(-a)b = -(ab)$. Similarly, $a(-b) = -(ab)$.
     (iii) Using (ii) twice and then (i) we obtain $(-a)(-b) = -[a(-b)] = -[-(ab)] = ab$.
     (iv) Using (ii) and (1.3), $(-1)a = -(1 \cdot a) = -a$.

Section 1.3, page 14

1 (a) $<$ (b) $>$ (c) $=$ **3** (a) $>$ (b) $<$ (c) $>$ **5** $-8 < -5$
7 $0 > -1$ **9** $x < 0$ **11** $3 < a < 5$ **13** $b \geq 2$ **15** $c \leq 1$
17 (a) 5   (b) -5 (c) 13   **19** (a) 0   (b) $4 - \pi$ (c) -1
21 (a) 4   (b) 6   (c) 6   (d) 10   **23** (a) 12   (b) 3   (c) 3   (d) 9
25 (a) No. For example, $2 - 5$ is not positive.
     (b) Yes. If a and b are positive, so is a/b.
27 $x - 5$ **29** $4 + x^2$

Section 1.4, page 20

1 $-27/125$ **3** 15   **5** $1/10$ **7** 1   **9** $27/625$ **11** $12a^8$
13 $1/2c^3$ **15** $48x^8$ **17** 16   **19** $4t^4u^2$ **21** $-10ab^4c$ **23** $2x^2$
25 $4a^2$ **27** $64a^6/b^3$ **29** $-8y^5/5u^5$ **31** $-a^{13}b^4/3$ **33** $25v^4/t^2$ **35** 1
37 $y^6/9x^4$ **39** $4/v^2$ **41** $3a + 2b$ **43** y^6/x^{12} **45** $1/a$
47 $-200s^8v^7/r^3p$ **49** 1

Section 1.5, page 27

1 9 **3** $-2\sqrt[5]{2}$ **5** $\sqrt[3]{4}/2$ **7** $8\sqrt{3}$ **9** $15\sqrt{5}$ **11** $3y^3/x^2$
13 $2a^2/b$ **15** $(3a^2/b)\sqrt[3]{2ab}$ **17** $\sqrt{3uv}/3u^2v$ **19** $3x^5/y^2$
21 $(2x/y^2)\sqrt[5]{x^2y^4}$ **23** $-3tv^2$ **25** $2pyz$ **27** $\sqrt[3]{9x}/3x$ **29** $(2x^2/y^2)\sqrt{3xy}$
31 $|a - b|$ **33** 1 **35** $(rs^6/4)\sqrt[4]{4r^2}$ **37** $3a\sqrt{b}$ **39** $(x^2y^2/z)\sqrt[7]{x^2y^4z^3}$
41 $2a^2b\sqrt{5b}$ **43** $a^2\sqrt[6]{b}$ **45** $\sqrt{ab}/ab$ **47** $(5x^2 + 3x - 1)\sqrt{x}$
49 (b) $|a/b| = \sqrt{(a/b)^2} = \sqrt{a^2/b^2} = \sqrt{a^2}/\sqrt{b^2} = |a|/|b|$

Section 1.6, page 32

1 $x^{3/4}$ **3** $(a + b)^{2/3}$ **5** $(x^2 + y^2)^{1/2}$
7 (a) $4x\sqrt{x}$ (b) $8x\sqrt{x}$ (c) $4 + x\sqrt{x}$ (d) $(4 + x)\sqrt{4 + x}$
9 16 **11** 0.000027 **13** $1/64$ **15** $15x^2$ **17** $18a^{5/6}$ **19** $1/64t^6$
21 $24c^2\sqrt[6]{c}$ **23** $8a^6p^3$ **25** p^2r^4/s^6 **27** $1/s^9$ **29** $u^{3/4}$ **31** 1
33 $(18bc^5/a^3)\sqrt{c}$ **35** $625x^8/y^4$ **37** $a + 1$ **39** $\sqrt{x}$ **41** $\sqrt{cd}$
43 $\sqrt[4]{a^2b}$ **45** $2\sqrt[6]{2}$ **47** $x\sqrt[6]{xy^5}$ **49** $\sqrt[6]{u^5v^4}/uv$ **51** $\sqrt[6]{b^5}$
53 1.7×10^{-24} **55** (a) 4.27×10^5 (b) 9.8×10^{-8} (c) 8.1×10^8
57 (a) 830,000 (b) 0.0000000000029 (c) 563,000,000 **59** 6.93
61 1.125×10^{24} **63** 3.6×10^9 **65** 5.87×10^{12} miles

Section 1.7, page 38

1 $5x^3 - x^2 - 6x + 6$ **3** $5x^4 - 2x^3 - 9x^2 + 17x - 4$ **5** $-12y^2 - 9$
7 $3a^4 - 4a + 5$ **9** $6x^3 - 5x^2 - 19x + 20$
11 $3r^4 + 4r^3 + 9r^2 + 2r + 12$ **13** $12x^5 + 18x^4 + 16x^3 - 5x^2 + 23x + 35$
15 $r^3 - t^3$ **17** $6x^5 - x^4 + 29x^3 - 2x^2 + 20x + 8$ **19** $4y^2 - 5x$
21 $3v^2 - 2u^2 + uv^2$ **23** $2x^2 - 5x - 3$ **25** $8s^2 - 38ts + 35t^2$
27 $15x^4 - 29x^2y - 14y^2$ **29** $36t^2 - 25v^2$ **31** $9r^2 + 60rs + 100s^2$
33 $16x^4 - 40x^2y^2 + 25y^4$ **35** $x^2 + 2 + (1/x^2)$ **37** $a - b$
39 $x^3 - 6x^2y + 12xy^2 - 8y^3$ **41** $27r^3 + 108r^2s + 144rs^2 + 64s^3$
43 $x^6 + 3x^4y^2 + 3x^2y^4 + y^6$ **45** $a - 3a^{2/3}b^{1/3} + 3a^{1/3}b^{2/3} - b$
47 $x^2 + 2xy + y^2 - z^2$ **49** $9x^2 + 12xy + 6xz + 4y^2 + 4yz + z^2$

Section 1.8, page 43

1 $s(r + 4t)$ **3** $3a^2b(b - 2)$ **5** $3xy^2(3x + 5y^2)$ **7** $(4x - 3)(x + 2)$
9 $(3c + 8)(2c + 3)$ **11** $(2r + 5t)(2r - 5t)$ **13** $(10x - 3y)(5x + 6y)$
15 $(4w^2 + 3s)(4w^2 - 3s)$ **17** $(6z + 5)^2$ **19** $(3x - y)(9x^2 + 3xy + y^2)$
21 $(3a + 4b)(9a^2 - 12ab + 16b^2)$ **23** $(2x^2 - 5)(4x^4 + 10x^2 + 25)$
25 $(3a + b)(2x - y)$ **27** $(5z + x^2)(8w - 7)$ **29** $(a^2 + b^2)(a - b)$
31 $(a - b)(a^2 + ab + b^2)(a + b)(a^2 - ab + b^2)$ **33** Irreducible
35 $6(x + 2)(x + 5)$ **37** $4(x - 3)^2$ **39** $3(5x + 4)^2$ **41** $9(2 + x)(2 - x)$
43 $(2x + 1)(xy - z)$ **45** $(3x^2 + 2y^2)(4z - 5w)$
47 $(y + 2)(y^2 - 2y + 4)(y - 1)(y^2 + y + 1)$
49 $(x + y - 3)(x^2 + 2xy + y^2 + 3x + 3y + 9)$
51 Irreducible **53** $(x^8 + 1)(x^4 + 1)(x^2 + 1)(x + 1)(x - 1)$

Section 1.9, page 49

1 $(x + 2)/(x + 3)$ **3** $3y/(5y + 1)$ **5** $-(a + 2)/(2a + 1)$
7 $(2x - 3)/x(5x - 2)$ **9** $34r/(3r - 1)(2r + 5)$ **11** $6x/(2x - 1)(x + 1)$

13 $3/2(4t^2 + 6t + 9)$ **15** $a/(a^2 + 4)(5a + 2)$ **17** $(6x - 7)/(3x + 1)^2$
19 $(3 - 2c)/c^3$ **21** $(5x^2 - 2x - 6)/(x - 1)^3$
23 $(18x^3 - 10x^2 - 66x + 63)/x^2(2x - 3)^2$ **25** $(p^2 + 2p + 4)/(p - 3)$
27 $-2/(2x + 2h + 3)(2x + 3)$ **29** $-(3x^2 + 3xh + h^2)/x^3(x + h)^3$
31 $11/(7x - 3)(2x + 1)$ **33** $a - b$ **35** $(x^2 + xy + y^2)/(x + y)$
37 $(2x^2 + 7x + 15)/(x^2 + 10x + 7)$ **39** $a - b$ **41** $(\sqrt{a} - \sqrt{b})/(a - b)$
43 $(y\sqrt{x} + x\sqrt{y})/xy$ **45** $c(1 + \sqrt{c})/(1 - c)$
47 $(a + b - \sqrt{c})/(a^2 + 2ab + b^2 - c)$ **49** $-1/(x + h)x$

Section 1.10, Review, page 51

1 (a) $-5/12$ (b) $39/20$ (c) $-13/56$ (d) $5/8$
3 (a) $x < 0$ (b) $1/3 < a < 1/2$ (c) $|x| \le 4$
5 (a) 5 (b) 5 (c) 7 **7** $18a^5b^5$ **9** $xy^5/9$ **11** $-p^8/2q$ **13** x^3z/y^{10}
15 b^6/a^2 **17** $s + r$ **19** x^8/y^2 **21** $\sqrt[3]{2}/2$ **23** $2x^2y\sqrt[3]{x}$
25 $(1 - \sqrt{t})/t$ **27** $2x/y^2$ **29** $x\sqrt[12]{x^5}\sqrt[6]{y^5}$ **31** $(\sqrt{a} + \sqrt{b})/(a - b)$
33 (a) 9.37×10^{10} (b) 4.02×10^{-6} **35** $x^4 + x^3 - x^2 + x - 2$
37 $-x^2 + 18x + 7$ **39** $3y^5 - 2y^4 - 8y^3 + 10y^2 - 3y - 12$ **41** $a^4 - b^4$
43 $6a^2 + 11ab - 35b^2$ **45** $169a^4 - 16b^2$ **47** $8a^3 + 12a^2b + 6ab^2 + b^3$
49 $81x^4 - 72x^2y^2 + 16y^4$ **51** $10w(6x + 7)$ **53** $(14x + 9)(2x - 1)$
55 $(2w + 3x)(y - 4z)$ **57** $8(x + 2y)(x^2 - 2xy + 4y^2)$
59 $(p^4 + q^4)(p^2 + q^2)(p + q)(p - q)$ **61** $(w^2 + 1)(w^4 - w^2 + 1)$
63 $(3x - 5)/(2x + 1)$ **65** $(3x + 2)/x(x - 2)$
67 $(5x^2 - 6x - 20)/x(x + 2)^2$ **69** $-(2x^2 + x + 3)/x(x + 1)(x + 3)$

CHAPTER 2

Section 2.1, page 58

1 $-7/4$ **3** $5\sqrt{2}/2$ **5** $5/2$ **7** $8/13$ **9** $-12/19$ **11** -6
13 $34/25$ **15** $10/9$ **17** 1 **19** $31/79$ **21** $9/38$ **23** $s = -20/39$
25 3 **27** $89/48$ **29** No solution   **31** No Solution   **33** $5/9$ **35** $-2/3$
37 No Solution   **45** $5/7$
47 Choose any a and b such that $b = 5a/3$. For example, take $a = 3, b = 5$, etc.
49 $x + 1 = x + 2$; 2

Section 2.2, page 67

1 $h = 2A/b$ **3** $h = 3V/\pi r^2$ **5** $R_2 = RR_1R_3/(R_1R_3 - RR_3 - RR_1)$
7 $P = S/(1 + rt)$ **9** $r = (3V + \pi h^3)/3\pi h^2$ **11** $r = (a - Sl)/(l - S)$
13 $m = Ft/(v_1 - v_2)$ **15** $b_1 = (2A - hb_2)/h$ **17** $v_1 = v_2 - at$
19 $n = -rI/(IR - E)$ **21** 43 quarters, 27 dimes   **23** 136, 137, 138, 139
25 -40 **27** 88   **29** \$12000 at 8.5%; \$18000 at 6%
31 \$1820 at 5%; \$1180 at 4.5%   **33** 64 seconds   **35** 3 cm
37 After an additional 50 games   **39** 20/3 liters   **41** 3 liters
43 225 miles   **45** 36 minutes   **47** 500/3 gm   **49** $d(b - a)/b$ gm

Section 2.3, page 74

1 $2/3, -5/2$ **3** $7/5, 1/4$ **5** $-5/4, -6$ **7** $3/4$ **9** $5/6, -2/3$
11 $3/2, -1$ **13** $-1 \pm \sqrt{7}$ **15** $(2 \pm \sqrt{14})/2$ **17** $5/2$

19 $(-9 \pm \sqrt{21})/10$ **21** $(3 \pm \sqrt{13})/2$ **23** $0, -4/9$ **25** $r = \sqrt{3\pi hV}/\pi h$
27 $d = \sqrt{gm_1 m_2 F}/F$ **29** $(-v_0 + \sqrt{v_0^2 + 2gs})/g$ **31** $y = (b/a)\sqrt{x^2 - a^2}$
33 $15, 17$ **35** 2 cm decrease   **37** 2.5 ft   **39** 3:15 P.M.   **41** 24 in., 76 in.
43 (a) After $25 - 5\sqrt{17}$ seconds and also after $25 + 5\sqrt{17}$ seconds (b) After 50 seconds (c) 10,000 feet
45 $-3/5$ **49** $x = (y \pm \sqrt{2y^2 - 1})/2;\ y = -2x \pm \sqrt{8x^2 + 1}$

Section 2.4, page 80

1 $-3, -3/2, 3/2$ **3** $0, -5/3, \pm\sqrt{2}/2$ **5** $0, 16$ **7** 1 **9** $-57/5$
11 $9/5$ **13** $\pm\sqrt{62}/2$ **15** 6 **17** -1 **19** $-5/4$
21 3 **23** $0, 4$ **25** $1/2, -1/2, 3, -3$
27 $\pm(1/6)\sqrt{30 + 6\sqrt{13}}, \pm(1/6)\sqrt{30 - 6\sqrt{13}}$ **29** $1/27, -27$ **31** $16/9, 25$
33 $3, 1/2$ **35** $-5/6, -3/2$ **37** $h = \sqrt{S^2 - \pi^2 r^4}/\pi r$
39 $x = (a/b)\sqrt{b^2 - y^2}$ **41** $y = (a^{2/3} - x^{2/3})^{3/2}$

Section 2.5, page 88

1 (a) $-1 < 1$ (b) $-11 < -9$ (c) $-3 < -2$ (d) $3 > 2$
3 $(2, 5)$ **5** $(-1, 3]$ **7** $[1, 4]$

9 $(-1, \infty)$ **11** $(-\infty, 2]$

13 $-1 < x < 7$ **15** $8 < x \le 9$ **17** $5 < x$ **19** $(17/5, \infty)$
21 $[-2, \infty)$ **23** $(5/2, \infty)$ **25** $(-\infty, 44/3]$ **27** $(-3, 1)$
29 $[1, 5]$ **31** $[2, 8/3)$ **33** $(6/38, \infty)$ **35** $(-2/3, \infty)$
37 $(-7/3, \infty)$ **39** $(1/4, \infty)$ **41** $(-\infty, 1)$ and $(1, \infty)$ **43** $140/9 \le c \le 80/3$
45 $20/9 \le x \le 4$
47 The result is false if $a < 0$ (consider, for example, $a = -2$ and $b = 2$).
49 If $a < b$ and $c < d$, then $b - a$ and $d - c$ are positive and hence $(b - a) + (d - c)$ is positive. Consequently, $(b + d) - (a + c)$ is positive; that is, $a + c < b + d$.

Section 2.6, page 94

1 $(-2, 2)$ **3** $(6, \infty) \cup (-\infty, -6)$ **5** $[-10, 10]$ **7** $(9.95, 10.05)$
9 $[-10, 2]$ **11** $(-2/3, 4)$ **13** $(-\infty, -6) \cup (2, \infty)$ **15** $(-\infty, 1] \cup [4, \infty)$
17 $(-2, 3)$ **19** $(-\infty, -5/2) \cup (1, \infty)$ **21** $[0, 10]$ **23** $(-\infty, -1/2) \cup (10/3, \infty)$
25 $(-3, 3)$ **27** $(4, \infty) \cup (-4, 0]$ **29** $[-5, -3/2) \cup (3/2, 5]$
31 $(-\infty, 1/2) \cup (7/5, \infty)$ **33** $(-\infty, -1) \cup (0, 1)$ **35** $[-2, 1] \cup [2, \infty)$
37 $(-\infty, -1] \cup (1, 2] \cup (3, \infty)$ **39** $[1/2, 4]$

Section 2.7, Review, page 96

1 $-5/6$ **3** -32 **5** All $x > 0$ **7** $(-2 \pm \sqrt{19})/3$ **9** $(1 \pm \sqrt{21})/2$
11 $125, -27$ **13** $1/4, 1/9$ **15** 2 **17** 5
19 The interval $(-11/4, 9/4)$ **21** $(-\infty, -3/10)$ **23** $(-\infty, 11/3] \cup [7, \infty)$
25 $(-\infty, -3/2) \cup (2/5, \infty)$ **27** $(-\infty, -3/2) \cup (2, 9)$ **29** $(1, \infty)$
31 $R = (S/\pi s) - r$ **33** $R = \sqrt[4]{8nVL/\pi P}$
35 $R_2 = (n - 1)R_1 R/(R_1 - R(n - 1))$ **37** $36 - 25\sqrt{2}$ cm ≈ 0.645 cm
39 64 mph   **41** 16 ml   **43** 9 in., 11 in.

CHAPTER 3

Section 3.1, page 104

1

3 The line bisecting quadrants I and III.

5 (a) The line parallel to the y-axis which intersects the x-axis at $(3, 0)$
(b) The line parallel to the x-axis which intersects the y-axis at $(0, 1)$
(c) All points to the right of, and on, the y-axis
(d) All points in quadrants I and III
(e) All points under the x-axis

7 (a) $\sqrt{29}$ (b) $(5, -1/2)$ **9** (a) $\sqrt{13}$ (b) $(-7/2, -1)$

11 (a) 4 (b) $(5, -3)$ **13** Area $= 28$ **17** $(13, -28)$

19 $d(A, P) = d(B, D)$

21 $\sqrt{x^2 + y^2} = 5$. A circle of radius 5 with center at the origin

23 $(0, 3 + \sqrt{11}), (0, 3 - \sqrt{11})$ **25** $(-2, -1)$ **27** $a > 4$ and $a < 2/5$

Section 3.2, page 112

1 **3** **5** **7** **9**

11 **13** **15** **17** **19**

21 **23** **25** **27**

29 Circle of radius 4, center at the origin

31 Circle of radius 1/3, center at the origin

33 Circle of radius 2, center at $(2, -1)$

35 Circle of radius 3, center at $(0, 3)$

37 $(x - 3)^2 + (y + 2)^2 = 16$ **39** $(x - 1/2)^2 + (y + 3/2)^2 = 4$

41 $x^2 + y^2 = 34$ **43** $(x + 4)^2 + (y - 2)^2 = 4$

45 $(x - 1)^2 + (y - 2)^2 = 34$

Section 3.3, page 119

1 3, 9, 4, 6   **3** $2, \sqrt{2} + 6, 12, 23$

5 (a) $5a - 2$ (b) $-5a - 2$ (c) $-5a + 2$ (d) $5a + 5h - 2$ (e) $5a + 5h - 4$ (f) 5

7 (a) $2a^2 - a + 3$ (b) $2a^2 + a + 3$ (c) $-2a^2 + a - 3$
(d) $2a^2 + 4ah + 2h^2 - a - h + 3$ (e) $2a^2 - a + 2h^2 - h + 6$
(f) $4a + 2h - 1$

9 (a) $3/a^2$ (b) $1/3a^2$ (c) $3a^4$ (d) $9a^4$ (e) $3a$ (f) $\sqrt{3a^2}$

11 (a) $2a/(a^2 + 1)$ (b) $(a^2 + 1)/2a$ (c) $2a^2/(a^4 + 1)$
(d) $4a^2/(a^4 + 2a^2 + 1)$ (e) $2\sqrt{a}/(a + 1)$ (f) $\sqrt{2a^3 + 2a}/(a^2 + 1)$

13 $[5/3, \infty)$ **15** $[-2, 2]$

17 All real numbers except $0, 3,$ and -3

19 All nonnegative real numbers except 4 and 3/2

21 $9/7, (a + 5)/7, \mathbb{R}$ **23** $19, a^2 + 3,$ all nonnegative real numbers

25 $\sqrt[3]{4}, \sqrt[3]{a}, \mathbb{R}$

27 Yes   **29** No   **31** Yes   **33** No   **35** Odd   **37** Even

39 Even   **41** Neither   **43** Neither

45 The equation $y = f(x)$ is not changed if $-x$ is substituted for x.

47 $r = C/2\pi; 6/\pi \approx 1.9$ inches   **49** $V = 4x^3 - 100x^2 + 600x$

51 $P = 4\sqrt{A}$ **53** $d = 2\sqrt{t^2 + 2500}$

55 $A = 20x$ if $x < 50$ and $A = 20x - 0.02x^2$ if $50 \le x \le 600$

57 Yes   **59** No   **61** Yes   **63** No

Section 3.4, page 130

1 Increasing on $\mathbb{R}$ **3** Decreasing on $\mathbb{R}$ **5** Increasing on $(-\infty, 0]$
Decreasing on $[0, \infty)$

7 Decreasing on $(-\infty, 0]$
Increasing on $[0, \infty)$ **9** Increasing on $[-4, \infty)$ **11** Increasing on $[0, \infty)$

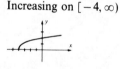

13 Decreasing on $(-\infty, 0)$
and on $(0, \infty)$ **15** Decreasing on $(-\infty, 2]$
Increasing on $[2, \infty)$ **17** Decreasing on $(-\infty, 0]$
Increasing on $[0, \infty)$

19 Neither increasing
nor decreasing   **21**   **23**   **25**

27   **29**   **31**   **33**   **35**

37 $f(x) = -\sqrt{1 - x^2}$

39 If $a < b$, then $f(a) > f(b)$; and if $b < a$, then $f(b) > f(a)$. Hence, if $a \neq b$, then $f(a) \neq f(b)$.

Section 3.5, page 139

1 $m = 8$ **3** $m = 1/11$ **5** $m = -1/4$ **9** $x - 3y - 11 = 0$

11 $x + 5y + 23 = 0$ **13** $4x + y - 7 = 0$ **15** $7x - 3y + 21 = 0$

17 (a) $y = -3$ (b) $x = 7$ **19** $y = -x$ **21** $13x - 8y + 132 = 0$

23 $8x - 15y - 13 = 0,\ x + 6y + 1 = 0,\ 10x - 3y - 11 = 0$

25 $m = 2/5,\ b = 2$ **27** $m = -3/5,\ b = 0$ **29** $m = 0,\ b = -2$

31 $m = -5/6,\ b = 10/3$ **33** $m = 0,\ b = 0$ **35** $k = -3$

37 **39**

41 $-2/5$ **43** $f(x) = -6x - 2$ **47** $\dfrac{x}{(3/2)} + \dfrac{y}{(-3)} = 1$ **49** $r < 1$ or $r > 2$

Section 3.6, page 147

1 $6x - 1,\ 6x + 3$ **3** $36x^2 - 5,\ 12x^2 - 15$ **5** $12x^2 - 8x + 1,\ 6x^2 + 4x - 1$

7 $x^3 - 1,\ x^3 - 3x^2 + 3x - 1$ **9** $x + 9 + 9\sqrt{x + 9},\ \sqrt{x^2 + 9x + 9}$

11 $x^2/(2 - 5x^2),\ 4x^2 - 20x + 25$ **13** $5,\ -5$

15 $1/x^4,\ 1/x^4$ **17** $x,\ x$ **19 and 21** Show $f(g(x)) = x = g(f(x))$

23 $f^{-1}(x) = (x + 3)/4$ **25** $f^{-1}(x) = (1 - 5x)/2x,\ x > 0$

27 $f^{-1}(x) = \sqrt{9 - x},\ x \leq 9$ **29** $f^{-1}(x) = \sqrt[3]{(x + 2)/5}$

31 $f^{-1}(x) = (x^2 + 5)/3,\ x \geq 0$ **33** $f^{-1}(x) = (x - 8)^3$ **35** $f^{-1}(x) = x$

37 (a) $f^{-1}(x) = (x - b)/a$ (b) No (not one-to-one) (c) Yes, it is its own inverse.

39 If g and h are both inverse functions of f, then $f(g(x)) = x = f(h(x))$ for all x. Since f is one-to-one this implies that $g(x) = h(x)$ for all x, that is, $g = h$.

Section 3.7, page 151

1 $a = kv,\ k = 2/5$ **3** $r = ks/t,\ k = -14$ **5** $y = kx^2/z^3,\ k = 27$

7 $c = ka^2b^3,\ k = -2/49$ **9** $F = km_1m_2/d^2,\ k = 225/8$

11 295 pounds per square foot **13** 20 pounds

15 $V = 20T/3P;\ 640/9$ **17** $50/9$ ohms **19** $3\sqrt{3}/2$ seconds

Section 3.8, Review, page 153

1 Area $= 10$ **3** The points in quadrants II and IV

7 $(x + 5)^2 + (y + 1)^2 = 81$ **9** $18x + 6y - 7 = 0$

11 **13** **15** **17** **19**

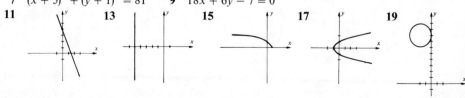

21 Decreasing on $\mathbb{R}$ **23** Decreasing on $[-1, \infty)$ **25**

27 (a) $1/2$ (b) $-\sqrt{2}/2$ (c) 0 (d) $-x/\sqrt{3-x}$ (e) $-x/\sqrt{x+3}$
(f) $x^2/\sqrt{x^2+3}$ (g) $x^2/(x+3)$
29 $18x^2 + 9x - 1, 6x^2 - 15x + 5$ **31** $f^{-1}(x) = (10 - x)/15$
33 $V = C^3/8\pi^2$

CHAPTER 4

Section 4.1, page 160

1

3

5

7 $3, -1/4$ **9** $2/3$ **11** $(-1 \pm \sqrt{41})/4$ **13** $(-9 \pm \sqrt{85})/2$
15 $2(x-4)^2 - 9$ **17** $-5(x+1)^2 + 8$ **19** $4(x - \frac{1}{2})^2 + 1$
21   **23**   **25**   **27**    **29**

31 $10, 2$ **33** $-1/2$

Section 4.2, page 167

1

3 $f(x) > 0$ if $x > 2$ **5** $f(x) > 0$ for all x
   $f(x) < 0$ if $x < 2$

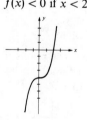

7 $f(x) > 0$ if $-3 < x < 0$ or $x > 3$
$f(x) < 0$ if $x < -3$ or $0 < x < 3$

9 $f(x) > 0$ if $x < -2$ or $0 < x < 1$
$f(x) < 0$ if $-2 < x < 0$ or $x > 1$

11 $f(x) > 0$ if $-4 < x < 1$ or $x > 5$
$f(x) < 0$ if $x < -4$ or $1 < x < 5$

13 $f(x) > 0$ if $x < -4$ or $x > 4$
$f(x) < 0$ if $-4 < x < 4$

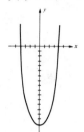

15 $f(x) > 0$ if $-1 < x < 1$
$f(x) < 0$ if $x < -1$ or $x > 1$

19 $k = -4/3$

17 $f(x) > 0$ if $x < -3$, $-1 < x < 0$, or $x > 2$
$f(x) < 0$ if $-3 < x < -1$ or $0 < x < 2$

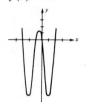

Section 4.3, page 179

1   **3** 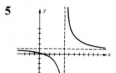   **5** 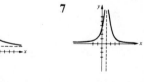   **7**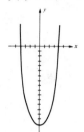

9 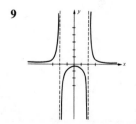   **11** 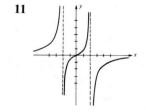   **13**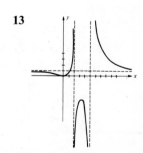

15   **17**    **19**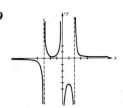

Section 4.4, page 188

1 $(1,2)$; 2   **3** $(-3,5)$; 4   **5** $(0,-3)$; $\sqrt{10}$ **7** $(1/2, 3/2)$; 1

9   **11**   **13**   **15**

17   **19**   **21**   **23**

25 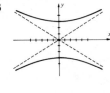   **27**    **29** 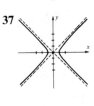   **31**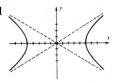

33 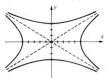   **35**   **37**   **39**

41 $x^2 + 4y^2 = 100$ **43** $2\sqrt{21}$ feet   **45** $A = 4a^2b^2/(a^2 + b^2)$
47 The graphs have the same asymptotes.

Section 4.5, Review, page 190

1   **3**   **5**   **7**

9    **11**   **13**   **15**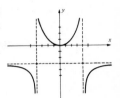

17    **19**

CHAPTER 5

Section 5.1, page 197

1   **3** 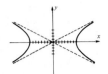   **5**   **7**   **9**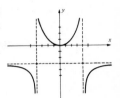

11 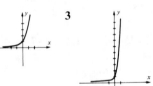   **13**   **15**   **17**   **19**

21 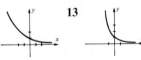   **23** 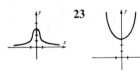   **25**

27 $-1/1600$ **29** $1010.00, $1020.10, $1061.52, $1126.83
31 a^x is not always real if $a < 0$.   **33** Reflection through the x-axis.

Calculator Exercises

1 (a) $1,061.36   (b) $1,126.49   (c) $1,346.86   (d) $1,814.01
3 (a) 5 years   (b) 8 years   (c) 12.5 years

5 (a) (b)   **7**

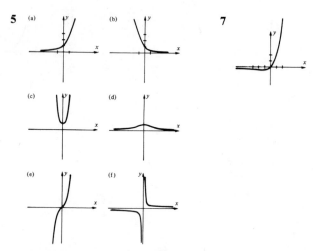

9   2.7048, 2.7169, 2.7181, 2.7183, 2.7183; $f(x)$ gets closer to e as x gets closer to 0.
11   7.44

Section 5.2, page 204

1 $\log_4 64 = 3$ **3** $\log_2 128 = 7$ **5** $\log_{10}(0.001) = -3$
7 $\log_t s = r$ **9** $10^3 = 1000$ **11** $3^{-5} = 1/243$
13 $7^0 = 1$ **15** $t^p = r$ **17** -2 **19**   2   **21**   5   **23**   1/3
25 -1 **27** -3 **29**   16   **31**   0.3   **33**   2.4   **35** -0.25
37   1.8   **39**   1.25   **41**   13   **43**   27   **45** $1/5, -1/5$ **47**   2, 3
49 $\sqrt[3]{7}$ **51**   7/2   **53**   No solution   **55**   1   **57** $2 + \sqrt{8}$
59 $2\log_a x + \log_a y - 3\log_a z$ **61** $(1/2)\log_a x + 2\log_a z - 4\log_a y$
63 $(2/3)\log_a x - (1/3)\log_a y - (5/3)\log_a z$
65 $(1/2)\log_a x + (1/4)\log_a y + (3/4)\log_a z$
67 $\log_a(x^2\sqrt[3]{x-2}/(2x+3)^5)$ **69** $\log_a x^4/y^{5/3}$

Section 5.3, page 210

1   **3**

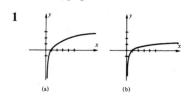   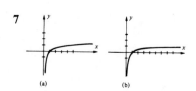

(a)   (b)   (a)   (b)

5   **7**

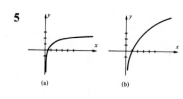   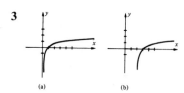

(a)   (b)   (a)   (b)

9

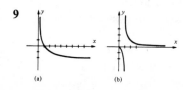

(a)   (b)

11 They are reflections of one another through the line $y = x$. They are inverse functions of one another.

13 $t = -1600 \log_2 (q/q_0)$ **15** $t = (1/k)(\ln Q - \ln Q_0)$

17 (a) 10   (b) 30   (c) 40

Calculator Exercises

1
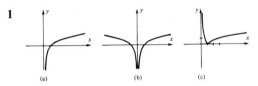

(a)   (b)   (c)

5 (a) 1.2619   (b) 0.3074   (c) 0.1781   (d) 2.5821   (e) -0.1203

Section 5.4, page 216

1 $2.5403, 7.5403 - 10, 0.5403$ **3** $9.7324 - 10, 2.7324, 5.7324$

5 $1.7796, 5.7796 - 10, 2.7796$ **7** $3.3044, 0.8261, 6.6956 - 10$

9 $9.9235 - 10, 0.0765, 9.9915 - 10$ **11** 4.0428   **13** 2.0878   **15** 3.3759

17 4,240   **19** 8.85   **21** 162,000   **23** 0.543   **25** 0.0677   **27** 189

29 0.0237   **31** (a) 2.2   (b) 5   (c) 8.3   **33** Acidic if pH < 7, basic if pH > 7

35 $\alpha \approx 54.5$

Section 5.5, page 219

1 $\log 7/\log 10$ **3** $\log 3/\log 4$ **5** $4 - (\log 5/\log 3)$ **7** $(\log 2 - \log 81)/\log 24$

9 $-\log 8/\log 2$ **11** 5   **13** $(2/3)\sqrt{1001/111}$ **15** 100, 1   **17** 10^{100}

19 10,000   **21** $x = \log(y \pm \sqrt{y^2 - 1})$ **23** $x = (1/2)\log(1 + y)/(1 - y)$

25 $x = \ln(y + \sqrt{y^2 + 1})$ **27** $x = (1/2)\ln(1 + y)/(1 - y)$

29 $t = -(L/R)\ln(1 - Ri/E)$ **31** (a) $I = I_0 10^{\alpha/10}$

Calculator Exercises

1 $x \approx 1.2091$ **3** $x \approx \pm 1.347$ **5** 139 months

Section 5.6, page 224

1 1.4062   **3** 3.7294   **5** $7.1000 - 10$ **7** 5.0913   **9** $9.8913 - 10$

11 2.5851   **13** $9.9760 - 10$ **15** 4.8234   **17** $8.6356 - 10$

19 0.4776   **21** 27.78   **23** 49,330   **25** 0.1467   **27** 0.006536   **29** 234.7

31 1.367   **33** 0.09445   **35** 109.900   **37** 0.001345   **39** 0.2442

Section 5.7, page 227

1 36,500   **3** 4.26   **5** 1,750,000   **7** 0.876   **9** 6.20   **11** 22.0

13 22.1   **15** 0.000724   **17** 1.37   **19** 1.99   **21** 90.27   **23** 0.9356

25 3.068   **27** 88.6 square units   **29** 1.84 seconds

Section 5.8, Review, page 229

1 -4 **3** 4   **5** 6

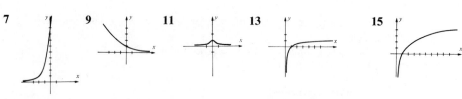

7　　　　**9**　　　　**11**　　　　**13**　　　　**15**

17 9　　**19** 1　　**21** $\log(16/3)/\log 2$　　**23** $\log(3/8)/\log(32/9)$
25 $4\log x + (2/3)\log y - (1/3)\log z$　　**27** $x = (1/2)\log(y+1)/(y-1)$
29 1.6796　　**31** 5.4780　　**33** 315.9　　**35** 0.06266　　**37** 25,800　　**39** 31.4

CHAPTER 6

Section 6.1, page 236

1 $(3,5), (-1,-3)$　　**3** $(1,0), (-3,2)$　　**5** $(0,0), (1/8, 1/128)$　　**7** $(3,-2)$
9 No solutions　　**11** $(-4,3), (5,0)$　　**13** $(-2,2)$
15 $((-6 - \sqrt{86})/10, (2 - 3\sqrt{86})/10), ((-6 + \sqrt{86})/10, (2 + 3\sqrt{86})/10)$
17 $(-4,0), (12/5, 16/5)$　　**19** $(0,1), (4,-3)$　　**21** $(\pm 2, 5), (\pm\sqrt{5}, 4)$
23 $(\sqrt{2}, \pm 2\sqrt{3}), (-\sqrt{2}, \pm 2\sqrt{3})$　　**25** $(2\sqrt{2}, \pm 2), (-2\sqrt{2}, \pm 2)$
27 $x = 3, y = -1, z = 2$　　**29** $(\log 3/\log 5, 7)$ or $(\log 2/\log 5, 2)$　　**31** $r = 3, s = -4$
33 $x = 35/22, y = 43/44$　　**35** 13, 9　　**37** 8 inches, 12 inches
39 $5 - \sqrt{19}, 5 + \sqrt{19}, 10$

Section 6.2, page 242

1 $(4,-2)$　　**3** $(8,0)$　　**5** $(-1, 3/2)$　　**7** $(76/53, 28/53)$　　**9** $(51/13, 96/13)$
11 $(8/7, -3\sqrt{6/7})$　　**13** No solution　　**15** All ordered pairs (m,n) such that
$3m - 4n = 2$
17 $(0,0)$　　**19** $(-22/7, -11/5)$　　**21** $a = 2, b = 4$　　**23** 137 adults, 313 children
25 68　　**27** 25 dimes, 18 quarters　　**29** \$6,650 at 6.5% and \$3,325 at 8%

Section 6.3, page 249

1 $(2,3,-1)$　　**3** $(-2,4,5)$　　**5** No solution　　**7** $(2/3, 31/21, 1/21)$

There are other forms for the answers in Exercises 9–15.

9 $(2c, -c, c)$ where c is any real number
11 $(0, -c, c)$ where c is any real number
13 $((12 - 9c)/7, (8c - 13)/14, c)$, where c is any real number
15 $((7c + 5)/10), (19c - 15)/10, c)$ where c is any real number
17 $(1,3,-1,2)$　　**19** $(1/11, 31/11, 3/11)$　　**21** $(-2,-3)$　　**23** No solution
25 $(2,1,2)$　　**27** 24　　**29** 17 l of 10%, 11 l of 30%, 22 l of 50%
31 $x^2 + y^2 - x + 3y - 6 = 0$

Section 6.4, page 255

29 $x = 2, y = -1, z = 3, s = 4, t = 1$

Section 6.5, page 259

1 $\begin{bmatrix} 9 & -1 \\ -2 & 5 \end{bmatrix}, \begin{bmatrix} 1 & -3 \\ 4 & 1 \end{bmatrix}, \begin{bmatrix} 10 & -4 \\ 2 & 6 \end{bmatrix}, \begin{bmatrix} -12 & -3 \\ 9 & -6 \end{bmatrix}$

3 $\begin{bmatrix} 9 & 0 \\ 1 & 5 \\ 3 & 4 \end{bmatrix}, \begin{bmatrix} 3 & -2 \\ 3 & -5 \\ -9 & 4 \end{bmatrix}, \begin{bmatrix} 12 & -2 \\ 4 & 0 \\ -6 & 8 \end{bmatrix}, \begin{bmatrix} -9 & -3 \\ 3 & -15 \\ -18 & 0 \end{bmatrix}$

5 $[11 \quad -3 \quad -3], [-3 \quad -3 \quad 7], [8 \quad -6 \quad 4], [-21 \quad 0 \quad 15]$

7 $\begin{bmatrix} -3 & 4 & 1 & 6 \\ 3 & 2 & 7 & -7 \end{bmatrix}, \begin{bmatrix} 3 & 4 & -1 & 0 \\ -1 & 2 & -7 & -3 \end{bmatrix}, \begin{bmatrix} 0 & 8 & 0 & 6 \\ 2 & 4 & 0 & -10 \end{bmatrix}, \begin{bmatrix} 9 & 0 & -3 & -9 \\ -6 & 0 & -21 & 6 \end{bmatrix}$

9 $\begin{bmatrix} 16 & 38 \\ 11 & -34 \end{bmatrix}, \begin{bmatrix} 4 & 38 \\ 23 & -22 \end{bmatrix}$ **11** $\begin{bmatrix} 3 & -14 & -3 \\ 16 & 2 & -2 \\ -7 & -29 & 9 \end{bmatrix}, \begin{bmatrix} 3 & -20 & -11 \\ 2 & 10 & -4 \\ 15 & -13 & 1 \end{bmatrix}$

13 $\begin{bmatrix} 4 & 8 \\ -18 & 11 \end{bmatrix}, \begin{bmatrix} 3 & -4 & 4 \\ -5 & 2 & 2 \\ -51 & 26 & 10 \end{bmatrix}$ **15** $\begin{bmatrix} 1 & 2 & 3 \\ 4 & 5 & 6 \\ 7 & 8 & 9 \end{bmatrix}, \begin{bmatrix} 1 & 2 & 3 \\ 4 & 5 & 6 \\ 7 & 8 & 9 \end{bmatrix}$

17 $[15], \begin{bmatrix} -3 & 7 & 2 \\ -12 & 28 & 8 \\ 15 & -35 & -10 \end{bmatrix}$ **19** $\begin{bmatrix} 4 \\ 12 \\ -1 \end{bmatrix}$ **21** $\begin{bmatrix} 18 & 0 & -2 \\ -40 & 10 & -10 \end{bmatrix}$

Section 6.6, page 265

1 $\frac{1}{10}\begin{bmatrix} 3 & 4 \\ -1 & 2 \end{bmatrix}$ **3** Does not exist   **5** $\frac{1}{8}\begin{bmatrix} 2 & 1 & 0 \\ -2 & 3 & 0 \\ 0 & 0 & 2 \end{bmatrix}$ **7** $\frac{1}{3}\begin{bmatrix} -4 & -5 & 3 \\ -4 & -8 & 3 \\ 1 & 2 & 0 \end{bmatrix}$

9 $\begin{bmatrix} \frac{1}{2} & 0 & 0 \\ 0 & \frac{1}{4} & 0 \\ 0 & 0 & \frac{1}{6} \end{bmatrix}$ **11** $\frac{1}{6}\begin{bmatrix} -8 & -7 & 4 & -9 \\ -8 & -4 & 4 & -6 \\ -4 & -5 & 2 & -3 \\ 6 & 3 & 0 & 3 \end{bmatrix}$ **13** $ab \neq 0; \begin{bmatrix} 1/a & 0 \\ 0 & 1/b \end{bmatrix}$

17 $(13/10, -1/10)$ **19** $(-25/3, -34/3, 7/3)$

Section 6.7, page 272

1 $M_{11} = -14, M_{21} = 7, M_{31} = 11, M_{12} = 10, M_{22} = -5, M_{32} = 4, M_{13} = 15, M_{23} = 34,$
$M_{33} = 6; A_{11} = -14, A_{21} = -7, A_{31} = 11, A_{12} = -10, A_{22} = -5, A_{32} = -4, A_{13} = 15,$
$A_{23} = -34, A_{33} = 6$
3 $M_{11} = 0, M_{12} = 5, M_{21} = -1, M_{22} = 7; A_{11} = 0, A_{12} = -5, A_{21} = 1, A_{22} = 7$
5 -83 **7** 5 **9** 2 **11** 0 **13** -125 **15** 48
17 -216 **19** $abcd$

Section 6.8, page 277

1 6.30 (i)   **3** 6.30 (iii)   **5** 6.30 (ii)   **7** 6.31   **9** 6.30 (ii)   **11** 6.29 or 6.31
13 6.30 (iii)   **15** -10 **17** -142 **19** -183 **21** 44   **23** 359

Section 6.10, page 288

1 The region above the graph of $2y = 3x - 6$
3 The region above and on the graph of $y = 1 - 2x$
5 The region under the graph of $y = x^2 - 2$
7 The region above and on the graph of $y = x^2 + 1$
9 The region above and on the graph of $y = 1/x^2$
11 The region under the graph of $3x + y = 3$ and also above the graph of $y = 4 - 2x$

13 The region under the graph of $y = x$ and also under the graph of $2x + 5y = 10$

15 The quadrilateral bounded by the graphs of $3x + y = 6$, $y - 2x = 1$, $x = -2$, and $y = 4$

17 The region above and on the line $x + y = 1$ which is also inside or on the circle $x^2 + y^2 = 4$

19 The region under or on the graph of $y = 1 - x^2$ and also under or on the line $x = 1 + y$

21 The region in the first quadrant between the graphs of $y = 2^x$ and $y = 3^x$

23 The region bounded by the graphs of $y = \log x$, $y + x = 1$, and $x = 10$

25 Let y denote the number of bats and x the number of balls. Then $x \geq 2$, $y \geq 3$, and $5x + 7y \leq 80$

27 If x and y denote the number of brand A and brand B, respectively, then $x \geq 20$, $y \geq 10$, $x \geq 2y$, and $x + y \leq 100$.

29 If x and y denote the amounts in the first and second accounts, respectively, then $x \geq 2000$, $y \geq 2000$, $y \geq 3x$, and $x + y \leq 15,000$.

Section 6.11, page 295

1 50 Double Fault and 30 Set Point   **3** 3.5 pounds of S and 1 pound of T

5 Send 25 from W_1 to X and 0 from W_1 to Y. Send 10 from W_2 to X and 60 from W_2 to Y.

7 45 acres of crop A and 50 acres of crop B

9 Minimum cost: 16 ounces X, 4 ounces Y, 0 ounces Z; maximum cost: 0 ounces X, 8 ounces Y, 12 ounces Z

Section 6.12, Review, page 297

1 $(19/23, -18/23)$ **3** $(-3, 5), (1, -3)$ **5** $(2\sqrt{3}, \pm\sqrt{2}), (-2\sqrt{3}, \pm\sqrt{2})$

7 $(14/17, 14/27)$ **9** $(6/11, -7/11, 1)$

11 $(-2c, -3c, c)$ where c is any real number

13 $(5c - 1, (-19c + 5)/2, c)$ where c is any real number

15 $(-1, 1/2, 1/3)$

17 The region inside the circle $x^2 + y^2 = 16$ which is also above the graph of $y = x^2$

19 The triangular region bounded by the graphs of $x - 2y = 2$, $y - 3x = 4$, and $2x + y = 4$

21 -6 **23** 48   **25** -84 **27** 120   **29** 0   **31** $a_{11}a_{22}a_{33}\cdots a_{nn}$

33 $\left(-\dfrac{1}{2}\right)\begin{bmatrix} 2 & 4 \\ 3 & 5 \end{bmatrix}$ **35** $\dfrac{1}{7}\begin{bmatrix} 2 & 1 & 0 & 0 \\ -1 & 3 & 0 & 0 \\ 0 & 0 & 3 & 2 \\ 0 & 0 & -5 & -1 \end{bmatrix}$ **37** $\begin{bmatrix} 4 & -5 & 6 \\ 4 & -11 & 5 \end{bmatrix}$

39 $\begin{bmatrix} 0 & 4 & -6 \\ 16 & 22 & 1 \\ 12 & 11 & 9 \end{bmatrix}$ **41** $\begin{bmatrix} -12 & 4 & -11 \\ 6 & -11 & 5 \end{bmatrix}$ **43** $\begin{bmatrix} a & 3a \\ 2b & 4b \end{bmatrix}$ **45** $\begin{bmatrix} 5 & 9 \\ 13 & 19 \end{bmatrix}$

CHAPTER 7

Section 7.1, page 304

1 $-2 + 6i$ **3** $-2 + 4i$ **5** $7 - 5i$ **7** $-4 - i$ **9** $4 + 7i$

11 $-6 + 3i$ **13** $-10 + 5i$ **15** $20 - 10i$ **17** 25   **19** $-72 - 36i$

21 $32 - 44i$ **23** 15   **25** 10   **27** $5 + 12i$ **29** $7 + 17i$ **31** 1

33 -1 **35** i **37** $x = -8/5, y = 3$ **39** $x = -1/2, y = -1/7$

Section 7.2, page 309

1 $(3/13) - (2/13)i$ **3** $(35/61) + (42/61)i$ **5** $(-1/5) - (11/10)i$
7 $(-7/13) + (17/13)i$ **9** $-7 - 21i$ **11** $(12/17) - (54/17)i$
13 $(42/5) - (9/5)i$ **15** $(-1/4) - (1/4)i$ **17** $-2 - i$ **19** $(1/125)i$ **21** 5
23 $\sqrt{85}$ **25** 8 **27** 1
The geometric representations in Exercises 29–37 are the following points:
29 $P(4, 2)$ **31** $P(3, -5)$ **33** $P(-3, 6)$ **35** $P(-6, 4)$ **37** $P(0, 2)$
39 $(26/29) - (7/29)i$ **41** $(-1/2)i$

Section 7.3, page 313

1 -5 **3** $-3 + 7i$ **5** 54 **7** $-2 + 2i$ **9** $-10\sqrt{10}i$
11 $(11/10) - (13/10)i$ **13** $(3 \pm \sqrt{31}i)/2$ **15** $-1 \pm 2i$ **17** $(-1 \pm \sqrt{47}i)/8$
19 $5, (-5 \pm 5\sqrt{3}i)/2$ **21** $2, -2, -1 \pm \sqrt{3}i, 1 \pm \sqrt{3}i$ **23** $\pm 2i, \pm(3/2)i$
25 $0, (-3 \pm \sqrt{7}i)/2$ **27** $x^2 - 8x + 17 = 0$ **29** $x^2 + 6x + 13 = 0$
31 $x^2 + 25 = 0$ **33** $x^2 - (10 + i)x + 10i = 0$ **35** $x^2 + 1 = 0$
37 $x^2 - 5ix - 6 = 0$

Section 7.4, Review, page 314

1 $-1 + 8i$ **3** $-28 + 6i$ **5** $40 + 56i$ **7** 97 **9** $16 + 15i$
11 $(7/13) + (17/13)i$ **13** $(5/82) + (2/41)i$ **15** $2\sqrt{34}$ **17** $-19 - 7i$ **19** $-i$
21 $(1 \pm \sqrt{14}i)/5$ **23** $\pm(\sqrt{14}/2)i, \pm(2\sqrt{3}/3)i$ **25** $x^2 - 14x + 50 = 0$
27 $x^2 + 2x + 17 = 0$ **29** $-3, (3/2) \pm (3\sqrt{3}/2)i$

CHAPTER 8

Section 8.1, page 320

1 $x^2 + x + 2, -2x + 13$ **3** $(5/2)x, (-9/2)x$ **5** $0, 7x^3 - 5x + 2$ **7** 95
9 -23 **11** 0 **13** $k = 17/12$ **15** $f(2) = 0$ **17** $f(c) > 0$
19 If $f(x) = x^n - y^n$, then $f(y) = 0$. If n is even, then $f(-y) = 0$.

Section 8.2, page 324

1 $2x^2 + x + 6, 7$ **3** $x^2 - 3x + 1, -8$ **5** $3x^4 - 6x^3 + 12x^2 - 18x + 36, -65$
7 $4x^3 + 2x^2 - 4x - 2, 0$ **9** $x^{n-1} + x^{n-2} + \cdots + x + 1, 0$ **11** $-23, 53$
13 3277 **15** $8 + 7\sqrt{3}$

Section 8.3, page 330

1 $(-1/15)x^3 + (19/5)x + 2$ **3** $-2x^3 + 12x^2 - 22x + 12$
5 $(3/10)x^3 - (33/10)x + 6$ **7** $x^4 - 2x^3 - 11x^2 + 12x + 36$
9 $(2/9)x^8 - (4/3)x^7 + (8/3)x^6 - (16/9)x^5$
11 -4 (multiplicity 3), $4/3$ (multiplicity 1)
13 0 (multiplicity 3), $5, -1$ (each of multiplicity 1)
15 $\pm 5/3$ (each of multiplicity 4), $\pm 4i$ (each of multiplicity 1)

17 -2 (multiplicity 3), 1 (multiplicity 2), 2 (multiplicity 1)
19 $(x + 3)^2(x + 2)(x - 1)$ **21** $(x - 1)^5(x + 1)$
25 $3/(x - 2) + 5/(x + 3)$ **27** $2/(x - 1) + 3/(x + 2) - 1/(x - 3)$
29 $2/(x - 1) + 5/(x - 1)^2$
31 $A/(2x - 1) + B/(x + 2) + C/(x + 2)^2$, where $A = -23/25$, $B = 24/25$, $C = 2/5$
33 $A = 1, B = 5, C = 0$
35 $A = 5, B = -3, C = 2, D = 0$

Section 8.4, page 336

1 $x^2 - 8x + 17$ **3** $x^3 - 10x^2 + 37x - 52$
5 $x^4 - 6x^3 + 30x^2 - 78x + 85$ **7** $x^5 - 2x^4 + 27x^3 - 50x^2 + 50x$
9   No. By (10.15), if i is a root, then $-i$ is also a root. Hence, the polynomial would have factors $x - 1$, $x + 1$, $x - i$, $x + i$ and, therefore, would be of degree greater than 3.
11 $4, -2, 3/2$ **13** $1/2, -2/3, -3$ **15** $-3/4, (-3 \pm 3\sqrt{7}i)/4$
17 $4, -3, \pm\sqrt{3}i$ **19** $1, -2, -4/3, 2/3$ **21** $3, -4, -2, 1/2, -1/2$
27   By (10.15), complex zeros occur in conjugate pairs.

Section 8.5, Review, page 338

1 $3x^2 + 2, -21x^2 + 5x - 9$ **3** $9/2, 53/2$ **5** -132
7 $6x^4 - 12x^3 + 24x^2 - 52x + 104. -200$
9 $(2/41)x^3 + (14/41)x^2 + (80/41)x + (68/41)$
11 $x^7 + 6x^6 + 9x^5$ **13**   1 (multiplicity 5), -3 (multiplicity 1)
15 $-3/(x + 1) + 4/(x - 2) + 1/x$ **19** $-1/2, 1/4, 3/2$

CHAPTER 9

Section 9.2, page 353

1 $a^6 + 6a^5b + 15a^4b^2 + 20a^3b^3 + 15a^2b^4 + 6ab^5 + b^6$
3 $a^8 - 8a^7b + 28a^6b^2 - 56a^5b^3 + 70a^4b^4 - 56a^3b^5 + 28a^2b^6 - 8ab^7 + b^8$
5 $81x^4 - 540x^3y + 1350x^2y^2 - 1500xy^3 + 625y^4$
7 $u^{10} + 20u^8v + 160u^6v^2 + 640u^4v^3 + 1280u^2v^4 + 1024v^5$
9 $r^{-12} - 12r^{-9} + 60r^{-6} - 160r^{-3} + 240 - 192r^3 + 64r^6$
11 $1 + 10x + 45x^2 + 120x^3 + 210x^4 + 252x^5 + 210x^6 + 120x^7 + 45x^8 + 10x^9 + x^{10}$
13 $(3^{25})c^{10} + 25(3^{24})c^{52/5} + 300(3^{23})c^{54/5} + 2300(3^{22})c^{56/5}$
15 $60(3^{14})b^{13} - (3^{15})b^{15}$ **17** $30618a^{10}b^2$
19 $840u^4v^6$ **21** $924x^2y^2$ **23** -61236 **25** $24x^2y^6$ **27** $210c^3d^2$
29 $1 + 0.2 + 0.018 + 0.00096 = 1.21896$. If $n = (1.02)^{10}$, then $\log n = 10\log(1.02) \approx 0.0860$ and $n \approx 1.2190$.

Section 9.3, page 360

1 $9, 6, 3, 0, -3; -12$ **3** $1/2, 4/5, 7/10, 10/17, 13/26; 22/65$
5 $9, 9, 9, 9, 9; 9$ **7** $1.9, 2.01, 1.999, 2.0001, 1.99999; 2.00000001$
9 $4, -9/4, 5/3, -11/8, 6/5; -15/16$ **11** $2, 0, 2, 0, 2; 0$
13 $2/3, 2/3, 8/11, 8/9, 32/27; 128/33$ **15** $4, 9, 25, 49, 121; 289$
17 $7, 19, 29, 37, 47; 79$ **19** $1, 2, 3, 4, 5; 8$
21 $2, 1, -2, -11, -38$
23 $-3, 3^2, 3^4, 3^8, 3^{16}$ **25** $5, 5, 10, 30, 120$ **27** $2, 2, 4, 4^3, 4^{12}$

29 $a_n = 2n + (1/24)(n - 1)(n - 2)(n - 3)(n - 4)(a - 10)$. There are many other answers.
31 -5 **33** 10   **35** 25   **37** $-17/15$ **39** 61   **41** 10,000
43 $(2n^3 + 12n^2 + 40n)/6$ **45** $(4n^3 - 12n^2 + 11n)/3$ **47** $\sum_{k=1}^{5}(4k - 3)$

49 $\sum_{k=1}^{4} k/(3k - 1)$ **51** $1 + \sum_{k=1}^{n}(-1)^k \dfrac{x^{2k}}{2k}$

53 $\sum_{k=1}^{7}(-1)^{k-1}/k$ **55** $\sum_{n=1}^{99} 1/n(n + 1)$

Section 9.4, page 364

1 $18, 38, 4n - 2$ **3** $1.8, 0.3, 3.3 - 0.3n$ **5** $5.4, 20.9, (3.1)n - (10.1)$
7 $\log 3^5, \log 3^{10}, \log 3^n$ **9** -8.5 **11** -9.8 **13** -105
15 30   **17** 530   **19** 423/2   **21** 60; 12,780   **23** 24
25 10/3, 14/3, 6, 22/3, 26/3   **27** $2, 2, 2^2, 2^6, 2^{24}$
29 23.25, 22.5, 21.75, 21, 20.25, 19.5, 18.75

Section 9.5, page 369

1 $1/2, 1/6; 8(1/2)^{n-1} = 2^{4-n}$ **3** $0.03, -0.00003; 300(-0.1)^{n-1}$
5 $3125, 5^8; 5^n$ **7** $81/4, -3^7/2^5; 4(-3/2)^{n-1}$ **9** $x^8, -x^{14}; (-1)^{n-1}x^{2n-2}$
11 $2^{4x+1}, 2^{7x+1}; 2^{nx-x+1}$ **13** 243/8   **15** $a_1 = 1/81, S_5 = 211/1296$
17 $(-3/2)(1 - 3^{10}) = 88,572$ **19** $(-1/3)(1 - 2^{-10})$ **21** $25/256\%$
23 $10,000(6/5)^t, 10,000(6/5)^{10}$ **25** 2/3   **27** 50/33   **29** $2 - \sqrt{2}$
31 23/99   **33** 2393/990   **35** 5141/999   **37** 16123/9999   **39** 30 meters

Section 9.6, page 376

1 (a) 60   (b) 125   **3** 64   **5** 336   **7** 24
9 (a) $(2.34)10^6$ (b) $(2.16)10^6$ **11** 151,200; 2880   **13** 1024
15 210   **17** 60480   **19** 120   **21** 6   **23** 40,320   **25** 120
27 (a) 60   (b) 125   **29** 9,000,000

Section 9.7, page 381

1 35   **3** 9   **5** n **7** 1   **9** 166,320   **11** 151,200
13 252   **15** 28, 56   **17** 8!5!4!3!   **19** 4,082,400

Section 9.8, Review, page 382

7 $x^{12} - 18x^{10}y + 135x^8y^2 - 540x^6y^3 + 1215x^4y^4 - 1458x^2y^5 + 729y^6$
9 $-(63/16)b^{12}c^{10}$
11 $5, -2, -1, -20/29, -7/19$ **13** $1, 1/2, 5/4, 7/8, 65/64$
15 10, 11/10, 21/11, 32/21, 53/32   **17** $9, 3, \sqrt{3}, \sqrt[4]{3}, \sqrt[8]{3}$ **19** 75   **21** 1000
23 $\sum_{k=1}^{5} 3k$ **25** $\sum_{k=0}^{4}(-1)^k 5(20 - k)$ **27** $13 + 10\sqrt{3}, 85 + 55\sqrt{3}$
29 $-31, 50$ **31** 64   **33** $4\sqrt{2}$ **35** 530   **37** 2041   **39** 5/7
41 $_{52}P_{13}; (_{13}P_5)(_{13}P_3)(_{13}P_3)(_{13}P_2)$ **43** 495,126

Index